GENETIC DAMAGE IN MAN CAUSED BY ENVIRONMENTAL AGENTS

ARRANGED BY
THE NORWEGIAN ACADEMY OF SCIENCE AND LETTERS
and
NORSK HYDRO A/S
in collaboration with
THE ACADEMIES OF SCIENCES of
DENMARK, FINLAND, ICELAND, and SWEDEN,
and sponsored also by
NORDISK KULTURFOND

ORGANIZING COMMITTEE

C. Overgaard Nielsen, Denmark
Diter von Wettstein, Denmark
Albert de la Chapelle, Finland
Esko Suomalainen, Finland
Kåre Berg, Norway
Anton Brøgger, Norway
Jan Rosenberg, Norway
Arne Semb-Johansen, Norway
Rune Grubb, Sweden
Åke Gustafsson, Sweden

CONFERENCE CHAIRMAN

Kåre Berg, Norway

GENETIC DAMAGE IN MAN CAUSED BY ENVIRONMENTAL AGENTS

Edited by

KÅRE BERG

*Professor of Medicine
Director, Institute of Medical Genetics, University of Oslo,
Director, Department of Medical Genetics, City of Oslo,
Norway*

ACADEMIC PRESS   New York   San Francisco   London   1979
A Subsidiary of Harcourt Brace Jovanovich, Publishers

ACADEMIC PRESS, INC.
111 Fifth Avenue, New York, New York 10003

United Kingdom Edition published by
ACADEMIC PRESS, INC. (LONDON) LTD.
24/28 Oval Road, London NW1 7DX

Library of Congress Cataloging in Publication Data

Genetic damage in man caused by environmental agents.
    Proceedings of a conference held in Oslo, Norway, May 11-13,
1977, sponsored by the Norwegian Academy of Science and Letters
and others.

    1. Human chromosome abnormalities—Congresses.
2. Environmentally induced diseases—Congresses.
3. Medical genetics—Congresses.   4. Occupational
diseases—Congresses.   I. Berg, Kare.   II. Norske
videnskaps-akademi i Oslo.
RB155.G375   616′ .042   79-414
ISBN 0-12-089550-1

Contents

SISTER CHROMATID EXCHANGE

EPIDEMIOLOGICAL APPROACHES

COST OF MUTATION

APPENDIX

Contributors

Numbers in parentheses indicate the pages on which the authors' contributions begin.

Anderson, Diana (383), Central Toxicology Laboratory, Imperial Chemical Industries Ltd., Cheshire, England

Beckman, Gunhild (205, 317), Department of Medical Genetics, University of Umeå, Umeå, Sweden

Beckman, Lars (205, 317), Department of Medical Genetics, University of Umeå, Umeå, Sweden

Berg, Kåre (1), Institute of Medical Genetics, University of Oslo, Oslo, Norway

Berlin, Maths (187), Department of Environmental Health, University of Lund, Lund, Sweden

Brøgger, Anton (87, 259), Genetics Laboratory, Norsk Hydro's Institute for Cancer Research, Oslo, Norway

de Carli, Luigi (301), Institute of General Biology and Zoology, University of Milan, Milan, Italy

Crimaudo, C. (301), Institute of General Biology and Zoology, University of Milan, Milan, Italy

Edwards, John H. (465), Infant Development Unit, Birmingham Maternity Hospital, Edgbaston, Birmingham, England

Eriksson, Kalervo (327), Department of Genetics, University of Helsinki, Finland

Fredga, Karl (187), Institute of Genetics, University of Lund, Lund, Sweden

German, James (65), Laboratory of Human Genetics, The New York Blood Bank, New York, N.Y., USA

Giorgi, R. (301), Institute of Genetics, University of Pavia, Pavia, Italy

Halkka, Olli (327), Department of Genetics, University of Helsinki, Helsinki, Finland

Hamilton, Howard B. (29), Department of Clinical Laboratories, Radiation Effects Research Foundation, Hiroshima, Japan

Hansteen, Inger-Lise (279), Telemark Central Hospital, Laboratory of Genetics, Porsgrunn, Norway

Harper, Elaine (371), Department of Clinical Genetics, Karolinska Sjukhuset, Stockholm, Sweden

Holm, Preben Bach (123), Department of Physiology, Carlsberg Laboratory, Copenhagen, Denmark

Hultén, Maj (143), Regional Cytogenetics Laboratory, East Birmingham Hospital, Birmingham, England

Infante, Peter F. (289), Industry-wide Studies Branch, Division of Surveillance, Hazard Evaluation and Field Studies, National Institute for Occupational Safety and Health, Center for Disease Control, Cincinnati, Ohio, USA

Kilian, D. Jack (101), Occupational Health and Medical Research, Dow Chemical USA, Freeport, Texas, USA

Klein, George (117), Department of Tumor Biology, Karolinska Institutet, Stockholm, Sweden

Knuutila, Sakari (269), Research Department, Rinnekoti Institution for the Mentally Retarded, Espoo, Finland

Lambert, Bo (371), Department of Clinical Genetics, Karolinska Sjukhuset, Stockholm, Sweden

Leer, Johan Chr. (419), Department of Molecular Biology, University of Aarhus, Aarhus, Denmark

Legator, Marvin S. (343), Department of Preventive Medicine and Community Health, Division of Environmental Toxicology and Epidemiology, University of Texas Medical Branch, Galveston, Texas, USA

Luciani, J. M. (143), Laboratory of Histology and Embryology II, Faculty of Medicine, Marseilles, France

Lundsteen, Claes (125), Section of Teratology, Department of Obstetrics and Gynecology, Rigshospitalet University of Copenhagen, Copenhagen, Denmark

Marcker, Kjeld A. (419), Department of Molecular Biology, University of Aarhus, Aarhus, Denmark

Meretoja, Tytti (213), Department of Industrial Hygiene and Toxicology, Institute for Occupational Health, Helsinki, Finland

Milunsky, Aubrey (441), Harvard Medical School, Genetics Division, Eunice Kennedy Shriver Center and Massachusetts General Hospital, Boston, Massachusetts, USA

Mitelman, Felix (363), Department of Clinical Genetics, University of Lund, Lund, Sweden

Mohrenweiser, Harvey (29), Department of Human Genetics, University of Michigan School of Medicine, Ann Arbor, Michigan, USA

Neel, James V. (29), Department of Human Genetics, University of Michigan School of Medicine, Ann Arbor, Michigan, USA

Nevstad, Nils Petter (247), Genetics Laboratory, Norsk Hydro's Institute for Cancer Research, Oslo, Norway

Nielsen, Ole F. (419), Department of Molecular Biology, University of Aarhus, Aarhus, Denmark

Nordenson, Ingrid (205, 317), Department of Medical Genetics, University of Umeå, Umeå, Sweden

Nordström, Stefan (205, 317), Department of Medical Genetics, University of Umeå, Umeå, Sweden

Norppa, Hannu (269), Department of Genetics, University of Helsinki, Helsinki, Finland

Nuzzo, F. (301), C.N.R. Laboratory of Biochemical Genetics and Evolution, Pavia, Italy

Oftedal, Per (335), Institute of General Genetics, University of Oslo, Oslo, Norway

Pero, Ronald W. (363), Wallenberg Laboratory, University of Lund, Lund, Sweden

Philip, John (125, 457), Section of Teratology, Department of Obstetrics and Gynecology, Rigshospitalet, University of Copenhagen, Copenhagen, Denmark

Picciano, Dante J. (101), Occupational Health and Medical Research, Dow Chemical USA, Freeport, Texas, USA

Ramel, Claes (405), Environmental Toxicology Unit, Wallenberg Laboratory, University of Stockholm, Stockholm, Sweden

Rannug, Ulf (405), Environmental Toxicology Unit, Wallenberg Laboratory, University of Stockholm, Stockholm, Sweden

Rasmussen, Søren Wilken (123), Department of Physiology, Carlsberg Laboratory, Copenhagen, Denmark.

Rinkus, Stephen J. (343), Department of Preventive Medicine and Community Health, Division of Environmental Toxicology and Epidemiology, University of Texas, Medical Branch, Galveston, Texas, USA

Reitalu, Juhan (187), Institute of Genetics, University of Lund, Lund, Sweden

Satoh, Chiyoka (29), Department of Clinical Laboratories, Radiation Effects Research Foundation, Hiroshima, Japan

Simoni, G. (301), Institute of General Biology and Zoology, University of Milan, Milan, Italy

Sorsa, Marja (431), Department of Genetics, University of Helsinki, Helsinki, Finland

Stamatoyannopoulos, George (49), Division of Medical Genetics, Department of Medicine, University of Washington, Seattle, Washington, USA

Tenchini, M. L. (301), Institute of General Biology and Zoology, University of Milan, Milan, Italy

Thune, Per (259), Department of Dermatology, University of Oslo, Oslo, Norway

Vainio, Harri (213), Department of Industrial Hygiene and Toxicology, Institute for Occupational Health, Helsinki, Finland

Waksvik, Helga (259), Genetics Laboratory, Norsk Hydro's Institute for Cancer Research, Oslo, Norway

Westergaard, Ole (419), Department of Molecular Biology, University of Aarhus, Aarhus, Denmark

Westermark, Tuomas (269), Research Department, Rinnekoti Institution for the Mentally Retarded, Espoo, Finland

Wolff, Sheldon (229), Laboratory of Radiobiology and Department of Anatomy, University of California, San Francisco, California, USA

Preface

The developments in modern society have drastically changed the chemical environment in industrialized countries. It has been realized for a long time that one price which we continuously pay for the wealth created by modern industry and for the modern way of living, is an increase in morbidity and mortality from damage caused by polluting agents of many kinds. This realization has led to a wide spectrum of measures aimed at preventing diseases caused by pollution, and many of these measures have had an excellent effect. They have all been aimed at preventing disease in people now alive.

More recently, we have realized that environmental agents may have effects which are far-reaching and difficult to control. Damage to the genetic material may not only incite tumor development in people now alive but can also cause genetic disease in future generations. Awareness of the latter possibility became pronounced when extensive damage to the chromosomes of people working in some categories of industry was observed a few years ago. The implications of these findings and of possible undetected genetic damage, for the health of future generations, are questions of prime importance. Other vital questions are: How can we create or improve potential methods for detecting damage to the units of heredity, the individual genes? Which findings would represent cause for serious concern and which ones may be trivial or caused by factors other than occupational exposure to mutagens? How can we best monitor people exposed to chemicals which could have a mutagenic effect? Which realistic measures may be useful in detecting an increase in mutational events at the population level? What action should be taken today, in cases of accidental, major exposure to environmental mutagens? How do we identify chemicals which may cause genetic damage in man when we know that there are wide variations, between and within species, in biotransformation of xenobiotics? To what extent, and how, should industry take advantage of knowledge available from the field of medical genetics in order to protect employees and their families against genetic damage? Is it possible to identify individuals who are particularly susceptible to damage, because of their genetic constitution? If so, to what extent could, or should such possibilities be utilized in industrial medicine?

Definite answers to these questions have not yet been found. Many people working in traditional industrial medicine feel that they lack the basic knowledge and technological competence to deal with these problems and that they have to draw upon experts in medical genetics. Most of those in the latter group, however, lack experience in handling problems of industrial medicine. There is a need to establish an interaction between these fields.

The above questions are among those addressed in this book. The conference on which this book is based was called out of concern for the genetic health of future generations. It was felt that a conference between experts, aimed at reviewing the methods which are at present available for revealing, monitoring and quantifying genetic damage in man would be most useful. The contributions comprise several examples of application of existing methods on exposed human populations. Thus, the emphasis of this particular conference was on genetic damage *in man*. A critical evaluation of existing and potential methods for revealing genetic damage in man, was considered to be of particular interest. To facilitate this, study groups were assigned to several important areas. The reports from the study groups appear in the Appendix.

It is hoped that this volume will serve a most useful purpose for workers in preventive medicine, medical genetics, toxicology, public health, and industrial medicine, and for other people who carry a responsibility for the health and welfare of people working or living in polluted areas, or for the health of future generations. Some sections of this book may be of interest to a number of people outside the specified disciplines, who have an interest in the various problems posed by damage to the human genome.

The enthusiastic collaboration in this venture, of the members of the Organizing Committee, the Medical Directors and Management of Norsk Hydro A/S, and the financial support from Nordisk Kulturfond, are gratefully acknowledged.

K.B.

INHERITED SUSCEPTIBILITY

INHERITED VARIATION IN SUSCEPTIBILITY AND
RESISTANCE TO ENVIRONMENTAL AGENTS

Kåre Berg

Institute of Medical Genetics
University of Oslo
Oslo, Norway

It has been known for many decades that not everybody who
is exposed to unfavourable environmental conditions contracts
an illness, even if some of those exposed become very severely
affected. Nevertheless, we owe much of the present awareness
in this area to work conducted by pharmacogeneticists during
the last 20 years or so. This work has revealed many examples
of genetic variation in response to drugs, some of which are
listed in Table I. Porphyria and malignant hyperthermia with
muscular rigidity are rare genetic disorders whereas the other
two examples in Table I represent variant genes with a frequency fulfilling the requirements to a genetic polymorphism. The
severe attacks provoked by barbiturates in porphyria patients,
and the deadly condition caused by Halothane in persons with
malignant hyperthermia, are very severe threats to the health
of these persons, who may be free of symptoms until exposed to
the drug. The polymorphism of N-acetyl-transferase, which is
determined by one gene for rapid acetylation and one for slow
acetylation, is relevant to many drugs which are acetylated
by this liver enzyme.

Pharmacogeneticists have demonstrated clearly the monogenic nature of several drug reactions, including those listed
in Table I, and the relatively rare ones caused by G-6-PD deficiency. More commonly, multiple genetic factors may influence the outcome of drug administration in man. Some quantitative enzyme variants which influence, directly or indirectly,
the metabolism of drugs in inbred mice (Nebert and Felton,
1976) are listed in Table II. There is no reason to doubt that
the same principles apply to man, for genetic variation in
many enzymes has been recognized in our species. It is also
reasonable to assume that the genetic variation in response to

TABLE I. *Examples of Atypical Reactions to Drugs in Certain Genetic Disorders or in People with Certain Normal Genetic Variants*

Genetic disorder/variant	Drug	Effect
Porphyria	Barbiturate	Increased porphyrin synthesis
Malignant hyperthermia	Halothane and succinyldicholine	Hyperthermia, rigidity, acidose, death
Serum cholin-esterase variants (several)	Succinyldicholine	Prolonged apnoea
N-acetyltrans-ferase variant (liver)	Isoniazid Sulfamethazine	Slow acetylation, neuropathy

TABLE II. *Quantitative Enzyme Variation Affecting, Directly or Indirectly, the Metabolism of Foreign Compounds, in Inbred Mice (Extracted from Nebert and Felton, 1976)*

β-glucuronidase
Liver catalase
Serum catalase
Phosphorylase kinase
Esterase
Alcohol dehydrogenase
Dipeptidase
Glutamate oxalate transaminase
β-galactosidase
Glutathione reductase

environmental compounds which has been observed in mice or rats, has some relevance to the problems of xenobiotics bio-transformation in man.

Genetic variation observed in experimental animals includes variation in chloroform susceptibility, barbital sleeping time, ethanol preference, cholesterol biosynthesis, cadmium induced testicular necrosis and serotonin toxicity (for review, see Nebert and Felton, 1976).

The realization that genetic variation may be a determining factor in an individual's reaction to a wide spectrum of xenobiotics has lead to the emergence of a sub-discipline of human and medical genetics, that of ecogenetics (Brewer, 1971; Motulsky, 1976). This paper will be limited to the central theme of human ecogenetics: the effect of existing genetic variability on response to environmental agents (Motulsky, 1976). Damage to the genetic material itself, caused by environmental agents is within the scope of this discussion as there is evidence for genetic variation in susceptibility to such damage.

Injury to the genetic material in germ cells causes mutations which may be transmitted to subsequent generations, whereas damage to the genetic material in somatic cells may incite carcinogenesis or the development of a benign tumor. Since the transfer to subsequent generations of new mutations caused by damage to germ cells can be extremely difficult to demonstrate in man, even in the offspring of atomic bomb survivors (Neel *et al.*, 1974), there is virtually no information available concerning possible genetic variability in germ cell mutagenesis. The present comments on damage to the genetic material will therefore be limited to data concerning tumors in man. If the widespread assumption that a majority of cancers are caused by environmental agents is correct, then genetic variability in tumor development may reflect genetic variability in response to carcinogenic factors.

CHROMOSOMAL DISORDERS, CHROMOSOMAL FRAGILITY AND CANCER

Cancers occur more frequently than expected by chance in patients with certain chromosomal anomalies (Table III). In patients with Down's syndrome, leukemia is 10 times more frequent than in the general population, and the risk of breast cancer in Klinefelter's syndrome is close to the risk in normal women (Mulvihill, 1975).

In addition to the greatly increased cancer risk in patients with such major chromosomal abnormalities, there are many observations linking carcinogenesis and chromosomes. Of particular interest is a small group of rare autosomal recessive disorders which exhibit chromosomal fragility and a high cancer risk. Bloom's syndrome, Fanconi's anemia and

TABLE III.   *Cancer in some Categories of Patients with Chromosomal Anomalies*

Syndrome	Type of malignancy
Down's	Leukemia
Klinefelter's	Breast Cancer
Turner's	Neural crest tumors; various

ataxia-telangiectasia all have an increased number of chromatid breaks and rearrangements in cultured cells. The relationship of chromosomal instability to cancer in these disorders have been particularly well examined by German (1972, 1973) and it seems likely that defective repair, following damage to the genetic material, may be a common denominator in these disorders as well as in xeroderma pigmentosum. Perhaps the most important information from these disorders in today's context is the suggestion that their cancer proneness may result from an inherited biochemical defect in the response to irradiation and chemical damage.

Homozygotes for the disorders are necessarily very rare, and it could be argued that these diseases are of little practical importance. The situation would, however, be quite different, if heterozygotes for these and perhaps certain other rare recessive diseases, also had an increased susceptibility to malignancy. This is suggested by results of *in vitro* studies on the susceptibility of cells from persons in Fanconi's anemia families to viral transformation. Miller and Todaro (1969) observed that heterozygotes for Fanconi's anemia had cells which were 5-40 times more susceptible than usual to transformation with the oncogenic virus SV 40 (Table IV). Heterozygotes are phenotypically normal and usually exhibit no chromosomal abnormalities, but these characteristics do not exclude an increased susceptibility of cells derived from heterozygotes to oncogenic viruses and thus to one kind of environmental agent. An excess of deaths from malignant disorders, particularly in younger persons, has been found in blood relatives of patients with ataxia-telangiectasia (Swift *et al.*, 1976).

There are also occasional families which, while they do not have any of these rare disorders, have presented findings suggestive of a genetic connection between chromosomal aberrations and neoplasms. Observations in one such family studied by Wurster-Hill *et al.* (1974) are summarized in Table V. Four out of five sibs in this sibship had chromosomal anomalies in

TABLE IV. *Increase in Susceptibility to SV Transformation in Fanconi's Anemia Families over the General Population in* in vitro *Cell Studies (from Miller and Todaro, 1969)*

Genotype	No. of persons	Increase in susceptibility (times)
Homozygous	5	10-50
Heterozygous	6	5-40

TABLE V. *Chromosome Aberrations in Four out of Five Siblings, two of whom Had Been Exposed to Organic Solvents (Extracted from Wurster-Hill et al., 1974)*

Individual	Chromosome anomalies in myeloid and lymphoid cells	Health condition	Exposure to organic solvents
LS (female)	Yes[x]	Chronic myelogeneous leukemia	Yes
IW (female)	Yes[x]	Myelofibrosis, breast cancer	Yes
BB (female)	Yes[xx]	Normal	No
RT (male)	Yes[x]	Lip carcinoma	No
WT (male)	No	Normal	No

x) *C group deletion*
xx) *Some aneuploidy, random trisomy, breakage*

both the myeloid and the lymphoid cells, and in three of them there was a C chromosome deletion, whereas the anomalies in the fourth were of a more random nature. Two of the patients

had a malignant blood disorder and both of them had a history
of pronounced exposure to organic solvents. There was no evidence of a chromosomal anomaly being transferred from one generation to the next in this family, which appears to represent
another example of a genetic predisposition to malignancy and
to chromosomal instability. Conclusions concerning the significance of the exposure to organic solvents must necessarily
be drawn with caution, but the fact that the two patients who
had been exposed both had a severe malignant disorder suggests
that this could be an example of industrial exposure inciting
malignant change in people with a high degree of genetic predisposition to such disorders.

Two tumor disorders are of particular interest because
they combine a characteristic chromosome abnormality with
evidence of monogenic inheritance. In bilateral retinoblastoma,
a deletion of chromosome 13 has been observed in a number of
cases and in Wilm's tumor patients, a small deletion in chromosome 8 may be found.

Finally, it is a distinct possibility that some of the
wide variation observed between individuals in response to
certain chromosome damaging agents (Brøgger, 1971) reflects
genetic variation among normal persons in susceptibility to
such injury.

The interesting hypothesis proposed by Rowley (1974), that
in human tumors, specific chromosome abnormalities may be related to particular etiological agents, deserves thorough
testing.

FAMILIAL CANCER

Kitchin and Ellsworth (1974) and others have found that
survivors of bilateral retinoblastoma have an increased risk
of developing a second primary tumor, particularly osteogenic
sarcoma. Some of these tumors arise far from the radiation
site. It has been suggested that this indicates an inborn
susceptibility to neoplasms. The idea is supported by
the fact that some relatives of retinoblastoma patients have
sarcoma but not retinoblastoma.

Retinoblastoma is of considerable interest also because
it was this tumor which Knudson (1971) first used to test his
two-step mutation model to explain the occurrence of heritable
and non-heritable tumors. A broad spectrum of information indicates that carcinogenesis is a multistage process. Knudson
proposed that   retinoblastoma is the consequence of two mutational events. According to this model, the first event may
be prezygotic or postzygotic; the second is always postzygotic.
Thus, the first mutation may be inherited. Carriers

of the postulated mutant gene would develop 0,1,2 or more
tumors in accordance with a Poisson distribution. The first
mutational event will not be hereditary when it is postzygotic and occurs in a somatic cell. For tumor development, two
different mutational events in one single cell are needed. The
non-inherited type occurs relatively late in life, as a single
tumor. Whenever an inherited mutation is present, only one
single additional event, in any cell, is needed to incite
tumor development from that cell. It follows that the heritable type will manifest itself at an early age, often with
multiple primary tumors. Studies on retinoblastoma, Wilm's
tumor, neuroblastoma, and pheochromocytoma have given results
in close agreement with the predictions from the model.

If Knudson's two-step mutation model is correct for a
number of human tumors, it follows that a significant number
of members of certain families will carry in all their cells
the result of the first mutational event, and require only
the second to develop a tumor. Hypothetically, such individuals should be more at risk for developing cancers than other
people. Whether or not this hypothesis is correct in all details, it may be of considerable practical importance to identify persons at risk from kindreds with familial neoplasms. It
may not be unreasonable to advise members of kindreds with
familial tumors to avoid working in industry where pollution
with chemicals is particularly pronounced.

Anderson (1975), reviewing familial cancer, pointed out
that most cancers in man occur in a heritable as well as a non-
heritable form. The characteristics of familial neoplasms according to Anderson (1975) are summarized in Table VI. The
most striking features are the early age at onset of the tumors and the frequent occurrence of multiple primary tumors.

There are several reports of various clinical and laboratory findings in non-affected members of families where
multiple malignancies or leukemia occur. Some findings in such
families are summarized in Table VII. Abnormalities related to
immunological functions are of particular significance, because it is known that a number of inherited immunodeficiency
states carry a greatly increased risk for malignant disorders.
Table VIII summarizes data from the Immunodeficiency - Cancer
Registry at the University of Minnesota (Kersey and Spector,
1975). The malignancy risk in these immunodeficiency states
varies between 2% in severe, combined system immunodeficiency
to 10% in ataxia-telangiectasia. Lymphoreticular tumors and
leukemia dominate in these disorders, but epithelial, mesen-
chymal and nervous system tumors may occur. The high risk of
malignant disorders in immunodeficiency states is in good
agreement with the well established facts that chemical carcinogens can also be immunosuppressive and that immunosuppres-

TABLE VI.   *Characteristics of Familial Neoplasms*
            (Adapted from Anderson, 1975).

---

1.   *Early age at onset*

2.   *Multiple primary tumors frequent*

3.   *Similarity between affected relatives with*
     respect to age at onset and disease pattern

TABLE VII.   *Some Findings in Occasional Families with*
             Multiple Cancers or Leukemia (Extracted from
             Kersey and Spector, 1975)

---

1.   *Abnormalities of cell mediated and humoral immunity*

2.   *Selective IgM deficiency*

3.   *Waldenström's macroglobulinemia*

4.   *Apparent HLA-linked immune response gene defect*

---

sive influences may further chemical carcinogenesis, and with the results of numerous studies on viral carcinogenesis in animals.

All these immunodeficiency disorders are rare, and thus the increased likelihood of these patients to develop tumors is relatively unimportant. It would, however, be of much more importance if heterozygous carriers of the genes for these disorders also had an increased cancer risk. Interesting observations have been made in relatives of immunodeficiency patients who could be heterozygous carriers of immunodeficiency genes. Gastric carcinoma has been reported in ataxia-telangiectasia families, and pancreatic carcinoma may be significantly more frequent in such families (Kersey and Spector, 1975). The possibility that heterozygous carriers of immunodeficiency genes are at risk to develop cancers remains interesting, but requires further studies.

Based on McKusick's catalogue of Mendelian traits in man, Mulvihill (1975) listed more than 150 examples of neoplasia as a single gene trait or as a feature or complication of other Mendelian disorders. If one were to include only disorders for which the single gene property may be considered as definitely proven, such a list would be considerably shorter. Table IX summarizes the numbers of disorders which have been

TABLE VIII. *Malignancy Risk in Patients with Inherited Immunodeficiency States According to the Immunodeficiency – Cancer Registry (Extracted from Kersey and Spector, 1975)*

Disorder	Malignancy risk (%)	Histological types of tumors [x)
Congenital X-linked immunodeficiency	6	Leukemia and lymphoreticular
Severe, combined system immunodeficiency	2	Lymphoreticular and leukemia
IgM deficiency	8	Lymphoreticular, epithelial and mesenchymal
Wiscott-Aldrich syndrome	8	Lymphoreticular and leukemia
Ataxia-telangiectasia	10	Lymphoreticular, leukemia, epithelial and nervous system
Common variable immunodeficiency	8	Lymphoreticular, epithelial and leukemia

[x) *Only tumor types occurring with a frequency of more than 5% are listed.*

definitely proven to be Mendelian and which feature neoplasia as a primary or secondary phenomenon, distributed on essentially the same categories of disorders as analyzed by Mulvihill (1975). The summary in Table IX, which is based entirely on McKusick's catalogue (McKusick, 1975), comprises more than sixty monogenic disorders. Thus, even after removing conditions with an uncertain inheritance, a wide spectrum of (albeit rare) monogenic disorders and types of cancer remains. Mulvihill (1975) has pointed out the practical importance of diagnosing not only the cancer but also the syndrome of which it is a manifestation. It is possible that close relatives of persons with one of these single gene entities may benefit not only from measures to detect cancer at an early stage, but also from advice about risks from exposure to environmental agents, when future knowledge provides a rational basis for such advice.

TABLE IX. *Review of Single Gene Traits (as listed in
McKusick's Catalogue) in Which Neoplasia is
Part of Disorder or a Frequent Complication.
Only Disorders with Proven Mode of Inheritance
have been counted*

Category	Number of disorders			
	Autosomal dominant	Autosomal recessive	X-linked	Total
Phacomatoses	3			3
Nervous system disorders	2			2
Endocrine disorders	4	1		5
Soft tissue disorders	6			6
Gastrointestinal tract disorders	5	2		7
Urogenital disorders		1	1	2
Pulmonary disorders	1			1
Vascular disorders	3			3
Skeletal disorders	2			2
Skin disorders	9	4	2	15
Hematological disorders		4		4
Immunodeficiencies		2	2	4
Various (multiple system) disorders	1	4	2	7
Total	36	18	7	61

BIOCHEMICAL DEFECTS OR VARIANTS IN PEOPLE WITH A HIGH CANCER RISK

The data summarized thus far, and much other information,
have made it an important task to try to identify biochemical
defects or variants in persons at particularly high risk for
cancer. As summarized in Table X, two different biochemical
abnormalities have been claimed to be present in people with
cancer of the urinary bladder. The defect in tryptophan metabolism was also reported in healthy family members and this

TABLE X. *Metabolic Abnormalities Reported in Cancer of the Urinary Bladder*

Disorder	Familial occurrence of disorder	Authors
Abnormal tryptophan metabolism	Yes	Leklem and Brown, 1976
Altered oxidative metabolism of nicotine	No data	Gorrod et al., 1974

may suggest that it occurred as an inherited biochemical abnormality rather than as a consequence of the disease. For the altered oxidative metabolism of nicotine, no family information is available.

Barclay *et al.* (1970) found a significantly reduced level of serum high density lipoprotein-2 (HDL$_2$) not only in women with advanced breast cancer, but also in people with a pronounced positive history of cancer in close relatives, compared to healthy people without a positive family history of cancer. Kwa and Wang (1977) reported an abnormally high luteal-phase evening peak of plasma prolactin in women with a family history of breast cancer.

At the present time, the significance of the above findings concerning cancer of the urinary bladder and breast cancer remains unclear. However, their potential importance is such that further studies should be conducted.

During the last several years there has been some excitement over the possibility that inherited variation in aryl hydrocarbon hydroxylase (AHH) inducibility may be significantly related to susceptibility to certain cancers. Polycyclic aromatic hydrocarbons are regarded as the main carcinogens of tobacco smoke. They are metabolically activated in the cell and this is mediated by AHH, a component of the mixed function oxidases. Formation of epoxide and its subsequent covalent binding to cellular macromolecules may represent a biochemical basis for carcinogenesis induced by polycyclic aromatic hydrocarbons. AHH is inducable by various agents, including the hydrocarbon substrates, barbiturates and others, and in the mouse, inducibility is under genetic control (Nebert *et al.*, 1972, 1973, 1975, Bürki *et al.*, 1975, Nebert and Felton, 1976). Kellermann and his co-workers (Kellermann *et al.*, 1973a) introduced a test in which the enzyme is induced in human lymphocytes using 3-methylcholanthrene. Table XI shows the distribution of inducubility of AHH, which Kellermann and his co-wor-

TABLE XI.   *Population Distribution of Postulated AHH*
Phenotypes (A = Low; B = High Inducibility)
(from Kellermann et al., 1973a)

	Phenotypes		
	AA	AB	BB
No. observed	86	59	16
No. expected [x]	82.8	65.3	12.9

[x] *From Hardy-Weinberg equilibrium*

TABLE XII.   *Result of Family Study Reported by Kellermann*
et al. (1973a) on their AHH Variants

Type	Mating no.	No. of children	% with phenotype		
			AA	AB	BB
AA x AA	17	39	100		
AA x AB	28	63	50	50	
AA x BB	4	10		100	
AB x AB	10	35	26	48	26
AB x BB	6	13		69	31
BB x BB	2	5			100
Total	67	165			

kers obtained by this method. Three classes of inducibility −
high, intermediate, and low − were observed and the authors
hypothesized that their trimodal distribution reflected genotypes at a single autosomal locus. Two alleles, A and B, at this
locus were thought to determine low and high inducibility, respectively. On the assumption that the upper mode reflected
homozygotes for the B gene, the lower mode homozygotes for the
A gene, and the intermediate mode the heterozygous condition,
Kellermann and co-workers found that the distribution of phenotypes in the population agreed well with that expected from
Hardy-Weinberg equilibrium (frequency of the A and B gene
taken as 0.72 and 0.28, respectively). Their hypothesis was
substantiated in a family study (Kellermann *et al.*, 1973a),
the results of which are summarized in Table XII. The distri-

bution of offspring of the different mating types was in extremely close agreement with the expectancy, assuming autosomal co-dominant or intermediate inheritance.

Kellermann *et al.*, (1973b) found that individuals with
bronchogenic carcinoma had high or intermediate inducibility
far more frequently than did healthy people (Table XIII). Thus,
it appeared that a simple polymorphism with a decisive influence on the risk for contracting lung cancer had been found
and there was a plausible connection between the biochemical
mechanism involved in the polymorphism and carcinogenesis. It
seemed to be a reasonable assumption that a fair proportion
of lung cancers might be avoided if people with an unfavourable phenotype within this polymorphism abstained from smoking
and other exposure to polycyclic hydrocarbons.

In the years following the reports by Kellermann and co-
workers several workers experienced very considerable difficulty in reproducing their results, and polygenic control of
AHH inducibility in man as well as in mice was favoured by
several authors (Kouri *et al.*, 1974, Coomes *et al.*, 1976,
Bürki *et al.*, 1975, Paigen *et al.*, 1977). The system in the
mouse has been further clarified by Nebert and Felton (1976)
who reported that matings between "responder" and "non-respon-
der" mice resulted in monogenic segregation of "responsiveness".
However, they also found over a dozen types of individual responses involving the AHH system. Intensive efforts to clarify
the genetics of AHH inducibility in man is greatly needed. At
the present time it seems plausible that there are several
forms of "AHH activity" in man, and that AHH induction is controlled by more than one locus. Evidence for genetic control
of AHH inducibility has appeared also from studies of twins
(Atlas *et al.*, 1976), but confirmation of a monogenic inheritance in man has not appeared (Paigen *et al.*, 1977). Evidence
for a unimodal population distribution of AHH inducibility
(rather than the trimodal distribution found by Kellermann and
co-workers) has led to the conclusion by several workers
(Kouri *et al.*, 1974) that polygenic rather than monogenic
control operates in man. It should be kept in mind, however,
that a unimodal distribution does in itself not exclude control by one locus. For example, human erythrocyte acid phosphatase activity is controlled by 3 alleles at one locus. Although each allele specifies one level of activity, the enzyme
activities of the resulting 6 genotypes are not sufficiently
separated to create a population distribution other than unimodal. The unimodal distribution of enzyme activity is thus
the result of six distinct, but overlapping distributions. It
is of interest that the distribution of AHH inducibility reported by Paigen *et al.* (1977) in progeny of lung cancer
patients and controls was such that it could well be composed
of several distinct but insufficiently separated distributions.

TABLE XIII. *Aryl Hydrocarbon Hydroxylase Inducibility*
             in Healthy People and in Patients with
             Bronchogenic Carcinoma (Kellermann et al.,
             1973b)

Category	AHH inducibility	
	High or intermediate	Low
Healthy people	55%	45%
Patients with bronchogenic carcinoma	96%	4%

Trell *et al.* (1976) also reported that they were able to separate the three phenotypes claimed by Kellermann *et al.* (1973a).
However, the data of Paigen *et al.* (1977) showed no difference
in AHH inducibility between progeny of lung cancer patients
and controls, as would have been predicted from the findings
of Kellermann and co-workers.

Studying hybrid cell clones segregating human chromosomes,
Brown and his co-workers (Brown *et al.*, 1976) obtained highly
suggestive evidence that AHH activity is determined by a locus,
or loci, linked to those controlling malate dehydrogenase and
isocitrate dehydrogenase. Both of these human enzymes had previously been assigned to chromosome 2. Thus, human genes required for AHH activity appear to have been assigned to chromosome 2. This finding should lead to renewed research, aimed
at identifying the effect of individual genes, in the control
of the variability of human AHH activity.

Trell *et al.* (1976) reported a highly significant connection between high AHH inducibility and laryngeal carcinoma
(Table XIV). Both bronchogenic carcinoma and laryngeal carcinoma are associated with heavy cigarette smoking, and it
seemed that both tumor forms had a much higher frequency of
high or intermediate inducibility than the general population.
However, Paigen *et al.* (1977) could not confirm that an association exists between high level of AHH inducibility and
lung cancer risk, and they concluded that the polymorphism
for AHH inducibility as measured at present, is probably not
important in determining susceptibility to lung cancer. Thus,
no definite conclusion can be drawn at the present time.
Paigen *et al.* (1977) point out, despite their findings, that
the concept of genetically determined differences in carcino-
gen-metabolizing enzymes affecting susceptibility to lung
cancer is still valid. Familial aggregation of lung cancer is

TABLE XIV. *Aryl Hydrocarbon Hydroxylase Inducibility in Laryngeal Carcinoma (from Trell et al., 1976)*

	AHH Inducibility		
	High	Intermediate	Low
No. observed	15	15	8
No. expected from population frequencies	4	17	17

$P < 0.001$

known to occur (Tokuhata and Lilienfeld, 1963) and a familial component common to lung cancer and chronic obstructive pulmonary disease has recently been reported (Cohen *et al.*, 1977, Lynch *et al.*, 1977). Thus, there is evidence for genetic differences in susceptibility to lung cancer which is independent of Kellermann's study.

ATHEROSCLEROTIC DISEASE

Table XV summarizes some of the evidence that genetic components are important in human atherosclerotic disease. The evidence for this comprises the existense of inherited hyperlipoproteinemias and an inherited, distinct, normal class of lipoprotein, the so-called Lp(a) lipoprotein (Berg, 1963, 1968). Presence of (high amounts of) this lipoprotein predisposes individuals to atherosclerotic disease (Berg *et al.*, 1974, Dahlén *et al.*, 1975, Berg, 1977, Frick *et al.*, 1977). A congenital thickening of arterial intimas may occur as a genetically determined trait. The relative risk for coronary heart disease in people with the inherited lipoprotein phenotype Lp(a+) over Lp(a-) is given i Table XVI.

There is an apparent discrepancy between the clear evidence that there are important genetic factors in atherosclerotic disease and the fact that morbidity and mortality from coronary heart disease has increased dramatically over a time period that is too short to allow for significant changes in gene frequency. The most plausible interpretation seems to be that environmental and nutritional factors, which are different from those operating 60 or 70 years ago, are now leading to atherosclerotic disease, particularly in those who have a genetically determined predisposition to such disease. Thus, if adequate preventive measures were available, there is little

TABLE XV.   *Some Hereditary and Environmental Components*
            in Atherosclerotic Disease

Genetic Components

Monogenic hyperlipidemias
Polygenic hyperlipidemias
Presence of Lp(a) lipoprotein
Polygenic hypertension
Familial effects independent of factors above
Arterial wall enzymes may transform promutagens to mutagens
Thickening of arterial intima

Environmental Components

Cigarette smoking
Nutritionally caused hyperlipidemia
Hypertension caused by nutritional/environmental factors
Environmental mutagens (if Benditt's hypothesis is correct)
Stress, etc.

TABLE XVI.   *Relative Risk for Coronary Heart Disease in*
             Lp(a+) over Lp(a-) Individuals

Population	No. of patients/ No. of controls	Relative risk
Finnish	*153/61*	*2.7*
Swedish	*58/103*	*2.0*
Norwegian	*188/1109*	*3.7*

Total (weighted) relative risk: 3.1

doubt that it would be worth identifying very early in life
all those individuals who have an inherited propensity to develop atherosclerotic disease. I have reviewed extensively the
evidence for hereditary and environmental components in the
etiology of atherosclerotic disease elsewhere (Berg, 1977).

One new development in this area is of major importance in
the present context. Benditt and Benditt (1973) have presented
evidence for a single cell origin of all smooth muscle cells
in a given atherosclerotic lesion. They were able to do this
from studying women heterozygous for X-linked enzyme variants,
in whom, because of the Lyonization, only one enzyme variant
is active in any one cell. Their evidence has been confirmed
by one group of workers (Pearson *et al.*, 1977) whereas another
has challenged their interpretation (Thomas *et al.*, 1976). A

recent study of plaque tissue nucleic acids showed conditions similar to those in benign tumors (Rounds *et al.*, 1976). Proliferation of smooth muscle cells is a key event in atherogenesis (Ross and Glomset, 1973). According to Benditt, the initial event is a somatic mutation in an arterial wall cell. Environmental as well as genetic factors may contribute to this. In genetically predisposed individuals, exposure to chemical mutagens, viruses, or ionizing radiation could cause a somatic cell mutation. The next step in the process would be cell proliferation. Chemical as well as physical injuries to the artery could initiate this, but it is also relevant that low density lipoprotein is known to stimulate growth of smooth muscle cells *in vitro* and that this lipoprotein is the main carrier of polycyclic aromatic hydrocarbons in the blood stream. A high level of low density lipoprotein may be genetically determined and could contribute to both of the first two steps in this sequence of events. Finally, lipoprotein deposition in the plaques occurs and degeneration and ulceration of plaques may lead to an acute stage of the disease and thrombus formation.

Thus, atherosclerotic disease is of interest in the present context, not only because of genetic variation in susceptibility, but also because of the possible role of somatic cell mutation as an inciting event. If Benditt's extremely interesting hypothesis is correct, one should find more evidence for somatic cell mutations in patients with atherosclerotic disease than in healthy persons, and atherosclerotic disease should be more common in people exposed to chemical mutagens than in those who are not so exposed. Some of the answers which will be needed ought to be forthcoming in the near future.

INHERITED α_1-ANTITRYPSIN DEFICIENCY AND DISEASE SUSCEPTIBILITY

Pulmonary emphysema is another incapacitating disorder occurring with an appreciable frequency. It was clearly documented in the mid-sixties (Eriksson, 1965) that genetic variation in the serum protein α_1-antitrypsin is associated with pulmonary emphysema. The variant allele referred to as the Z-allele determines a much lower level of α_1-antitrypsin than does its normal counterpart, although the protease inhibitory capacity of whatever amount of α_1-antitrypsin is present appears to be normal. The homozygotes for this disorder are highly prone to develop pulmonary emphysema; 80% will finally develop this disorder (Kueppers 1976a). However, non-smokers develop dyspnoea on the average 9 years later than smokers. The main cause for the lung damage may be granulocytic proteases which may be liberated during an inflammatory response or during pinocytosis of dust (Kueppers, 1976a). Although somewhat controversial, there is suggestive evidence that the heterozygotes are also predisposed to emphysema. If this were the case, a significant proportion of an industrial population may be at risk of developing this disorder. Screening for α_1-antitrypsin deficiency is a simple procedure and there may be good reasons to examine the serum of people who will be exposed to a high degree of air pollution with respect to this genetic variation. Smoking and intermediate α_1-antitrypsin levels may interact additively to produce emphysema in MZ heterozygotes (Kueppers, 1976a, Larsson *et al.*, 1977).

It is still unknown to what extent isoelectric focusing, which has recently permitted the identification of new, common, genetic variants of α_1-antitrypsin (Kueppers, 1976b) may become useful in attempts to identify further groups of people at risk for lung disease caused by smoking and environmental pollution.

OTHER GENETIC MARKERS

The α_1-antitrypsin variation is a normal polymorphism, similar to many others, such as the acetylator polymorphism mentioned previously.

One disease relationship has been clearly established for the latter polymorphism, namely that of systemic lupus erythematosus occurring in patients treated with hydralazine (Perry *et al.*, 1970). Patients on hydralazine treatment developing systemic lupus erythematosis are slow acetylators much more frequently than expected by chance alone. It therefore became important to learn if spontaneous systemic lupus erythematosus, that is, disease occurring without any known drug treatment, is associated with acetylator phenotypes. This was indeed found to be the case and it may therefore be hypothesized that spontaneous systemic lupus erythematosus may have unidentified aromatic amine-type compounds as inciting agent (Reidenberg and Martin, 1974).

A disorder of considerable interest for certain branches of industry is asbestosis. As shown in Table XVII, two studies have shown a higher frequency of HLA-B27 antigen in asbestosis patients than in controls, but a third series failed to show any such relationship and the combined information from the three series is not statistically significant.

Associations of this kind are of considerable interest because the HLA system has turned out to exhibit association with a number of disorders, many of which have an altered im-

TABLE XVII.   *HLA-B27 Antigen in Asbestosis Patients and Controls*

Series of	Per cent positives	
	Patients	Controls
Merchant et al., 1975	32	11
Matej and Lange, 1976	23	10
Evans et al., 1977	5	8
Combined	17	10

Combined result is not significant.

munological responsiveness (Dausset and Svejgaard, 1977). It is a distinct possibility that certain HLA-haplotypes may confer susceptibility or resistance to asbestosis development.

Table XVIII shows the result of a study on the frequency of the B-lymphocyte group-2 antigen in asthma patients and controls (Rachelefsky *et al.*, 1976). There is a highly significant excess of positives among the patients. Thus, in this lung disorder, a different kind of lymphocyte antigen shows a very strong association.

A final example of a normal genetic marker which may become helpful in identifying persons at risk of contracting a particular kind of disease is illustrated in Table XIX. Atkin (1977) has recently reported that the chromosome 1 heteromorphism which occurs in a fair number of healthy persons is considerably increased in frequency in patients with various malignant diseases. He hypothesizes that this heteromorphism may be a marker for a high cancer risk group. Although it is too early to evaluate fully the significance of this information, it serves to illustrate the importance of the new techniques for fine structure studies of human chromosomes. If confirmed, the heteromorphism may be one additional way of identifying high risk individuals.

INHERITED RESISTANCE TO ENVIRONMENTAL AGENTS

From what has already been said, it follows that there are not only examples of particular genetic susceptibility to environmental agents, but also of genetically determined resistance. At face value Kellermann's data on AHH inducibility suggest that people with low inducibility have an inherited resistance to some of the environmental agents causing lung

TABLE XVIII.   *B-Lymphocyte Group-2 Specificity in Asthma*
                Patients and Controls (from Rachelefsky et
                al., *1976)*

Category	No. of persons B-lymphocyte group-2 antigen		Total
	present	absent	
Asthma patients	26	4	30
Controls	26	83	109
Total	52	87	139

P < 0.001

TABLE XIX.   *Chromosome 1 Heteromorphism as a Possible*
              Marker for a High Cancer Risk Group, according
              to Atkin (1977)

	Number of persons Heteromorphism		Total
	present	absent	
Patients with malignant disease	32	26	58
Controls	17	38	55
Total	49	64	113

0.01 < P < 0.02

cancer. Table XX summarizes a comparison (Evans *et al.*, 1977) of asbestos workers without demonstrable pulmonary changes by radiography (category 0), with those who had demonstrable changes (categories 1 and 2), with respect to the HLA antigen BW5. There was a highly significant excess of positives among those who did not have radiologically demonstrable changes although they had been exposed to asbestos to the same extent as had those with such changes. The preliminary evidence suggests that the lymphocyte antigen HLA-BW5 may confer a certain degree of resistance to asbestosis.

TABLE XX. *HLA-BW5 Antigen in Exposed Asbestos Workers with or without Pulmonary Changes Demonstrable by Radiography (from Evans et al., 1977)*

ILO-UC perfusion category	No. of persons HLA-BW5		Total
	positive	negative	
0	9	23	32
1 and 2	1	41	42
Total	10	64	74

$P < 0.01$

A third example of inherited resistance to diseases which is at least environmentally influenced, is shown in Table XXI. Glueck *et al.*, (1976) have studied people in kindreds with either high level of high density lipoprotein (HDL) or low level of low density lipoprotein (LDL). It is clear from Table XXI that both of these hereditary traits confer a greatly increased life expectancy in both sexes. Persons with one or the other of these "longevity traits" presumably have some degree of protection against atherogenic factors of a nutritional or environmental nature.

CONCLUDING REMARKS

Practical application of preventive medicine based on the concept of genetic variation in susceptibility to environmental agents can be done in selected cases. For instance, it seems reasonable to advise homozygotes for the α_1-antitrypsin Z allele, to avoid air pollution and not to smoke. Persons who belong to kindreds with familial clustering of cases of multiple primary tumors may be advised to avoid industrial exposure to mutagenic or carcinogenic chemicals. Most of the areas I have discussed, however, represent fields which demand much more research. In a fair proportion of these areas, it should be possible to conduct effective research in a rather short time. There is every reason to stimulate further study in such areas as the identification of individual human genotypes in the AHH system, and the resolution of the theory that some HLA-types confer either susceptibility or resistance to environmentally caused diseases, such as asbestosis. The possibility that normal chromosomal markers may be helpful in iden-

TABLE XXI.   *Familial Hyper-HDL-emia and Hypo-LDL-emia as*
"Longevity Syndromes" Expressed as Expectation
of Life at Age 35 (from Glueck et al.*, 1976)*

Kindreds with	Increase (years) in life expectancy over general population		Significance level of increase
	Males	Females	
Hyper-HDL-emia	5	7	P < 0.02
Hypo-LDL-emia	9	12	P < 0.002

tifying high risk groups should be examined closely and work
which may confirm or refute Benditt's hypothesis of a connection between somatic cell mutation and atherosclerotic disease
should be encouraged.

Furthermore, between species variation in xenobiotics bio-
transformations makes it imperative to extract as much information as possible in this matter from man himself because
data from other species will, in many cases, not be valid.
Considerable intraspecies variation will be a problem even
after data from a limited number of individuals have become
available. These facts may limit the usefulness of mutagenicity testing based on incorporation of liver microsomes in
bacterial test sytems. Clearly, there is no substitute for man
in the study of environmental damage to our species, and genetic variability in susceptibility and resistance to environmental agents can only be approached by genetic methods in
man.

It would seem to be an important task for preventive industrial medicine to make use in the future of any practical method geneticists may develop to identify those with a genetic
susceptibility so that they can be protected as mush as possible from an unfavourable environment. This would help to
secure for each individual a type of environment which suits
his or her particular genetic make-up in an optimal way.

Although development along these lines may in theory lead
to prevention òf a significant part of occupational diseases,
this approach is not without problems. Some aspects of preemployment screening may not be to the individual's advantage.
It has been emphasized by Motulsky (1976) that considerable
harm could be done to many black U.S. men if suggestions by
well-meaning physicians that carriers of the sickling trait
should be barred from military services, were followed. Cri-

tical voices have been heard about the "blame the victim" mentality which could result from too much emphasis on genetic susceptibility to diseases caused by environmental agents (Powledge, 1976). Despite the existence of such a danger, it would be unjustified to discourage research in this area since protection against injury will be facilitated once a susceptible subpopulation can be identified. There is little doubt that the greatest potential of medical genetics, at present, is its growing ability to identify populations at risk, so that efforts at prevention can be concentrated on those individuals most susceptible to disease.

REFERENCES

Anderson, D.E. (1975). *In* "Persons at High Risk of Cancer" (J.F. Fraumeni, ed.), p. 39. Academic Press, New York.

Atkin, N.B. (1977). *Brit. Med. J. 1,* 358.

Atlas, S.A., Vesell, E.S., and Nebert, D.W. (1976). *Cancer Res. 36,* 4619.

Barclay, M., Skipski, V.P., Terebus-Kekish, O., Greene, E.M., Kaufman, R.J., and Stock, C.C. (1970). *Cancer Res. 30,* 2420.

Benditt, E.P., and Benditt, J.M. (1973). *Proc. Natl. Acad. Sci. (USA) 70,* 1753.

Berg, K. (1963). *Acta path. microbiol. scand. 59,* 369.

Berg, K. (1968). *Ser. Haematol. 1,* 111.

Berg, K. (1977). *In* "The Biochemistry of Atherosclerosis" (A. Scanu, R. Wissler, and G. Getz eds.), in press, Marcel Dekker, New York.

Berg, K. Dahlén, G., and Frick, M.H. (1974). *Clin. Genet. 6,* 230.

Brewer, G.J. (1971). *Amer. J. hum. Genet. 23,* 92.

Brown, S., Wiebel, F.J., Gelboin, H.V., and Minna, J.D. (1976). *Proc. Natl. Acad. Sci. (USA) 73,* 4628.

Brögger, A. (1971). *In* "Excerpta Medica International Congress Series No. 233", p. 35. 4th International Congress of Human Genetics, Paris, Sept. 6-11.

Bürki, K., Liebelt, A.G., and Bresnick, E. (1975). *Biochem. Genet. 13,* 417.

Cohen, B.H., Diamond, E.L., Graves, C.G., Kreiss, P., Levy, D.A., Menkes, H.A., Permutt, S., Quaskey, S., and Tockman, M.S. (1977). *Lancet ii,* 523.

Coomes, M.L., Mason, W.A., Muijsson, I.E., Cantrell, E.T., Anderson, D.E., and Busbee, D.L. (1976). *Biochem. Genet. 14,* 671.

Dahlén, G., Frick, M.H., Berg, K., Valle, M., and Wiljasalo,
    M. (1975). *Clin. Genet. 8,* 183.
Dausset, J., and Svejgaard, A., eds. (1977). "HLA and Disease",
    Munksgaard, Copenhagen.
Eriksson, S. (1965). *Acta med. scand. 177,* Suppl. 432.
Evans, C.C., Levinsöhn, H.C., and Evans, J.M. (1977). *Brit.
    med. J. 1,* 603.
Frick, M.H., Dahlén, G., Berg, K., Valle, M., and Hekali, P.
    (1977). *Chest,* in press.
German, J. (1972). *In* "Progress in Medical Genetics" (A.G.
    Steinberg and A.G. Bearn, eds.), Vol VIII. Grune & Strat-
    ton, New York.
German, J. (1973). *Hosp. Pract. 8,* 93.
Glueck, C.J., Gartside, P., Fallat, R.W., Sielski, J., and
    Steiner, P.M. (1976). *J. Lab. clin. Med. 88,* 941.
Gorrod, J.W., Jenner, P., Keysell, G.R., and Mikhael, B.R.
    (1974). *J. nat. Cancer Inst. 52,* 1421.
Kellermann, G., Luyten-Kellermann, M., and Shaw, C.R. (1973a).
    Amer. J. hum. Genet. 25, 327.
Kellermann, G., Shaw, C.R., and Luyten-Kellermann, M. (1973b).
    New Engl. J. Med. 289, 934.
Kersey, J.H., and Spector, B.D. (1975). *In* "Persons at High
    Risk of Cancer" (J.F. Fraumeni, ed.), p. 55. Academic
    Press, New York.
Kitchin, F.D., and Ellsworth, R.M. (1974). *J. Med. Genet. 11,*
    244.
Knudson, A.G. (1971). *Proc. Natl. Acad. Sci. (USA) 68,* 820.
Kouri, R.E., Ratrie, H., Atlas, S.A., Niwa, A., and Nebert,
    D.W. (1974). *Life Sci. 15,* 1585.
Kueppers, F. (1976a). *In* "Proceedings of the Fifth Internatio-
    nal Congress of Human Genetics", Excerpta Medica, Amster-
    dam, in press.
Kueppers, F. (1976b). *Amer. J. hum. Genet. 28,* 370.
Kwa, H.G., and Wang, D.Y. (1977). *Int. J. Cancer 20,* 12.
Larsson, C., Eriksson, S., and Dirksen, H. (1977). *Brit. Med.
    J. 2,* 922.
Leklem, J.E., and Brown, R.R. (1976). *J. nat. Cancer Inst. 56,*
    1101.
Lynch, H.T., Guirgis, H.A., and Harris, R.E. (1977). *Lancet ii,*
    815.
Matej, Ḥ., and Lange, A. (1976). *In* "First International Sym-
    posium on HLA and Disease", p. 256. Inserm, Paris.
McKusick, V.A. (1975). "Mendelian Inheritance in Man", 4th Ed.,
    Johns Hopkins University Press, Baltimore.
Merchant, J.A., Klouda, P.T., Soutar, C.A., Parkes, W.R.,
    Lawler, S.D., and Turner-Warwick, M. (1975). *Brit. Med. J.
    1,* 189.

Miller, R.W., and Todaro, G.J. (1969). *Lancet i*, 81.

Motulsky, A.G. (1976). *In* "Proceedings of the Fifth International Congress of Human Genetics". Excerpta Medica, Amsterdam, in press.

Mulvihill, J.J. (1975). *In* "Persons at High Risk of Cancer" (J.F. Fraumeni, ed.), p. 3. Academic Press, New York.

Nebert, D.W., and Felton, J.S. (1976). *Fed. Proc. 35*, 1133.

Nebert, D.W., Goujon, F.M., and Gielen, J.E. (1972). *Nature New Biol. 236*, 107.

Nebert, D.W., Heidema, J.K., Strobel, H.W., and Coon, M.J. (1973). *J. biol. Chem. 248*, 7631.

Nebert, D.W., Robinson, J.R., Niwa, A., Kumaki, K., and Poland, A.P. (1975). *J. Cell Physiol. 85*, 393.

Neel, J.V., Kato, H., and Schull, W.J. (1974). *Genetics 76*, 311.

Paigen, B., Gurtoo, H.L., Minowada, J., Houten, L., Vincent, R., Paigen, K., Parker, N.B., Ward, E., and Haynes, N.T. (1977). *New Engl. J. Med. 297*, 346.

Pearson, T.A., Kramer, E.C., Solez, K., and Heptinstall, R.H. (1977). *Amer. J. Path. 86*, 657.

Perry, H.M., Tan, E.M., Carmody, S., and Sakamoto, A. (1970). *J. Lab. clin. Med. 76*, 114.

Powledge, T. (1976). *New Scientist 71*, 486.

Rachelefsky, G., Terasaki, P.I., Park, M.S., Katz, R., Siegel, S., and Saito, S. (1976). *Lancet ii*, 1042.

Reidenberg, M.M., and Martin, J.M. (1974). *Drug Metab. Dispos. 2*, 71.

Ross, R., and Glomset, J.A. (1973). *Science 180*, 1332.

Rounds, D.E., Booher, J., and Guerrero, R.R. (1976). *Atherosclerosis 25*, 183.

Rowley, J.D. (1974). *J. nat. Cancer Inst. 52*, 315.

Swift, M., Sholman, L., Perry, M., and Chase, C.C. (1976). *Cancer Res. 36*, 209.

Thomas, W.A., Florentin, R.A., Reiner, J.M., Lee, W.M., and Lee, K.T. (1976). *Exp. mol. Path. 24*, 244.

Tokuhata, G.K., and Lilienfeld, A.M. (1963). *J. nat. Cancer Inst. 30*, 289.

Trell, E., Korsgaard, R., Hood, B., Kitzing, P., Nordén, G., and Simonsson, B.G. (1976). *Lancet ii*, 140.

Wurster-Hill, D.H., Cornwell, G.G. III, and McIntyre, O.R. (1974). *Cancer 33*, 72.

POINT MUTATIONS IN MAN

A CONSIDERATION OF TWO BIOCHEMICAL APPROACHES TO MONITORING
HUMAN POPULATIONS FOR A CHANGE IN GERM CELL MUTATION RATES

James V. Neel

Department of Human Genetics
University of Michigan School of Medicine
Ann Arbor, Michigan
and
Radiation Effects Research Foundation
Hiroshima

Harvey Mohrenweiser

Department of Human Genetics
University of Michigan School of Medicine
Ann Arbor, Michigan

Chiyoka Satoh
Howard B. Hamilton

Department of Clinical Laboratories
Radiation Effects Research Foundation
Hiroshima

This presentation will assume that there is no need at
this meeting to justify an intense interest in the possibility
of monitoring human populations for mutation rates. Rather,
we will consider whether such monitoring is feasible at the
present time. The word "feasible" is a relative term. What
is not feasible with a small effort may often be feasible
with a larger effort. It is inevitable that under these
circumstances we are drawn into "cost-benefit" analysis.
Before we proceed to consider two protocols for monitoring
human populations, and become involved in discussion of "cost",
it might be well to say just a few words about "benefit".

Opinions among geneticists concerning the mutagenic effects of current and future exposures to a variety of environmental contaminants vary widely, but it is the more pessimis-
tic opinions that tend to be heard. Whereas the earlier concerns were primarily directed towards the intrusion of ionizing radiation into our lives, more recently in the U.S. the
emphasis has been on the possible impact of chemical mutagens
(Committee 17, 1975). It must be admitted that there is room
for considerable uncertainty concerning the nature of the problem, and this uncertainty has created a great deal of concern in the minds of informed lay-persons, concern which
understandably often finds expression in resistance to developments which virtually promise to increase future exposures
to potential mutagens. It is important at this juncture to
recognize that there is no such thing as a "free lunch" in
Nature--if we introduce mutagens at any level of exposure into
the environment, the result, as the matter is understood today, will be an increase in mutation rates. The question is
whether with industrialization we have increased mutation
rates by 1%--which most geneticists would find tolerable--or
by 100% --which would disturb most geneticists. Elswhere we
have presented calculations, under carefully stated assumptions, of the magnitude of the program required to detect a
50% increase in mutation rates (Neel, 1971; see below). It is
a very large program by current standards. To detect as little
as a 10% increase requires an almost inconceivably large
program. Now, there are two possible outcomes to a monitoring
program. Either, at some stated statistical level, we detect
no increase, or we do detect an increase. In the case of the
latter, the question of the appropriate governmental response
has yet to be addressed. In the case of the former, the alar-
mist can maintain that a program resulting in the exclusion
of an increase of 50% or greater really does not meet the
public need, which is to say, he might not be satisfied with
any program which it is feasible to mount at this time. Thus
at present the "benefit" from the "cost" of a monitoring program is not nearly so clearcut and non-controversial as in
many other undertakings. In the final analysis the necessary
judgements will be reached by the political rather than the
scientific process.

In what follows, we will briefly present two different approaches to the monitoring problem. These are both illustrated by studies now in progress, out of which will presumably
emerge the experience on which future, improved programs will
be based. The technology in this field is evolving very rapidly. An approach which may be appropriate for one country may,
for logistical and sociological reasons, not be appropriate

for another country. Thus, we present these two protocols as preliminary proposals, certain to undergo modification in the future, and with no thought that they should be taken as *the* models.

PROTOCOL ONE: CONTINUOUS MONITORING BASED ON PLACENTAL CORD BLOOD SAMPLES

Both of the protocols which we present stem from the conviction that recent technical developments enable geneticists to carry the study of mutation to the level of protein phenotypes. The first, the effort which we have recently initiated at the University of Michigan, is designed to provide baseline data on the mutation rate of the loci encoding for some 25 proteins present in erythrocytes or blood serum. It is not at present a monitoring effort *per se*, since it is on far too small a scale, but clearly the experience gained in this study could be of great value in designing a monitoring program aimed to detect changing mutation rates. Furthermore, in addition to contributing useful data on the design of a more comprehensive effort, the data to be accumulated on spontaneous rates will be basic to estimating how large a monitoring program is required to detect an increase at some specified level.

Sampling and Sample Processing

The present protocol calls for the collection of placental cord blood samples from all infants delivered at the University Hospital and venous blood samples from both their parents. The Obstetric Service of the University Hospital is relatively small, encompassing some 1500 deliveries annually, with a disproportionate representation of high-risk pregnancies, including unmarried juveniles. We limit our program to samples for which informed consent for the use of their blood for research purposes has been obtained from both the parents, and from the parents on behalf of their child. During the first 7 months of the program we have completed the trio of samples in 65% of deliveries. An analysis of the cause of failure is given in Table I.

*Table I. An analysis of the results of the attempt to
collect blood samples from placenta of new
born infant plus mother plus father during
the period July, 1976 – February, 1977 at
Women's Hospital, The University of Michigan*

Successes		*545*
		(65%)
Failures		
Refusal by one or both parents	*125*	
Father unavailable	*106*	
"System failure"	*86*	
Miscellaneous other reasons		
for failure	*6*	
Reason for failure unknown	*15*	
	———	
		338
		(35%)
	Total	*883*

Note the role played by "father unavailable," so characteristic of the young unmarried mother. Note, also, the refusal
rate of 15 percent, probably higher in an academic town like
Ann Arbor than in most American cities. We believe that under
more favorable circumstances, the collection rate could be
80 percent, a belief we propose to test in the near future
with the extension of the study to several other obstetric
services.

    Cells and plasma from complete trios are processed into
1 cc. aliquots for storage at $-80^{\circ}C$ and in liquid N_2. The
child's specimen stored at $-80^{\circ}C$ is analyzed first. In the
event of a variant being discovered when the sample is processed, the full battery of tests is applied to both the parental samples. With the observed average variant rate (exclusive of polymorphisms) of approximately 2/1000 tests and a
battery of 25 systems, about 1 in 20 children will be found
to have at least one variant. Thus in 19 out of 20 instances,
the collection and processing of parental blood will have been
unnecessary. Nevertheless, we feel this approach is less time
consuming than attempting to contact the parents one or two
months after delivery in the event of the child being found to

have a variant, and the approach certainly results in a higher
proportion of successful family studies than would result by
delaying the collection of parental blood samples until the
presence of a variant in a child was established. Furthermore,
with the rate at which additional systems suitable for this
type of study are becoming available, and given the labor involved in processing the samples, we believe we have in storage an important resource against future developments.

Laboratory Strategy

The backbone of the effort to detect variants of these
proteins is currently electrophoresis, utilizing starch gel
or acrylamide as the supporting medium. Recently, we have been
relying increasingly on polyacrylamide gels in view of the
many techniques which increase the resolving ability of electrophoresis using such gels (Latner, 1975; Leaback, 1974).
Several of these techniques, including the utilization of
acrylamide gels of differing pore size and electrophoresis
with several buffer systems, have been employed by Coyne
(1976), Johnson (1976), and Singh *et al.* (1976) to detect
variant protein species not previously demonstrable. In
Table II is a listing of the proteins which are currently
being studied, or for which methods of electrophoretic analysis are currently being developed in our laboratory.
Additionally we will soon be initiating for these purposes an
effort to utilize those enzymes of the leukocyte not present
in the erythrocyte. The erythrocyte and plasma proteins generally include those with which our laboratory has had extensive experience by virtue of its studies of Amerindians.

In order for a variant enzyme to be detectable with electrophoretic techniques, the structural alteration must result
in a net charge change and the protein must retain functional
properties. An unknown percentage of amino acid substitutions
will result in loss of function, and it is to be expected on
theoretical grounds that about 2/3 of the substitutions involve no net charge change. Thus most of the amino acid substitutions are not detected by electrophoresis. Also most
genetic deletions will result in loss of functional protein.
Accordingly, electrophoresis probably detects less than one
quarter of specific locus mutations.

The extensive lists of metabolic errors associated with
loss of enzyme activity which have been compiled by Grimes
and de Gruchy (1974), Raivio and Seegmiller (1972), and
Kirkman (1972) are indicative of the range and prevalence of
mutations resulting in loss of activity. These reports plus

the studies of the first degree relatives of persons with various deficiency states involving such enzymes as glucose phosphate isomerase (Van Biervliet *et al.*, 1975; Vives-Couons *et al.*, 1975), pyruvate kinase (Kahn *et al.*, 1977) and iduronidase (Dulaney *et al.*, 1976), indicate that the heterozygous individual is distinguishable from the normal population.

TABLE II.   *A partial list of proteins of the erythrocyte and blood serum which can now be satisfactorily screened for electrophoretic variants. Relatively little has thus far been done with the proteins of the leukocyte; this appears to be an important field for investigation. Proteins indicated by a (1) are not present in sufficient quantities in cord blood for reliable studies; those indicated by a (2) are currently included in the program for screening cord bloods described in the text.*

Blood serum

Albumin
α_1-Antitrypsin
Ceruloplasmin [2]
Haptoglobin   [1]
Pseudocholinesterase
Transferrins [2]
Group specific component [2]

Erythrocyte

Acid phosphatase [2]
Adenosine deaminase [2]
Adenylate kinase
Aldolase
Carbonic anhydrase I [1]
Carbonic anhydrase II [1]
Catalase
Diaphorase
Diphosphoglycerate mutase
Enolase
Esterase A [2]
Esterase D [2]
Galactose-1-phosphate uridyl transferase [2]
Glucose-6-phosphate dehydrogenase [2]

Erythrocyte (Continued)

Glutamate dehydrogenase
Glutamic pyruvate transaminase
Glutathione reductase
Glyceraldehyde-3-phosphate dehydrogenase
Glyoxalate [2]
Hemoglobin A [2]
Hemoglobin A$_2$ [2]
Hexokinase [2]
Isocitrate dehydrogenase [2]
Lactate dehydrogenase [2]
Malic dehydrogenase [2]
Nucleoside phosphorylase [2]
Peptidase A [2]
Peptidase B [2]
6-Phosphoglucose dehydrogenase [2]
Phosphoglucomutase 1 [2]
Phosphoglucomutase 2 [2]
Phosphohexose isomerase [2]
Pyruvate kinase
Triose phosphate isomerase [2]
Glutamic oxalacetic transaminase [2]

In none of the reports quoted above could the variant enzyme
be detected by electrophoresis in the heterozygous individual,
which is as expected if the protein did not retain enzymatic
activity. This type of genetic damage is often observed in
the offspring of irradiation-treated mice (Glucksohn-Waelsch
et al., 1974; Erickson *et al.*,1974; Russell *et al.*,1976;
Mays *et al.*, 1977), a finding paralleling the earlier interpretation on other grounds that most of the X-ray induced
mutations of the mouse are due to gene inactivations or small
deletions (Russell, 1971).

The erythrocyte enzymes we are currently studying for
genetically-controlled deficiencies in activity are listed in
Table III. These enzymes were chosen because they have levels
of activity which are readily detectable by conventional
methods, because the enzyme protein is almost exclusively the
product of a single locus, and because our preliminary estimates suggest a between-individual coefficient of variation of
less than 15%. If individuals with a "null" allele at any of
these loci have a level of enzyme activity of approximately
50% of normal, they should be quite distinct from the normal
population in enzyme activity. Genetic diseases associated
with enzyme deficiencies have been reported for each of these
enzymes (Grimes and de Gruchy, 1974). The frequencies of
these diseases are still very poorly defined, but even if
they are as rare as 10^{-6}, heterozygote frequencies should approximate 2×10^{-3}.

TABLE III.   *A list of erythrocyte enzymes which should be*
amenable to the search for mutants which when
heterozygous result in half-normal levels of
enzyme activity.

Adenylate kinase	*Glucose-6-phosphate dehydro-*
Pyruvate kinase	*genase*
Triose phosphate isomerase	*Glutathione peroxidase*
Phosphoglucokinase	*Lactate dehydrogenase*
Phosphoglucose isomerase	*Glutathione reductase*

A technique which is often capable of distinguishing between proteins with similar catalytic capabilities but small
structural differences is heat stability. Many examples of
electrophoretic alterations being associated with changes in
the heat stability of a protein have been noted (Kozak and
Jensen, 1974; Sieginbeek *et al.*, 1976; Van Biervliet *et al.*,
1975). Genetic variants which cannot be detected by electrophoretic analysis may be demonstrated by differences in heat
stability profiles (Thorig *et al.*, 1975; Singh *et al.*, 1974;
Singh *et al.*, 1975). Studies utilizing heat stability as the
detection method have an advantage over electrophoresis in
that alterations of heat stability do not require that the
mutational event results in a net charge change. Our initial
efforts in utilizing heat stability as a method for detecting
variant enzymes will involve those listed in Table III. Variants of many of these enzymes have been reported to exhibit
altered heat stability profiles.

The instrumentation to be utilized for the studies based
upon kinetic measurement of enzyme activity is the Miniature
Centrifugal Fast Analyzer developed by Oak Ridge National
Laboratory (Burtis *et al.*, 1973). Enzyme assays with this
instrumentation usually require hemolysate from less than 1 µl
of packed cells. This analyser has excellent operating characteristics and the coefficient of variation for within-samp-
le repeatability is less than 2% for most of the 9 erythrocyte enzymes currently being examined. An additional asset of
this instrumentation is the availability of programs for its
PDP8E computer, facilitating data acquistion and manipulation,
including the calculation of K_m, K_i and V_{max}.

One of the practical problems in a study of this type is
the occurrence in a particular system of a specific electrophoretic variant in relatively high frequency--say 3-5/1000
determinations. If all such variants are truly identical, then
the probability is very high that they all have descen-
ded from a common ancestor rather than that some of them arose
in the preceding generation from a mutational event. Detailed
family studies of each variant are not apt to be rewarding
with respect to mutation. If, on the other hand, this electrophoretic class shelters a variety of variants due to different amino acid substitutions resulting in the same change in
the net charge of the molecule, then the probability that any
one variant arose in the preceding generation from a mutation
is greater (but still small), and family studies should be
pursued vigorously. Recent studies on electrophoretic variants in *Drosophila* and *Colias* have revealed the feasibility
of demonstrating heterogeneity in electrophoretic classes by a
combination of electrophoretic techniques and the study of

physical and kinetic properties (cf. especially Bernstein *et al.*, 1973; Johnson, 1976; Singh *et al.*, 1976; McDowell and Prakash, 1976). It is our plan to devote considerable effort to the search for heterogeneity in electrophoretic classes; we expect the Miniature Fast Analyzer to be especially useful in this context.

Data Storage and Retrieval

In any program designed to detect a difference in mutation rate between two groups of - say 50%, the volume of the determinations to be processed will by our current standards be enormous (see section on Sample Size below). It is imperative that data storage and retrieval be made as efficient as possible. In our operation this challenge is under the supervision of Dr. Charles Sing and Mr. Ronald Griffith. The laboratories supporting this project interface through terminals with the Departmental DEC 11/70 computer. The on-line time-sharing capabilities of this machine allows multiple members of the lab team simultaneous access to the data base. Identifying information concerning each sample is entered into the computer shortly after sample collection. As laboratory determinations become available, they are fed directly from the lab books into the computer. All data initially go into "temporary storage". After verification of the results against the laboratory records, plus any further studies necessary to remove ambiguities in the findings, the data are transferred into "permanent storage". The programming permits one at any moment either to review all the information on a given individual or group of individuals as it exists at any point in time, or to retrieve all of the data on any particular system. We hope in the future to place the Miniature Fast Analyzer and other suitable equipment "on-line" with our current computer configuration. While much of this is standard, current computer strategy, the fact is that it has not often been applied to the management of laboratory data of this type. In our case, it not only eliminates the laborious hand transcription, collation, and coding of the data which has characterized past operations, but also provides a vastly more flexible retrieval of the data according to specific needs than was previously possible.

Some Considerations of Sample Size

A procotol of this sort if implemented on a sufficient scale and extended over a sufficient time span could detect secular changes in mutation rates. The degree of implementation necessary depends on the baseline rate of mutation of this type (still very poorly defined), the magnitude of the change one would not wish to see go undetected, and the time span on which one is satisfied to operate. No two statisticians will proceed in quite the same way with their calculations of the number of observations necessary to achieve the stated objectives. On the assumption that the initial detectable mutation rate was 0.5×10^{-5}/gene/generation, we have previously calculated that to demonstrate a 50% increase in mutation rate with the confidence implied by an α value of 0.05 and a β value of 0.20 would require two samples of approximately 300,000 persons each typed for some 20 different proteins (Neel, 1971). If one is content to operate on a 5-years cycle, this requires the analysis of 60,000 samples per year plus the necessary follow-up studies. Based on the standards of prior biomedical research, this requires a most unusual effort and organization. Clearly an undertaking of this magnitude requires the most serious consideration. But while it may be dismissed as a larger undertaking than the circumstances warrant, it is already clear that it cannot be discounted on technical grounds alone.

PROTOCOL TWO: DISCONTINUOUS MONITORING, INVOLVING PERIODIC
CONTRASTS OF HIGH-RISK AND LOW-RISK GROUPS

An alternative approach to continuous monitoring is, whenever a group suspect for an increased mutation rate emerges, to contrast the rates in such a group with the rates in a suitable control group. This constitutes a sort of 'worst case' analysis, from which one can presumably extrapolate to the more usual circumstance.

The most important opportunity for this approach to date has undoubtedly been the situation existing in the wake of the atomic bombings at Hiroshima and Nagasaki. Shortly after World War II an extensive study was undertaken of the children born to parents exposed to a greater or lesser extent to the radiation spectrum of the atomic bombs, as well as the children born to a suitable control group. This study employed

what we may term the morphological approach, involving such
potential indicators of an increased mutation rate in the parents as the frequency of congenital defects, birth weight,
infant and childhood death rates, physical growth and development, and the sex ratio. In 1954, after an extensive analysis
(Neel and Schull, 1956), the detailed examinations of infants
at birth and nine months later were discontinued but data
continued to be collected on two indices, namely, the survival
and the sex-ratio of the children born to the survivors of the
bombings and to the controls, and these continuing studies
have been the subject of several reports (Neel and Schull,
1956; Neel, 1963; Schull, Neel and Hashizuma, 1966; Neel,
Kato and Schull, 1974). A great deal of very laborious investigation can be summarized with the statement that although
clear-cut results of the parental exposure have not been established, there are very borderline findings with respect to
the sex-ratio and survival which are compatible with the
findings in experimental systems.

The study on survival has involved the establishment of
two cohorts of children, one born to what are termed 'proximally exposed' parents, whose mean conjoint exposure to whole
body radiation has been estimated at 115 *rem*, and one born to
'distally exposed' parents, whose mean conjoint exposure is
estimated at less than 1 *rem*. Each cohort is comprised of
some 17,000 children. Periodically (on a 5 years cycle) the
survival of each child in the cohort is determined.

This population is clearly one of great interest, with the
results of the studies not only of immediate pertinence to
those concerned, but to a variety of scientific and governmental groups. Surely it is the responsibility of the scientific community to ensure that the maximum useful information
can be extracted from this unfortunate experience, the discussion being reopened from time to time as the techniques
available to the study of the problem evolve. This thought led
to the initiation of cytogenetic studies on the two groups of
children in 1967 (Awa *et al.*, 1968, 1973; Awa, 1975). Although
the desirability of such studies had been recognized when the
first follow-up studies were initiated in 1946, the necessary
techniques were of course not at hand. More recently, under
the auspices of the Radiation Effects Research Foundation, an
organization jointly sponsored by the U. S. and Japanese governments, an effort has been launched to extend the study of
the potential genetic effects of the bombs to the biochemical
level. This effort has been preceded by a three-year pilot
study which has demonstrated that the rare protein variants on
which one would base the quest for mutational events, have
about the same frequency in Japanese as in Caucasian populations, a frequency whose magnitude makes a study of this type

feasible (Ferrell *et al.*, in press; Ueda *et al.*, in press; Satoh *et al.*, in press; Tanis *et al.*, in press; Neel *et al.*, in press). The new study should be regarded as additive to our previous studies in Japan, the ultimate objective being a synthesis of all these studies.

The proteins being investigated for evidence of mutational events and the techniques by which they are being and will be scrutinized are essentially similar to those included in the description of Protocol One. This is scarcely surprising in view of the close working relationship between the group at the University of Michigan and the group in Hiroshima and Nagasaki.

Although the battery of indicators of mutation is thus far quite similar in the two undertakings, the practical problems are quite different. Whereas in programs of the first type the potential study population is quite large, in programs of the second type, as in Japan, the study population will usually be quite limited. It is of the utmost importance that maximum participation be achieved, and that the data be collected in such a fashion that the results of various studies can be combined. In Japan there are of course some losses from the cohorts through death. A much more important source of loss is emigration from Hiroshima and Nagasaki. Japan now has a relatively high internal mobility, and the oldest children of survivors conceived since the bombings are now entering their 30's and so are well into the job market. It has not seemed feasible to pursue these children beyond the confines of the two cities (although we do attempt to contact them when they return on festival occasions!). Finally, these 'children' are now independent adults, not all of whom wish to participate in a study which reminds them of a very traumatic event in their family history.

A further set of practical difficulties arises when a variant is encountered, which, with a battery of 25 protein indicators and the variant frequencies in Japan (2/1000 determinations, exclusive of polymorphisms), is the case for approximately 1 in each 20 subjects. We are finding that for almost one-quarter of persons with a variant, one or both parents are deceased, which of course in any strict sense invalidates the use of the family in a mutational context. Furthermore, if both parents are alive, they may not wish to participate in such a study.

It is still too early for reliable figures concerning sample attrition, but it is clear it will be relatively heavy. While the situation in Japan perhaps presents a rather extreme example of sample attrition, a similar problem will arise whenever and wherever an attempt is made to contact the children born to a cohort of particular interest. Under these circumstances, consideration must be given to maximizing the information to be gained from each sample through an increase in the number of systems studied, but this puts added stresses on the laboratory (to be discussed later). Furthermore, most of these populations of unusual interest will fortunately be relatively small. It will be highly desirable to be able to pool the results of different studies, a goal that immediately raises questions of standardization.

Some Considerations of the Power of the Study

Since (it is to be hoped) the cohort of children born to proximally exposed survivors of the atomic bombings will constitute the largest identifiable sample of humans whose parents have been exposed to a well defined mutagen in the world's history, we may inquire as to what might be demonstrated with such a sample (R.E.R.F. RP 4-75). We have made three assumptions: 1) as before, the detectable spontaneous mutation rate is 0.5×10^{-5}/locus/generation, 2) the average conjoint parental dose is 115 *rem* of atomic radiation, and 3) it will be possible to study 10,000 children (and their parents when necessary) from each cohort. Under these conditions, if 115 *rem* is the doubling dose for mutation of the type this study will detect, there are 12 chances in 100 of obtaining a between-group difference using an $\alpha=0.05$ test. This probability increases to 35 percent if the doubling dose is 57 *rem* and to 59 percent if it is 38 *rem*. It is clear that on these assumptions, even this extensive study may well fail to yield a significant result. The probability of significance is of course increased if the spontaneous rate is higher and decreased if it is lower.

STUDIES ON SOMATIC CELL MUTATION RATES

It is clear that studies on germinal mutation rates in man will need to be laborious and extensive if they are to serve a monitoring function. Studies on *in vivo* somatic cell mutation rates in control and exposed subjects offer an appealing short-cut, since now for any enzyme system, each person offers, as in a blood sample, the possibility of a million or more locus tests, rather than the number defined by his or her reproductive performance. It is to be hoped that efforts to develop such tests will proceed vigorously. Our failure in

this presentation to devote more time to this development -
which we are pursuing on a small scale - stems solely from
limitations of time. However, at some stage in developing
our monitoring strategy, before we can rely on guidance by an
in vivo somatic cell approach, there must be one study that
simultaneously measures, on a large scale, somatic and germinal mutation rates, so that we can establish the necessary
conversion factor. A mutation in a germ cell must pass a
transmission test which mutation in a somatic cell need not.
Once the implications of somatic cell mutation rates for future generations have been established, then monitoring through
observations on somatic cell rates can be expected largely to
replace germ cell monitoring.

SOME DECISIONS WHICH A MONITORING PROGRAM BASED ON BIOCHEMICAL
VARIANTS MUST FACE

We have presented in bare outline two different programs
designed to serve as prototypes for a particular approach to
monitoring. It is obvious that a variety of technical and
philosophical problems remain to be solved. Some of these
have already been alluded to. Let us briefly consider some
of the issues more commonly raised in discussions of monitoring programs (see also Neel, 1971, 1974, in press).

The Problem of Discrepancies between Stated and True Parentage

Any study of mutation of this type must deal with discrepancies between stated and actual parentage. Such discrepancies may arise not only from what is usually termed 'non-pa-
ternity' - a poor expression - but also, in a program using
cord blood samples, from mislabelling of samples in the delivery room, and, in a study of an identified cohort, from the
presence of adopted children whose status is not revealed by
the parents. We propose to investigate the accuracy of parental assignment in all instances of apparant mutation, through
the use of stored samples. Let us assume that it will be
possible to bring to bear the standard battery of genetic
tests on the detection of what we will call "parentage discrepancies" and that this battery can detect 90% of such discrepancies. Then if "parentage discrepancies" characterize 5% of
the children under study, and if for any system the frequency
of the electrophoretic class of which the putative mutant is
a member is 10^{-3}, then the probability that the presence of a

variant in a child of normal parents may be attributed to an undetected "parentage discrepancy" is 10^{-1} x 5 x 10^{-2} x 10^{-3} = 0.5 x 10^{-5}. If this is also the frequency of spontaneous mutation, then half of all the apparent mutations in the control sample would be the result of these discrepancies. While a 'dilution' of the mutation rate of this magnitude is disappointing, allowance can be made for it in any analysis. Incidentally, it seems generally to have been overlooked that the same problem arises in the study of mutation resulting in "sentinel phenotypes" (*i.e.*, dominantly-inherited syndromes), in which it has *not* been customary to conduct parentage checks.

If mutation rates are more like 10^{-6}/locus/generation, then there is a real danger that a purported study of mutation is in fact one of undetected errors in the assignment of parentage and confusion in the labelling of samples. With respect to the latter, one would wish to reexamine all the samples from the parents of children born on the same day as an apparent mutant, to check on labelling error. With respect to the former, in addition to striving to increase the power of the exclusion tests, we believe we can also begin to develop probability functions for the chance that the alleged father will have a particular genetic constitution - given the genetic findings in the mother and child. This is a type of Vaterschaftsbestimmung. The extreme case is represented by the situation in which the child and father both possess a particular rare variant. This is of course not proof of paternity but it certainly is circumstantial evidence.

Despite strict observance of the doctrine of informed consent in such studies, it is clear these studies may result in very sensitive unanticipated by-products. Clearly the precautions to preserve anonymity of participation must be extreme.

Fewer Observation on more People or more Observations on Fewer People?

A dilemma already emerging in these studies is what constitutes the optimum number of observations to undertake on each participant in the study. Given the expenditure of effort in obtaining a specimen and preparing it for analysis, it is obviously desirable to extract as much information as possible from that sample. On the other hand, every additional system which must be maintained at the requisite level of technical excellence imposes an added burden on the laboratory, and, depending on the size of the operation, there is surely a trade-off which maximizes return on effort. Where,

however, sample size is clearly limited, and the sample is of
exceptional interest, as in the previously described situation
in Japan, there seems no alternative to a vigorous effort to
increase the number of observations per sample.

Detection vs Characterization

There is inherent in monitoring studies a genuine scientific conflict of interest. The monitoring function must place
maximum emphasis on the efficient detection of variants in
subjects and their parents. The precise characterization of
the variants, and their comparison with one another, and with
the variants present in other populations - so dear to the
heart of the biochemical and population geneticist - is really
not essential to monitoring. On the other hand, herein lies
much of the scientific interest. As a practical matter - to
maintain the interest and enthusiasm of competent scientists
in the program - some compromise must be struck between these
two functions. This compromise will not be easily reached.

Assessing the impact of mutants of this type

Concern over a possible increase in mutation rates stems
from the belief that such an increase would be detrimental to
human health. But to reach any cost-benefit analysis, one must
be able to quantify the issue. In this connection, it must be
admitted that whereas the biochemical approach appears to be
the most precise available, genetically speaking, the health
significance of this kind of mutation is far less clear than
that of the dominant syndromes whose frequency in the population is maintained by mutation: for example, multiple neurofibromatosis, retinoblastoma, aniridia, Marfan's syndrome. It
is an article of faith that in time we will understand the
biomedical significance of these enzyme mutants much better
than at present. As indicated earlier, already some of them,
accompanied by severely decreased levels of enzyme activity,
have been implicated in human disease. Furthermore, it is a
reasonable working hypothesis that a doubling in the mutation
rate leading to enzymatic variants would be accompanied by a
doubling in the rate leading to more serious entities.

Cost of Monitoring Programs of this Type

Sooner or later in discussions of this type we come to questions of cost. Both of the programs we have described are in the developmental stage, so that unit costs are highly speculative. However, it is clear that a program which seeks to follow up a high-risk group will, relatively speaking, be far more expensive than a program dealing with consecutive births. This is because of the extra effort in locating the members of the study population. But it will also be more informative per person studied. We hope that within another year we will have initial cost estimates for the cord blood program. However, in considering these costs, which will seem high to some, we must consider the costs of *not* monitoring. Rather than be faced with uncertainty concerning the genetic impact of industrilization and its attendant pollution, surely it is worth a considerable expenditure to have the facts at our disposal when the need arises to make critical decisions concerning the directions of our society.

Organization of a Collaborative Study

Finally, let us point out the obvious, that a monitoring program of this type eminently lends itself to an international collaborative effort. The problem is clearly so large that it should not be entrusted to any one laboratory or any one country. Perhaps in the informal discussions which will characterize this Conference we can begin to explore how such a collaborative effort could be organized.

REFERENCES

Awa, A.A. (1975). *J. Rad. Res.*, suppl. to Vol. 16, 78.
Awa, A.A., Bloom, A.D., Yoshida, M.C., Neriishi, S. and
    Archer, P. (1968). *Nature 218*, 367.
Awa, A.A., Neriishi, S., Honda, T., Satuni, T. and Hamilton,
    H.B. (1973). *In* "Excerpta Medica International Congress
    Series No. 297", p. 92. 4th International Conference on
    Birth Defects, Vienna, Austria, Sept. 2-8.
Bernstein, S.C., Throckmorton, L.H. and Hubby, J.L. (1973).
    Proc. Natl. Acad. Sci. (USA) 70, 3928.
Burtis, C.A., Johnson, W.F., Mailen, J.C., Overton, J.B., Tif-
    fany, T.O. and Watsky, M.B. (1973). *Clin. Chem. 19*, 895.
Committee 17. (1975). *Science 187*, 503.

Coyne, J.A. (1976). *Genetics 84*, 593.

Dulaney, J.T., Milunsky, A. and Moser, H.W. (1976). *Clin. Chim. Acta 99*, 305.

Erickson, R.P., Eicher, E.M. and Gluecksohn-Waelsch, S. (1974). *Nature 248*, 416.

Ferrell, R.E., Ueda, N., Satoh, C., Tanis, R.J., Neel, J.V., Hamilton, H.B., Inamizu, T. and Baba, K. *Ann. Hum. Genet.*, in press.

Gluecksohn-Waelsch, S., Schiffman, M.B., Thorndike, J. and Cori, C.F. (1974). *Proc. Natl. Acad. Sci. (USA) 71*, 825.

Grimes, A.J. and de Gruchy, G.C. (1974). *In* "Blood and Its Disorders" (R.M. Hardesty and D.J. Weatherall, eds.). pp. 473-525. Blackwell, London.

Johnson, G.B. (1976). *Genetics 83*, 149.

Kahn, A., Vives-Corron, J.L., Marie, J., Galand, C. and Boivin, P. (1977). *Clin. Chim. Acta 75*, 71.

Kirkman, H.N. (1972). *Progr. Med. Genet. 8*, 125.

Kozak, L.P. and Jensen, J.T. (1974). *J. Biol. Chem. 249*, 7775.

Latner, A.L. (1975). *Adv. Clin. Chem. 17*, 193.

Leaback, D.H. (1974). *Chem. in Britain 10*, 376.

Mays, J., McAninch, J., Feuers, R., Burkhart, J., Mohrenweiser, H. and Casciano, D. (1977). *In* "Proceedings of the 8th Environmental Mutagen Society Meeting", Colorado Springs, Co.

McDowell, R.E. and Prakash, S. (1976). *Proc. Natl. Acad. Sci. (USA) 73*, 4150.

Neel, J.V. (1973). "Changing Perspectives on the Genetic Effects of Radiation", pp. viii and 97, C.C. Thomas, Springfield.

Neel, J.V. (1971). *Perspectives in Biol. 14*, 522.

Neel, J.V. (1974). *Mutation Res. 26*, 319.

Neel, J.V. and Schull, W.J. (1956). "The Effect of Exposure to the Atomic Bombs on Pregnancy Termination in Hiroshima and Nagasaki.", pp. xvi and 241, *Nat. Acad. Sci.* National Research Publication 461, pp. xvi and 241.

Neel, J.V., Kato, H., and Schull, W.J. (1974). *Genetics 76*, 311.

Neel, J.V. *In* "Report of Study Group on Environmental Monitoring: Analytical Studies for the U.S. Environmental Protection Agency." Appendix A. National Research Council - National Academy of Sciences Report.

Neel, J.V., Ueda, N., Satoh, C., Ferrell, R.E., Tanis, R.J., and Hamilton, H.B. *Ann. Hum. Genet.*, in press.

Radiation Effects Research Foundation (1976). Research Plan for RERF Studies of the Potential Genetic Effects of Atomic Radiation. Research Protocol 4-75, p. 30.

Raivio, K.O., and Seegmiller, J.E. (1972). *Ann. Rev. Biochem.*
41, 543.
Russell, L.B. (1971). *Mutation Res. 11*, 107.
Russell, L.B., Russel, W.L., Popp, R.A., Vaughan, C. and
Jacobson, K.B. (1976). *Proc. Natl. Acad. Sci. (USA) 73*,
2843.
Satoh, C., Ferrell, R.E., Tanis, R.J., Ueda, N., Kishimoto,
S., Neel, J.V., Hamilton, H.B. and Baba, K. *Ann. Hum.
Genet.*, in press.
Schull, W.J., Neel, J.V. and Hashizume, A. (1966). *Amer. J.
Hum. Genet. 18*, 328.
Siegenbeek, von Heukelom, L.H., Boom, A., Bartstra, H.A. and
Staal, G.E.J. (1976). *Clin. Chim. Acta 72*, 109.
Singh, R., Hubby, J.L. and Lewontin, R.C. (1974). *Proc. Natl.
Acad. Sci. (USA) 71*, 1808.
Singh, R.S., Hubby, J.L. and Throckmorton, L.H. (1975). *Genet-
ics 80*, 637.
Singh, R.S., Lewontin, R.C. and Felton, A.A. (1976). *Genetics
84*, 609.
Tanis, R.J., Ueda, N., Satoh, C., Ferrell, R.E., Kishimoto,
S., Neel, J.V., Hamilton, H.B. and Ohno, N. *Ann. Hum.
Genet.* in press.
Thorig, G.E.W., Schoone, A.A. and Scharloo, W. (1975).
Biochem. Genet. 13, 721.
Ueda, N., Satoh, C., Tanis, R.J., Ferrell, R.E., Kishimoto,
S., Neel, J.V., Hamilton, H.B. and Baba, K. *Ann. Hum.
Genet.* in press.
Van Biervliet, J.P.G.M., van Milligen-Boersma, L. and Staal,
G.E.J. (1975). *Clin. Chim. Acta 65*, 157.
Van Biervliet, J.P.G.M., Xlug, A., Bartstra, H.A., Rotteveel,
J.J., deVaan, G.A.M. and Staal, G.E.J. (1975). *Hum.
Genet. 30*, 35.
Vives-Corrons, J.L., Roxman, C., Kahn, A., Carrera, A. and
Triginer, J. (1975). *Hum. Genet. 29*, 291.

POSSIBILITIES FOR DEMONSTRATING POINT MUTATIONS
IN SOMATIC CELLS, AS ILLUSTRATED BY STUDIES OF
MUTANT HEMOGLOBINS

George Stamatoyannopoulos[1]

Division of Medical Genetics
Department of Medicine
University of Washington
Seattle, Washington

INTRODUCTION

The studies and concepts discussed in this paper are
based on two premises: (1) that there is a demonstrated need
to monitor human populations for spontaneous or induced mutations, and (2) that monitoring for gametic mutations is
complicated by the monumental difficulties of surveillance of
large population samples. Somatic cell methods are among the
approaches proposed to meet this need, (Drake *et al.*, 1975)
and *in vitro* approaches are already in use. This paper dis-
cusses the possibility of using somatic cells for detection
of mutations occurring *in vivo*. In contrast to the methods of
utilizing tissue culture, this approach is based on the
screening of large cohorts of somatic cells for detection of
those that carry specific point mutations.

Spontaneous mutations are expected to occur in somatic
cells and mutagens are expected to exert their effect in proliferating cells of every tissue. In self-renewing tissues,
cells with genetic damage that is not associated with selective disadvantage will continue to proliferate, establishing
mutant clones. When mutation occurs in differentiating cells

[1]*Supported by contract NO1-ES-4-
2151 from the National Institute of Environmental Health
Sciences*

the size of the clones is expected to be small and the effect of the mutation temporary. When pluripotent stem cells of a tissue are affected by mutation, mutant clones that may last throughout life will be established. Their presence will be reflected in the population of terminally differentiated cells of the tissue, with the appearance of cells that are heterozygous for the phenotype caused by the mutation. Since the terminally differentiated cells of certain renewing tissues (like blood, sperm, epithelium) are easy to obtain, availability of methods to detect specific mutations in such cells could permit measurements of mutation frequencies and calculations of mutation rates at specific loci.

Availability of an efficient, sensitive, specific and fully validated method for detection of somatic mutations *in vivo* could offer several possibilities for mutation research. In contrast to the gametic methods, study of only a few individuals would be required to test the mutagenic effects of environmental contaminants. Population monitoring for the effects of known mutagens would require longitudinal studies of only a few persons. Since the effects of mutagens will be studied in man, current problems of relevance of screening for mutagenicity to human health will not arise. In addition, such questions as age- or sex-related differences in metabolism or the possibility of polymorphic variation in biotransformation of mutagens could be investigated with population and family studies. Furthermore, single cell systems for the detection of mutations might provide information of direct relevance to gametic mutations if screening of spermatozoa for specific mutations were utilized. This attractive picture, however, is presently dimmed by methodological problems and theoretical restrictions which are discussed in the subsequent sections of this report.

Previous Attempts To Demonstrate Somatic Mutations in Vivo

Although somatic mutations have been implicated in such processes as malignant transformation, atherosclerosis, aging, and the development of antibody diversity, only a few attempts have been made to detect them *in vivo*. Early attempts were made by Atwood (1958) and Atwood and Scheinberg (1958; 1959). They observed that in individuals who were genetically of the AB or A blood group type, a proportion of erythrocytes (usually around 1 in 10^3) failed to demonstrate the agglutinogen A. These cells were originally considered to represent somatic mutants and, under this assumption, a somatic mutation rate of 7×10^{-6} per cell division was calculated (Atwood and Scheinberg, 1958). The mutational origin of the abnormal red cells was subsequently tested with measurement of the fre-

quencies of non-A or non-B erythrocytes in AB heterozygous and
in B homozygous individuals. If they were mutants, frequencies
in B homozygotes were expected to equal the square of the frequency in heterozygotes. Although the homozygotes had fewer
abnormal cells than did the AB heterozygous individuals, the
frequency of these cells was 50 to 200 times higher than expected under the somatic mutation hypothesis (Atwood and
Pepper, 1961). Loss of A or B agglutinogen in a rare red cell
thus appears to be an epigenetic event.

Other attempts were made by Sutton (1972; 1974). He
utilized the cellular phenotype of the hemoglobin mutation
hereditary persistence of fetal hemoglobin (HPFH), which is
associated with continuation of synthesis of large amounts
of fetal hemoglobin (10 to 30 per cent of the intracellular
hemoglobin) in all the circulating red cells of the adult HPFH
gene carriers (reviewed by Weatherall *et al.*, 1974). This
abnormality is usually due to β and δ globin gene deletions.
Rare red cells containing Hb F are also found in persons who
carry normal hemoglobin genes and these were postulated to
represent somatic HPFH mutants (Sutton, 1974). More recent
work, however, renders this possibility highly unlikely. Every
normal adult has 0.5 to 5.0 per cent Hb F-containing cells in
the blood (Boyer *et al.*, 1975; Wood *et al.*, 1975) and these
frequencies are raised in various physiological states,
anemias and hemopoietic malignancies. Recent work with hemopoietic cells in culture (Papayannopoulou *et al.*, 1976a;
1977a) suggests that the Hb F-containing cells in adult bloods
are differentiation-related phenocopies of the HPFH mutation
rather than real mutants.

In another attempt (Sutton, 1972) the target cell was the
polymorphonuclear white cell while the marker was a kinetic
abnormality of certain G6PD deficiency mutants, namely the
ability to utilize efficiently the analogue substrates
galactose-6-phosphate and 2-deoxy-glucose-6-phosphate. Rare
white cells (about 1 in 10^3) with a high utilization of
2-deoxy-glucose-6-P were detected using cytochemical methods
and the possibility of their mutational origin was raised. It
is, however, difficult to accept this interpretation since
several factors, including variation in metabolic or phago-
cytic activity of the polymorphonuclear leukocytes, might be
responsible for the appearance of this rare cytochemical
finding.

Markers of in Vivo Somatic Mutations

Screening of somatic cells to detect those with somatic
mutations does not permit proof of the mutational origin of
cells with abnormal biochemical phenotypes by transmission

criteria. Even when cells with proliferative potentials (e.g., lymphocytes) are to be used, cloning of the abnormal cells may prove impossible since the experimental manipulations necessary for detection of the abnormality will most probably destroy the proliferative capacity of the cells. Since the transmission criterion cannot be used for exclusion of phenocopies, cellular phenotypes that are likely to be produced by epigenetic events cannot be used for screening for mutations of somatic cells *in vivo*.

This restriction excludes from use in single cell methods all the systems that detect protein deficiencies. It is known that the activity of several enzymes depends on cell age and other factors and that phenotypes interpretable as heterozygosity for deficiency can be produced epigenetically. Examples are the decreased G6PD activity in old red cells, the development of glutathione reductase deficiency when the dietary supply of riboflavin is reduced (Beutler, 1972), the biosynthetic findings suggestive of heterozygous thalassemia in acquired conditions (Bradley and Ranney, 1973), etc. It is reasonable, therefore, to assume that microenvironmental influences of various types can lead to appearance of phenotypes that mimic heterozygous protein deficiency in a small fraction of terminally differentiated cells; with the single-cell screening methods, these cells will be recognized as "somatic mutants".

It appears, thus, that the only methods that can be usefully applied to the detection of mutants by the screening of cohorts of cells are ones that depict the appearance, in a cell, of a protein with mutant primary structure and in the quantity expected for a genetically heterozygous cell. This is the rationale underlying the use of variant human hemoglobins in our attempt to detect mutations in somatic cells. The handicap of the approach is obvious: cells containing structural mutations are expected to be much less frequent than the deficiency mutants since the latter can be produced by several types of mutational events (gene deletions, frameshifts, initiation and termination mutants, mutants affecting mRNA stability and mRNA functional mutants), while the former are mostly due to single base replacements.

Possibilities Offered by Variant Hemoglobins

Among the several biochemical genetic systems in man, the hemoglobins offer the best possibilities for use in a single-cell detection system. Ninety-five per cent of the protein in erythrocytes is hemoglobin and in the case of most globin gene structural mutations, 20 to 50 per cent of the protein is abnormal. Over 250 variant hemoglobins have been structurally characterized and can be utilized for mutation research. Most show single aminoacid substitutions, though double substitutions, aminoacid deletions, aminoacid insertions and variants with elongated and shortened polypeptide chains have been observed (reviewed by Stamatoyannopoulos and Nute, 1974). Many of these mutants must exist, in very low frequencies, among the circulating red cells of every genetically normal person. The question is whether or not these mutations can be identified at the cellular level and if this can be done, whether or not rare mutant red cells can be detected in blood samples from normal persons.

Ideally, single-cell electrophoretic methods should be used for this purpose. Although single red-cell electrophoresis has been done previously (Rosenberg, 1970) and automated methods could conceivably be devised for efficient electrophoretic screening of millions of cells, the currently available methodology is impractical for mutation screening. As an alternative, we chose to test the possibility of immunochemical screening of red cells by developing antibodies against abnormal hemoglobins, using the fluorescent conjugates of these antibodies for detection of the mutant proteins in individual erythrocytes.

The common hemoglobins S, C and E (because of their availability) and four other variants, all due to single aminoacid substitutions representing transitions or transversions in the α or β globin genes are used in the study. In addition, an effort is being made to produce antibodies that recognize the abnormal sequences of three hemoglobins with elongated chains, Hb Constant Spring (Clegg *et al.*, 1971) Hb Wayne (Seid-Akhavan *et al.*, 1976) and Hb Cranston (Bunn *et al.*, 1975). These latter hemoglobins were incorporated into the study because of the interesting nature of the molecular events that gave rise to the abnormal structures.

Hb Constant Spring is due to a base substitution affecting the α-chain termination codon; as a result, the α-globin chain is translated beyond its normal length, having 172 instead of the normal 141 aminoacids. We reason that if specific antibodies to this normally untranslated part of the α-globin mRNA become available, all mutations at the duplicate α globin loci due to (missense) base substitutions at the termination codon or to deletions including the termination codon (provided they are not molecularly lethal and do not lead to frameshifts) would result in production of whole or part of the Constant Spring carboxylterminal sequence and, hence, would be detectable at the cellular level by the antibodies recognizing that sequence.

The molecular lesions underlying Hb Wayne is a single base

deletion, while in Hb Cranston insertion of two bases, both
near the 3' end of the α(Wayne) or β(Cranston) globin gene;
these variants represent two of the three examples of frameshift mutations in man. In Hb Wayne, a novel octopeptide is
added at the α-chain carboxylterminal end. In Hb Cranston the
mutation adds a novel carboxylterminal decatripeptide. We
reason that antibodies specific for these extra sequences of
the frameshift mutants will permit detection, at the cellular
level, of every frameshift mutation (that does not lead to
cellular lethality or premature chain termination) caused by
one or 1 + 3n base deletions or by two or 2 + 3n base insertions at the normally translated part of the α- or β-globin
genes.

Detection of Mutant Hemoglobins in Single Erythrocytes

Hemoglobins are poor antigens and antibody titers in the
sera of immunized animals are low. Since, after a vigorous
purification process, recovery of specific anti-mutant Hb
antibodies is very small, horses are used for immunization and
large amounts of sera (3,000 to 4,000 ml) are purified using
affinity chromatography (Papayannopoulou *et al.*, 1976b; 1977b).
Conjugation of the purified antibodies with fluorochromes
(usually fluorescein isothyocyanate) and reaction of the
antibody-FITC with fixed red cells in blood smears permits
the unequivocal identification of red cells containing the
mutant hemoglobin.

Figures 1 and 2 illustrate the use of anti-mutant Hb antibodies for detection of abnormal hemoglobins in single cells.
Cells from a person heterozygous for Hb S and Hb A are
strongly labeled by the fluorescent antibody (Fig. 1) while
normal red cells fail to bind the reagent. Similar results
have been obtained with other antibodies capable of recognizing abnormal hemoglobins that differ from normal by a
single aminoacid substitution.

Can the immunofluorescent identification of abnormal
hemoglobins in single cells be used for detection of *rare*
erythrocytes with a mutant phenotype? This question was
answered in experiments using anti-Hb S-FITC labeling of
artificial mixtures containing normal erythrocytes and AS
red cells, the latter present in frequencies ranging from
1×10^{-2} to 2×10^{-6}. Smears prepared from artificial mixtures
were labeled with anti-Hb S-FITC and the total numbers of
cells as well as numbers of fluorescing cells were counted.
The findings of an experiment are illustrated in figure 3.
There is an excellent correlation between observed frequencies
of cells binding the fluorescent anti Hb S and the expected

frequency of AS cells as determined from the dilution factors; small deviations can readily be attributed to the expected variation in preparation of dilutions and in cell counting.

These observations suggest that detection of rare mutant cells is possible with screening of red cell populations in fixed blood smears labeled with fluorescent anti-mutant Hb antibodies. The findings also indicate that the mutant AS cells in normal blood cannot be more frequent than 1 in 10^5 (otherwise the observed values in Fig. 3 would not correlate with the expected frequencies) and that cellular phenocopies of the $Hb\beta^A \rightarrow Hb\beta^S$ mutation, as depicted by the fluorescent antibody method, must also be rarer than 1 in 10^5 red cell.

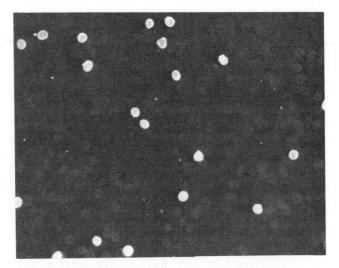

FIGURE 1. *Microphotograph illustrating the identification of abnormal hemoglobins in single erythrocytes using specific anti-mutant hemoglobin antibodies. The preparation is an artificial mixture of red cells from a person heterozygous for Hb S and an individual with normal hemoglobin genotype (ratio of AA to AS cells is 9:1). The blood smear has been labeled with anti-Hb S antibody conjugated to fluorescein isothyocyanate (FITC). Note that only about 10 per cent of the cells (corresponding to the proportion of the AS cells in the artificial mixture) bind the fluorescent antibody.*

Studies in Vivo and Interpretations

A highly purified anti-Hb S antibody (apparently recognizing only the structural alteration introduced by the replacement of glutamic acid by valine in position β6) was
utilized in these experiments. The fluorescein isothyocyanate-
conjugated antibody was used to search for red cells binding
the anti Hb S-FITC in blood samples from normal individuals.
Smears prepared from freshly drawn blood samples of adult
(20 to 35 years old) male and female volunteers were coded,
fixed, labeled with anti-Hb S-FITC and screened for fluorescing cells. Control smears containing 1 in 10^2 AS cells
were used for testing the labeling efficiency while coded
controls with 1 in 10^5 AS cells were used to test screening
efficiency. The coded preparations were screened under a
fluorescent microscope using a method of overlapping fields;
filters eliminated background fluorescence and permitted
visualization of only the strongly fluorescing cells. The
erythrocytic nature of cells detected by this approach was
subsequently verified by cytochemical stains.

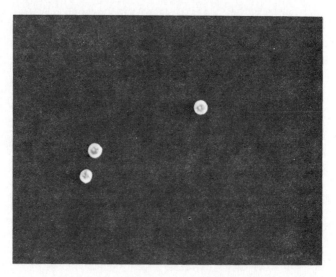

FIGURE 2.   *This preparation, an artificial mixture of*
cells from a normal individual and a person
heterozygous for Hb C (ratio AA:AC cells,
1:20), has been labeled with anti-Hb C-FITC.
Only about 5% of cells, i.e. the ones con-
taining Hb C, react with the fluorescent
antibody.

Blood smears of 15 individuals have been tested with this procedure and red cells binding the anti-Hb S-FITC in a manner similar to that of AS cells were observed at frequencies of 3×10^{-7} to 4×10^{-8} (average frequency of 1.1×10^{-7}). If these cells are examples of somatic β^A to β^S mutation, the total frequency of red cells containing *any one* of the expected abnormal hemoglobins due to base substitutions at the β-hemoglobin locus will approximate 1 in 10^4. If these cells are the progeny of mutant stem cells, a mutation rate per base replication of 3×10^{-10} (see next section for the rationale underlying this calculation) might apply to the β-hemoglobin locus of the human pluripotent hemopoietic stem cells. The question, however, is whether these cells showing

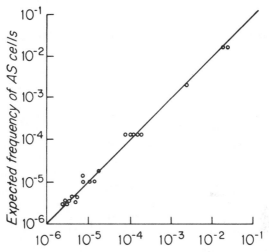

FIGURE 3. *Correlation between the observed frequency of cells binding the anti-Hb S-FITC and the expected proportion of AS cells in studies of artificial mixtures of AS and AA red cells. The total number of cells counted for each comparison was at least 20 times higher than the dilution factor (e.g. 2×10^6 cells were counted in artificial mixtures containing 1×10^{-5} cells).*

an AS immunofluorescent phenotype are *mutants* or epigenetic alterations of experimentally produced phenocopies of the $\beta^A \to \beta^S$ mutation, i.e. *artifacts*.

Although the antibodies used in the study are highly purified and specific, several possibilities for artifactual appearance of cells binding the anti-Hb S-FITC can be proposed. Single-cell screening for mutants stretches cellular identification procedures beyond their levels of reliability; any conceivable epigenetic cellular alteration, including phenomena that we cannot foresee, can occur with a frequency of 10^{-7}. The rare variant cellular phenotypes must therefore be considered as artifacts unless proof of their mutational origin can be provided. This view applies not only to the specific question of the nature of the immunochemically recognized "S cells", but it extends to any other approach to be devised for monitoring of mutation using single-cell surveillance, with the possible exception of single-cell electrophoretic screening for structural mutations. How can one test the mutational or artifactual origin of these rare variant cells?

An answer was first sought with an indirect approach, i.e. determination of the frequencies of cells reacting with anti-Hb S-FITC in situations where such cells were expected to be elevated, i.e. in persons exposed to mutagens. It was soon realized, however, that this approach was unrealistic because it is practically impossible to formulate predictions regarding frequencies of induced mutations in hemopoietic stem cells. Other problems relate to the availability of persons (other than those with hemopoietic malignancies) who have been exposed to high concentrations of mutagens that are likely to produce base substitutions. We are attempting, therefore, to find an answer through the following experiment.

The anti-Hb S and the anti-Hb C antibodies are purified to monospecificity in order to be used, together with anti-Hb E antibodies, for the immunofluorescent identification of "S cells", "C cells" and "E cells" in the blood. They will be used to determine the frequency of "S cells" and "E cells" in persons who are homozygous for the Hb C mutation (Hb C/Hb C patients) and the frequency of "C cells" and "E cells" in individuals homozygous for Hb S (Hb S/Hb S patients). "E cell-", "C cell-" and "S cell-" frequencies will be determined in control patients with hemolytic anemia of non-hemoglobinopathic etiology, matched to the Hb S/Hb S and Hb C/Hb C patients for age, sex and rate of erythropoiesis. If "E cells" but not "S cells" are found in the bloods of Hb C/Hb C patients and if "C cells" are not found in the bloods of Hb S/Hb S patients, we shall consider that these cellular phenotypes are produced by somatic mutations. If

"S cells" are found in the Hb C/Hb C patients, and if there are "C cells" in the Hb S/Hb S patients, we shall conclude that most probably these cellular phenotypes are artifacts.

The reasons for drawing the above conclusions are based on the mutational events responsible for production of the three abnormal hemoglobins. Hb E (the built-in control of the experiment) has a glu→lys substitution at β22, while Hb S has a glu→val, and Hb C a glu→lys substitution, both of which occur at β6. In the Hb S and Hb C homozygotes, β22 glu→lys can be derived by a single (G→A) base replacement in the codon corresponding to β22 (glutamic GAX, lysine AAX). In these patients, however, mutations from Hb S to Hb C (i.e. val→lys) or the reverse (i.e. lys→val) would require two (GU $\rightleftarrows$ AA) base substitutions at the codon for the structural site β6 (valine GUX, lysine AAX). If the frequency of "E cells" in these Hb S/Hb S or Hb C/Hb C persons with stimulated erythropoiesis is at the range of 10^{-6} to 10^{-7} the frequency of "C cells" or "S cells" must be as low as 10^{-12}, i.e. much below the level detectable by the fluorescent cell approach.

Somatic Methods for Monitoring Mutation in Vivo: How Useful?

So far I have discussed the rationale for monitoring somatic mutation using single-cell screening and the problems of verification of the nature of rare abnormal phenotypes, once they are observed. I should like now to consider another problem inherent in these approaches, i.e. the possibility that the *frequencies* of the spontaneously occurring mutations among differentiated cells are so high that they render detection of induced mutations, in most situations, an unrealistic goal. To clarify matters I shall refer briefly to certain aspects of hemopoietic cell proliferation (reviewed by Lajtha, 1968).

The erythrocytes in the circulation derive from erythroblasts (3 to 4 divisions) that are produced in turn through a process of differentiation and amplification of committed erythroid stem cells (5 to 10 divisions); the progenitors of these cells are the pluripotent stem cells of the hemopoietic system. The pluripotent stem cell pool is capable of self-renewal while there is no (or very limited) self-renewal within the committed stem cell compartment. The size of the pluripotent stem cell pool in man can be calculated indirectly. Models of hemopoiesis (supported by experiments in the mouse) suggest that a proportion of pluripotent stem cells (about 1/10 to 1/20 of the total pool) leave the pool daily and enter the committed stem cell compartments of the formed elements of blood. The size of the pluripotent stem cell pool

is kept constant by proliferation of an equivalent fraction
of cells of the pluripotent compartment. Thus, although the
majority of pluripotent stem cells are, at any given time,
out of cycle, the average frequency of division of each stem
cell is about once per ten to twenty days. Generation time of
the pluripotent stem cells is, in general, shorter in the
fetus than in the adult.

If this model of hemopoiesis is correct, the red cells of
a 30-year old adult must derive from a pluripotent stem cell
pool the cells of which have already undergone an average of
1,000 divisions. If the occurrence of spontaneous mutations
at a given locus relates to the number of divisions a proliferating cell undergoes, the frequency of mutant red cells
in the circulation of the 30-year old adult must be at least
10^3-fold higher than the spontaneous mutation rate per cell
generation per locus. The implication is that methods determining mutation *frequencies* in somatic cells will be insensitive in recording induced mutagenesis *in vivo* unless the
putative mutagens cause huge increases in mutation rates or
the exposure to mutagenic contaminants lasts for very long
periods of time (several years).

It seems, thus, rather unlikely that several of the
attractive prospects of the *in vivo* somatic cell monitoring
systems outlined in the introduction will be realized, even
if genetically sound and fully validated systems become
available. On the other hand, such *validated* systems might
prove useful in comparisons between human populations, and in
situations of very prolonged industrial exposures to strong
mutagenic contaminants.

CONCLUDING REMARKS

Studies with variant hemoglobins show that it is possible
to detect variant hemoglobins in individual erythrocytes and
that immunochemical methods can be used to identify AS cells
in low frequencies under controlled experimental conditions
(such as artificial mixtures of AS and normal red blood cells).
Rare erythrocytes binding the fluorescent anti-Hb S antibodies have been found in the bloods of genetically normal
individuals but it is not yet proved that they are mutants
rather than artifacts. We take the position that rare "mutant"
cell phenotypes should be considered as artifacts until acceptable *genetic* evidence of their mutational origin is obtained.

If the soundness and validity of single-cell methods is
proved genetically, methodological developments such as
automated detection systems and use of multiple genetic

markers will be required for practical application in mutation monitoring. Studies of special human populations (e.g. survivors of atomic bombings) or of animal models will be required in order to determine possible correlations of frequencies of somatic and gametic mutations.

The number of spontaneous somatic mutations accumulated *in vivo* is expected to exceed the number of new mutants produced by the average exposures to mutagenic contaminants, by several orders of magnitude. It is, therefore, predicted that the *in vivo* measurement of mutation frequencies in populations of terminally differentiated cells will be a rather insensitive index of induced mutagenesis.

REFERENCES

Atwood, K.C. (1958). *Proc. Nat. Acad. Sci. USA. 44*, 1054.
Atwood, K.C., and Scheinberg, S.L. (1958). *J. Cell. Comp. Phys. 52*, (1) 97.
Atwood, K.C., Scheinberg, S.L. (1959). *Science. 129*, 963.
Atwood, K.C., and Petter, F.J. (1961). *Science. 134*, 2100.
Beutler, E.E. (1972). *Biochimie. 54*, 759.
Boyer, S.H., Belding, T.K., Margolet, L., Noyes, A.N., Burke, P.J., and Bell. W.R. (1975). *The Johns Hopkins Med. J. 137*, 105.
Bradley, T.B., and Ranney, H.M. (1973). *Progr. in Hematol. 8*, 77.
Bunn, H.F., Schmidt, G.J., Haney, D.N., and Dluhy, R.G. (1975). *Proc. Nat. Acad. Sci. USA. 72*, 3609.
Clegg, J.B., Weatherall, D.J. and Milner, P.F. (1971). *Nature. 234*, 337.
Drake, W.T., *et al.* (1975). *Science. 187*, 503.
Lajtha, L.G. (1968). *In Vitro. 4*, 14.
Papayannopoulou, Th., Brice, M., and Stamatoyannopoulos, G. (1976a). *Proc. Nat. Acad. Sci. USA. 73*, 2033.
Papayannopoulou, Th., McGuire, T.C., Lim, G., Garzel, E., Nute, P.E., and Stamatoyannopoulos, G. (1976b). *Br. J. Haematol. 34*, 25.
Papayannopoulou, Th., Brice, M., and Stamatoyannopoulos, G. (1977a). *Proc. Nat. Acad. Sci. USA.* (in press).
Papayannopoulou, Th., Lim, G., McGuire, T.C., Ahern, V., Nute, P.E., and Stamatoyannopoulos, G. (1977b). *Am. J. Hematol.* (in press).
Rosenberg, M. (1970). *Proc. Nat. Acad. Sci. USA. 67*, 32.

Seid-Akhaven, M., Winter, W.P., Abramson, R.K., and
    Rucknagel, D.L. (1976). *Proc. Nat. Acad. Sci. USA. 73*,
    882.
Stamatoyannopoulos, G., and Nute, P.E. (1974). *Clinics in
    Haematol. 3*, 251.
Sutton, H.E. (1972). *In* "Mutagenic Effects of Environmental
    Contaminants" (Sutton, H.E., and M.I. Harris, eds.),
    p. 121. Academic Press, New York.
Sutton, H.E. (1974). *In* "Birth Defects. Proceedings of the
    Fourth International Conference" (Motulsky, A.G., and
    W. Lenz, eds.), p. 212.
Weatherall, D.J., Pembrey, M.E., and Pritchard, J. (1974).
    Clinics in Haematol. 3, 467.
Wood, W.G., Stamatoyannopoulos, G., Lim, G., and Nute, P.E.
    (1975). *Blood. 46*, 671.

CHROMOSOME DAMAGE

CLINICAL IMPLICATION OF
CHROMOSOME BREAKAGE

James German

Laboratory of Human Genetics
The New York Blood Center
New York, N.Y.

I have been asked to discuss the clinical implications of
chromosome breakage. To do this, I shall first attempt to answer the questions: What does one perceive in the phenotype and
how is man's health and well-being affected when structural rearrangement occurs in chromosomes of various cell types at
different stages of development? To answer these questions, it
seems natural to divide the effects that are to be considered
into those on the germ line of cells and those on somatic
cells. To facilitate this discussion I have made a drawing representing our life cycle (Fig. 1) that will help focus attention on these two classes of cells at various times during development and growth when chromosome mutation in them can be of
importance clinically.

At the phenotypic level, the effects of mutation in various cells will fall roughly into two classes, (a) those that
result, on the one hand, in abnormal embryonic development,
mental retardation, and subfertility and (b) those associated
with cancer. Because of the large number of published works
concerning karyotype-phenotype correlations in human cytogenetics and because the literature on chromosome change in cancer
is extensive, I see little point in reviewing those matters
here. However, I shall punctuate my talk with examples from our
own laboratory's files which show some clinical consequences
of chromosome breakage. Suffice it to say that chromosome imbalance is capable of disturbing cellular function, with clinical consequences ranging from mild mental retardation and
anatomical malformation to profound retardation and anomaly in-
compatable with intrauterine life, and from mild subfertility
to the complete absence of germ cells from the gonads. Less
well understood is the role of chromosome mutation in the con-

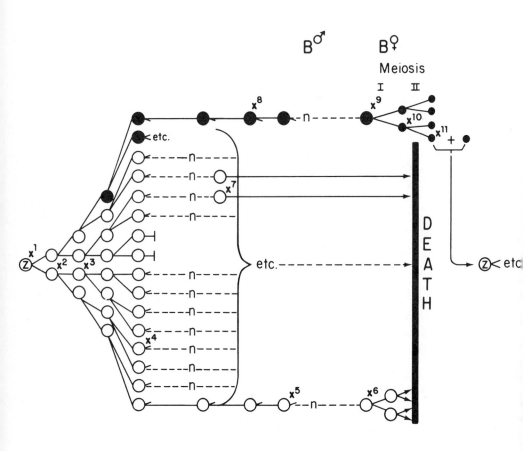

Soma:
O— *Programmed cell death*   — *Germ line*
O— *Living, ,non-dividing cells*   Ⓩ— *Zygote*
O-- *Proliferating cells*   B *Birth*

FIGURE 1. *Representation of man's life cycle, karyogamy
to karyogamy. The Xs show some times when damage to germ line
of somatic cells might be clinically significant.*

version of a normal cell to cancer, and of the evolution of a cancer once established; however, much evidence supports the idea that chromosome mutation does play a role there also.

My objective is to put my assigned subject into perspective and to ask what we have learned during the past two decades about human chromosome mutation with respect to the type of cells affected in the human body, and the times they are affected. And, broadly, what are the implications? What do we know about the cause of clinically important chromosome mutations relative to either developmental defects or cancer?

Figure 1 represents the human life cycle, from karyogamy (the coming together of the male and female pronuclei in the fertilized egg) to karyogamy. The germ line, presumed to be segregated early in cleavage, is represented as black cells, whereas the soma is represented by the remaining cells, the clear ones. In division, a cell, although it itself exists no more, avoids death, but ultimately the soma ends in death of the organism. In contrast the germ line, by giving rise to gametes which can participate in fertilization, can achieve immortality (unless the species under consideration should become extinct). The drawing emphasizes the adage - a hen is just the egg's way of making another egg. I have placed numbered Xs near various cells in the drawing and will discuss some of the theoretical consequences of chromosome mutation at those points. For most of the points, we now have clinical observations that mutation can occur.

At this time, I must mention that chromosome breakage can have several consequences, including death of the cell. (The breakage shown in Fig. 2 probably would be lethal.) Cellular death at the right time certainly is important in normal organogenesis, but extensive unscheduled death induced by some exogenous agent also could be important, possibly as a cause of abnormal embryonic development. Death of a cell is of no importance genetically, but changes short of death can be. These possibly important changes could produce deletion or duplication of various segments of chromatin and be transmissable from cell to cell thereafter. The mutation might take various forms such as translocation or inversion. Although the changes I am thinking of now are those that would be visible to the cytogeneticist, it must be assumed that many affect regions that are too small to be seen even with banding techniques and perhaps extend all the way down to frame-shift mutations.

The Time, the Cell, and the Consequences

Referring now to Fig. 1, and to breakage in different cells at different times, I will duscuss the germ line first and then the soma.

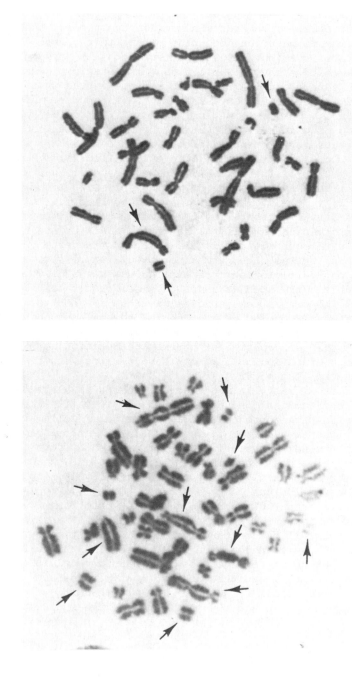

FIGURE 2. *Chromosome breakage and rearrangement (arrows) in PHA-stimulated lymphocyte metaphases which had received γ-radiation in* G_0.

Germ Line. The times during embryonic, fetal, and post-natal life when significant damage can occur in proliferating germ line cells are somewhat different in the male and the female. In the male, stem cells engage in ordinary mitotic division cycles throughout prenatal and postnatal life. At any time, a mutation might be introduced and the affected cell could expand into a clone of cells in the gonad. The sector of gonad occupied by the mutant clone would be larger the earlier in life the mutation occurred (X^8 in Fig. 1). For the female, the same is true about clones, but the mutation would have to be introduced before birth because stem cells do not proliferate significantly thereafter; from birth on, each cell that is a potential gamete has entered meiosis and is suspended in prophase, where it remains until stimulated during a monthly cycle to mature into a fertilizable egg. Spermatogonia and oogonia that carry a mutant chromosome complement might or might not produce viable gametes; the types of rearrangement in these gametes vary, and this variation depends on the meiotic segregation pattern. Thus, both chromosome and point mutation during embryonic life in either the male or female theoretically could result in a mutant clone of stem cells. Multiple gametes carrying the same mutation would emerge from these cells so that phenotypically nonmutant parents might have more than one offspring with the same mutant phenotype. For the male only, mutations in postnatal life can affect more than a single germ cell and its meiotic progeny in the same way.

The time during which cells actually in meiosis can be damaged also differs in the male and female (X^9, X^{10}, X^{11}). In the male, active sperm production occurs from puberty on, so that cells in some stage of the meiotic division, including spermatids and mature spermatozoa ready to engage in fertilization, are subject to the effects of a mutagen. In the female, the idea has often appealed to geneticists that a population of cells biding their time in meiotic prophase from birth up to their respective months of ovulation constitutes a flock of sitting ducks, ready targets for mutagens that may come along during the 50 years they might wait. Interestingly enough, however, gene mutation correlates better with increased paternal rather than maternal age (Stern, 1973). No maternal age effect has been recognized for *de novo* structural chromosome rearrangements. (Many, however, have suggested that the striking maternal age effect in the birth of trisomic babies may be the consequence of the prolonged period the oocyte spends in meiotic prophase, but this is only conjectural. Chromosome nondisjunction is not the subject of my paper anyway). For both male and female, mutation presumably can affect a primary gametocyte (X^9), a secondary gametocyte (X^{10}), or one of the haploid sperms or eggs (X^{11}) during the period immediately before it participates in fertilization.

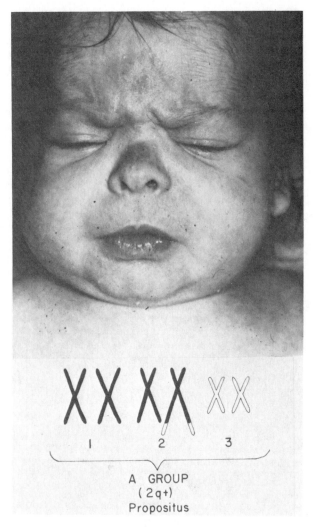

FIGURE 3a. Malformed child (IV. 3 in Figure 3b) with chromosome imbalance as result of fertilization by a genetically unbalanced gamete. The child's father (III.7) and grandfather (II.5) were transmitting in balanced form a translocation between chromosome Nos. 2 and 3. The child's chromosome complement contains only one of the aberrant chromosomes, thus this imbalance.

Soma. When we consider damage to the genetic material of somatic cells, our first concerns are (a) unfavourable effects on embryonic development and (b) development of cancer. A possible third concern is (c) aging, because the important theories of aging invoke mutation; however, if chromosome breakage is of

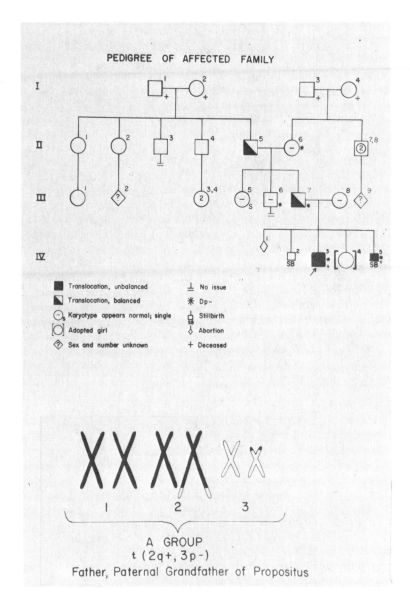

FIGURE 3b. *Pedigree of child in Figure 3a.*

importance here (at x^7, for example), it remains to be discovered. I have already implied that damage to the genetic material resulting in cellular death, *i.e.* that which is non-heritable, is unimportant with respect to cancer, but it could be
important with respect to embryonic malformations and aging.

Many teratogenic agents are known to be mutagenic and oncogenic. (That is not the case for all of them by any means; in
contrast, agents that are mutagenic usually are also clastogenic and oncogenic.) Doubtless, exogenous agents can interfere with embryonic development in several ways other than by
directly damaging genetic material, but some probably do act
that way. An exogenous teratogen could destroy some cells or
affect them significantly in some way relative to organogenesis, and at the same time induce mutations in other cells, the
mutations being unimportant in the coincident teratogenesis.
Such double and coincident action of a teratogen might be the
explanation for the cancer proneness Dr. Robert Miller (Miller,
1968) has shown for a number of malformation entities.

Now, to the first concern, embryonic maldevelopment: The
zygote itself may be genetically unbalanced because a rearranged chromosome was introduced by one of the fertilizing
gametes (Fig. 3). In this case, the cellular constitution of
the entire embryo presumably would have the same genetic imbalance. (Some postulate that mosaicism might ensue when an
aberrant chromosome which enters the zygote proves "unstable"
and is lost from some embryonic cells by non-disjunction;
however, not much evidence for this has been forthcoming.) On
the other hand, the zygote might begin with a normal chromosome makeup and itself be affected by a chromosone breaking
event (X^1 in Fig. 1), and, depending on the nature of the rearrangement, the cellular composition of the embryo could become either uniform or mosaic (Figs. 4(a) and 5). The rearrangement following the breakage could result in cells with two
different mutant genomes. However, if one was inviable or at a
proliferative disadvantage, evidence of only one mutant genome
might be found later by the cytogeneticist examining an aborted embryo or malformed baby. Rearrangements at X^1 (Fig. 1)
resulting in a mosaicism, one cellular component of which has
a normal and the other a mutated genome, can also be postulated. Similar events could occur in cells during the early cleavage phases of embryonic development (X^2 and X^3), and in this
case one could expect coexistence in the mosaicism of the
mutant and normal cells (Fig. 6), perhaps mosaicisms even of
three cell types (Fig. 4(b)). Some chromosome mutations that
occur in the preimplantation embryo, for example at X^2, could
not only be associated with mosaicism but could also be heritable if the cell affected happens to be one that will give rise
to somatic as well as germ line cells. Also, it is interesting
to consider the possibility that at this time in prenatal life
mutations may have occurred in a parent of a child with a
structurally rearranged chromosome imbalance in which child
the mutation seemingly would appear *"de novo"*. Many times
mosaicism had not been sought diligently in the parents of
such children; also, mosaicism can never be ruled out.

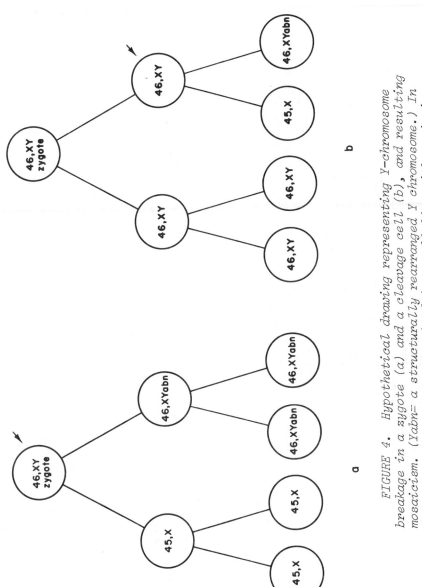

FIGURE 4. Hypothetical drawing representing Y-chromosome breakage in a zygote (a) and a cleavage cell (b), and resulting mosaicism. (Yabn= a structurally rearranged Y chromosome.) In (a), the rearrangement resulted in a cell line with a missing Y (45,X) and a cell line with an abnormal Y (46,XYabn). In (b), the mosaicism includes a third, normal, component (46,XY).

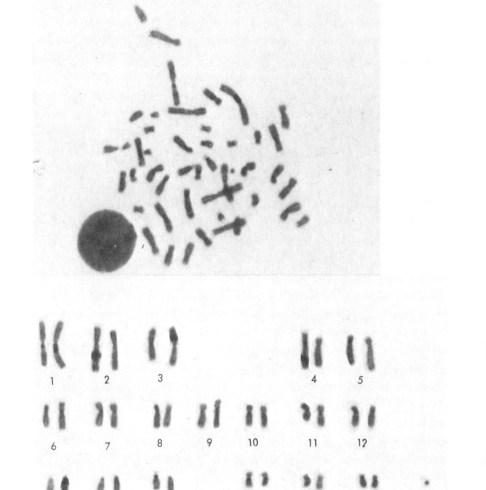

FIGURE 5a. *Mosaicism in a girl with Turner's syndrome, as represented in Fig 4(a). First cell: 46,X+ring, i.e. with a normal X and an abnormal (ring) derivative of an X or a Y.*

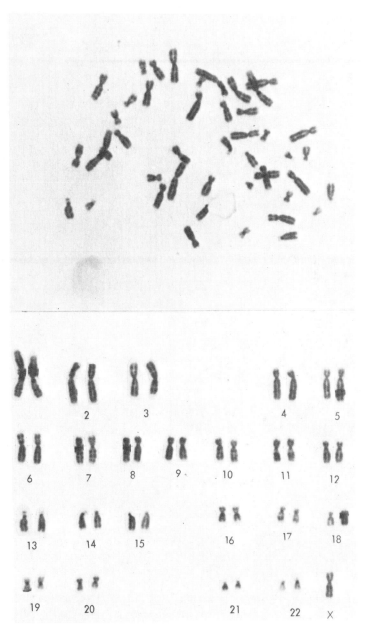

FIGURE 5b. Second cell: 45,X.

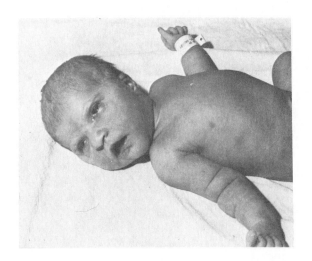

46,XX/46,XX,18q−

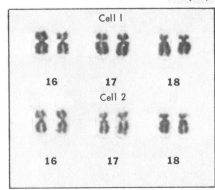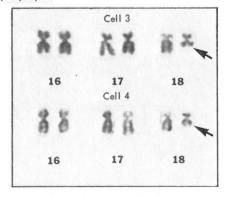

FIGURE 6. Mosaicism affecting an autosome: 46,XX/46,XX,18q−.
The zygote probably was normal (46,XX), the mutation (18q−)
and the mosaicism being post-fertilization events.

The mutational events to which I have just been referring
would occur in the preimplantation embryo. I would like to
emphasize that this is a period of human life that, in my
opinion, has been unduly neglected by experimental cytogeneticists and teratologists. It is a period when the rapidly
dividing embryo is suspended in maternal fluids of the
Fallopian tube or uterine cavity. This period should be a

particularly vulnerable stage of embryonic life, and 10-15% of
the wastage of human conceptuses has been estimated to occur
at this time (Hertig, 1967). Mosaicism in humans is reported
commonly, certainly more often in relation to sex chromosomal
rearrangement but also in relation to autosomal, normal cells
proliferating alongside cells with a structural rearrangement
(Fig. 6). Mosaicisms in which a normal cellular component is
present can best be explained by mutational events during preimplantation embryonic life. (This comment constitutes a plea
for the cytogenetic study of the preimplantation embryo. For
my more extensive presentation of this idea, see German, 1970.)

Chromosome mutation in a cell later in embryonic life
(X^4, X^5) has not been shown to be associated with malformations. Human teratogens apparently cause malformations mainly
through other mechanisms. However, cancer, especially that
present at birth or becoming clinically apparent during childhood, is the matter for concern when we consider mutations
that occur at this point of intrauterine life. About half of
acute leukemias and the majority of solid malignant tumors
display a mutated chromosome complement, with structural rearrangements, extra or missing intact chromosomes, or both
(*e.g.*, Fig. 7). Theoretically, the mutation could occur any
time during life, pre- or postnatal (X^6). Clones of mutant
cells that have a relatively poor or only slightly advanta-
geous growth potential might remain occult for years, perhaps
later undergoing secondary and tertiary changes and coinciden-
tally acquiring greater proliferative advantage. The studies
of Dr. Alice Stewart (Stewart *et al.*, 1958) correlating intrauterine X-irradiation with childhood leukemia pertain here.
Mutations in the chromosome complement can be induced by ionizing radiation in cells, the progeny of which are not malignant, at least not in the usual sense. Such mutant clones have
also been found repeatedly in individuals with one of the
"chromosome breakage syndromes" (German, 1972). It is in line
with present thinking to consider that at least a proportion
of such mutant clones already possess a low grade of proliferative advantage and are potentially capable of evolution to a
frank malignancy through additional changes, mutational or
otherwise; *i.e.*, are premalignant. I have summarized my views
earlier on the significance of such mutant clones in cancer-
prone populations (German, 1972), and evidence that points to
chromosome mutation whether environmentally or genetically
induced being of importance in the etiology of cancer.

Implications with Respect to Etiology

During recent years much concern has been expressed about
environmental situations associated with chromosome breakage,

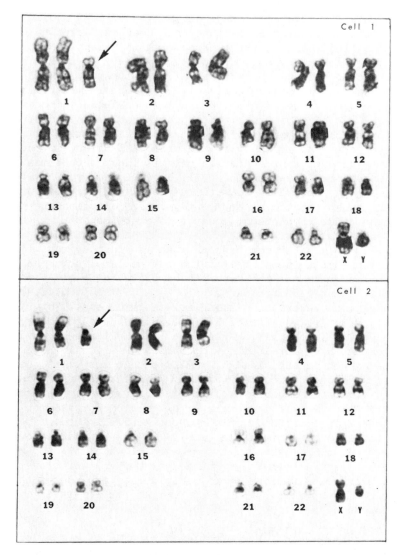

FIGURE 7. *Two cells from a mutant clone which was con-stituting about half the bone marrow of an anemic middle-aged man. Note the extra (47th) chromosome, derived through chromo-some rearrangement from a No. 1 and possibly some other chro-mosome. The man had an undiagnosed condition which was con-sidered "preluekemic". (Unpublished study of R.S.K. Chaganti, Memorial Hospital, New York.)*

both radiation and chemical, yet no epidemiological data exist, as far as I know, associating any of the environmental chromo-some breaking situations with the conception or birth of increased numbers of humans with chromosome imbalance, the result of chromosome breakage and rearrangement. This does not, of course, mean that there is no association. Situations potentially informative either have not shown increases or signs of increases (*e.g.*, the irradiated Japanese survivors of the atomic blasts who subsequently conceived have shown neither an increased number of malformed offspring nor an increased abortion rate) or else simply have not been studied (*e.g.*, the abortuses of anesthetists or of vinyl chloride workers have not been examined cytogenetically as far as I know). The clinical follow-ups made of known mutagen-exposed populations are, in this respect, nothing to brag about. Experimentally, X-irradiation of the rat testis has been used to produce chromosome breakage and increase the emergence of heritable chromosome rearrangements. Some of these rearrangements, even though transmitted balanced, are capable of giving rise to genetically unbalanced gametes and to abnormal embryos. So, animal studies show clearly the potential significance for man of chromosome mutating events. Little doubt exists that chromosome breakage may have clinical significance, certainly in relation to developmental malformations and probably in relation to cancer. But what proportion of it is environmentally induced? Correlations between exposure to a mutagen and emergence of cancer do exist, but even there the etiological significance of chromosome breakage remains uncertain. The agent could simultaneously cause chromosome mutation and - but by a different mechanism - cancer. Studies showing cancer proneness in the genetically-determined chromosome breakage syndromes, perhaps more than those of cancer proneness after exposure to certain environmental agents, point to chromosome mutation as a key event in cancer etiology.

Designing studies and surveys aimed at obtaining evidence that environmental agents actually are responsible for chromosome mutations that are genetically and clinically significant in relation to developmental defects would be difficult, time-consuming, and expensive. These studies would have to be long-term, because of such considerations as the following: Mutation in the germ line would have to occur in a female while she was herself *in utero* if a sizable clone of germ cells carrying the same rearrangement is to exist. Because few pregnancies occur in humans, clustering of conceptuses with different rearrangements in one woman would not be expected because the vast majority of rearrangements would be lethal for the embryo very early in development. Because spontaneous abortion is common even under what are considered regular circumstances of present-day life, the effects of an environmental agent that causes chromosome mutation in the follicular eggs of many women in a population would be detected only by a careful surveillance program. Central registries like the

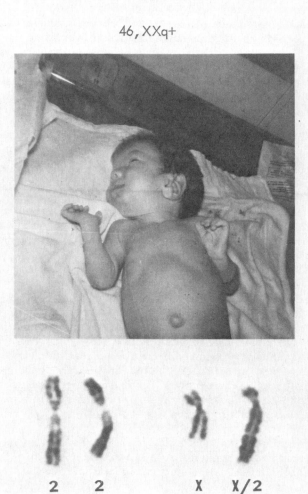

FIGURE 8. Malformed baby with partial trisomy for the long arm of chromosome No. 2 and partial monosomy of the end of the long arm of an X. This X/autosome (X/2) translocation had come about through chromosome breakage (See Fig. 9).

one maintained in New York State might, in due time, pick up an increase in liveborns with *de novo* structural rearrangements, but only many months to years after the time some exogenous mutagen might have been active.

Clinical case histories sometimes disclose events of possible etiological significance in association with the conception of a child with a *de novo* chromosome mutation. But, just

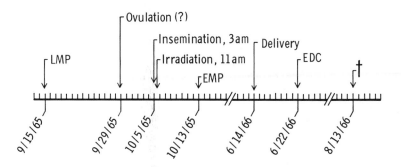

FIGURE 9. *Periconceptual events in the case of the child shown in Fig. 8. LMP, last menstrual period; EMP, expected mentrual period; EDC, expected date of confinement,* † *, death of baby.*

how useful have such clues been? (Efforts to obtain such histories are always in order, of course.) Two examples taken from our laboratory's files show how difficult it is to derive really helpful information in this way. In the first case, a malformed baby with an X/autosome translocation producing partial trisomy of the distal long arm of chromosome No. 2 (Fig. 8) was born to a woman in her 40s who had, on the very morning following the only sexual intercourse that could possibly have led to the fertilization, held her firstborn child (a boy with Wilm's tumor) while several diagnostic X-ray films were taken of him (Fig. 9). This periconceptual history of X-irradiation in this case is provocative, but what can we do with such a history? In the second example, a malformed child with a chromosome rearrangement was under study and while taking the family history, I learned that the father had for many months worked in an office located immediately above an industrial area in which, as he said, there was "radiation, consisting of high reactive neutrons". This bit of historical information caused considerable interest in our laboratory, to be quelled only when we showed that the mother was the carrier of a balanced translocation and that the unbalanced genome of her child was not the result of exogenous radiation but of an ordinary segregational event during meiosis (Fig. 10).

Considering further the difficulty of learning the clinical significance of chromosome mutation in relation to developmental defects, particularly that environmentally induced, I must emphasize at once that the proportion of wasted human conceptuses due just to chromosome rearrangements actually is not large, only a few percent (Boué *et al.*, 1975). (Chromosome imbalance of all types, especially triploidy, trisomy, and mono-

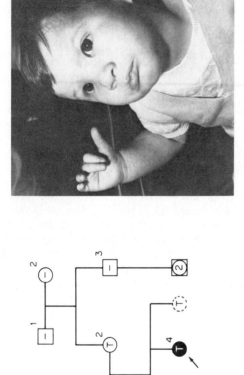

FIGURE 10. "De novo" balanced translocation (5p−; 21q+) in the phenotypically normal mother of a malformed baby with the cri du chat syndrome. The child's chromosome complement was unbalanced because he inherited the 5p− chromosome but not the 21q+ (German et al., 1976).

somy, of course probably is the major cause of wastage of
fertilized human eggs (Witschi, 1970; Boué *et al.*, 1975)). The
other thing often overlooked is that in fact a huge wastage of
fertilized human eggs takes place in ordinary families. A diagram prepared by the late Emil Witschi, mainly from Hertig's
data (Hertig, 1967), shows this very well (Fig. 11). He estimated that 16% of eggs coming in contact with sperm simply
fail to cleave. Of those cleaving, 15% will fail to survive the
preimplantation state, and 27% will fail during the ensuing
few weeks. Of the fertilized eggs that begin the course,
Witsci deduced that only about a third develop into liveborn
babies, and it is generally known that 1-2% of those liveborn
will have a malformation or defect of some kind. Our current
questions are: What proportion of the embryonic wastage and of
defective liveborns is attributable to environmentally induced
chromosome breakage, and what environmental agents will increase this? It will certainly be difficult to search for a
correlation between the intake of some low-grade chromosome
mutagen (for example caffeine) used by a large segment of most
populations and the births of babies with *de novo* chromosome
mutations, themselves rare events. Even when some unusual environmental event affects a population that can be submitted
to careful study, the massive background of embryonic wastage
already existing in man makes the task formidable.

I repeat, I know of no evidence that exposure of a population to a chromosome breaking agent increases the conceptions
of embryos, or the birth of babies, with structural alteration
of the chromosomes.

DISCUSSION

That this conference was called indicates the existence
today of a widespread commitment of many to become stewards
of the human genetic material. My negative review of the
damage known to have been produced by some environmental
agent and significant in relation to either sporadic or
inherited structural change in human chromosomes of importance in abnormal human development would seem to fail to
justify the widespread concern that exists. However, when
consideration is given to the great lag periods which might
be involved in the long-lived human between a mutation and
its visible effect, and the possibility that some agents of
importance may be quite ubiquitous, it seems unwise, for the
present, either to abandon the problem or to dismiss our
concerns. It seems unreasonable knowingly to expose human
genetic material unnecessarily to any mutating agent,
whether that material is of germ line or soma. Experiments
in other species support this attitude. The argument that
some might advance, that increased mutation rate is desir-

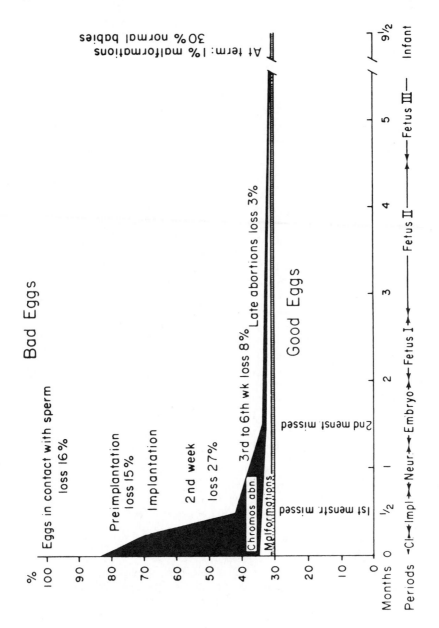

FIGURE 11. Drawing prepared by Emil Witschi (1969) to
show that much wastage of fertilized human eggs occurs. The
graph represents postovulatory survival and development in man.
On the horizontal axis are given ages and developmental
periods. (Cl.= cleavage; Impl.= implantation; Neur.= neurula.)
The black area represents the proportion of embryos with chro-

mosome abnormality. "Under assumed favourable conditions 30%
of eggs develop to normal babies ("good eggs"); 1% become
live-born infants with cognitive defects (hatched), of which
about 1 in 25 with chromosome anomalies; 69% perish, resorbed
or aborted; this class, summarily designated as "Malforma-
tions", may be chromosomally normal (white) or anomalous
(black)". (Quoted from Witschi's legend to the figure.)

able in order to hurry evolution along, rarely appeals to
the physician engaged in clinical cytogenetics and essentially never to the parents of a child with mental retardation
and multiple congenital anomalies, who would more readily agree
that most mutations are detrimental. At the moment, it is impossible to estimate the proportion of all chromosome rearrangements, those being transmitted in balanced form plus those
occurring sporadically – both types are usually brought to
light first by the birth of an abnormal child – which is environmentally induced, much less whether the proportion is increasing. We know just about as much about what could be called
spontaneous chromosome rearrangements, *i.e.*, those which would
be the product of some non error-free chromosomal mechanism.

Admittedly, studies that would provide an answer to the
question of the clinical significance of environmentally induced chromosome mutation important in relation to abnormal
development are enormously difficult to devise and carry out.
(Accidents, events of wartime, and so-called experiments of
nature can sometimes be studied to advantage.) But, it seems
desirable to encourage and support surveys, not limit the investigation just to the development of more techniques for
detecting chromosome-breaking agents, and to the detection of
agents themselves! If the proportion of mutations environmentally induced turns out to be trivial, which is possible, this
is well worth knowing. Certainly the number of human embryos
aborted spontaneously because of chromosome imbalance of all
types is huge, but which type (trisomy, monosomy, structural
rearrangement, etc.), if any, is environmentally induced or
even significantly affected by the environment? (We do not
know.) We do know that the proportion of all chromosome imbalance that derives from *structural* rearrangement (*i.e.*, that
coming from "breakage") actually is very low, and that it is
responsible for only a small percentage of all spontaneous
abortions and defective liveborns.

The stand against mutagens is much firmer in relation to
cancer, because in contrast to what some might think, we long
ago identified some "causes" of cancer, *i.e.*, certain chemicals
and radiations. Mutation in a somatic cell, often one visible
to the cytogeneticist, appears ever more convincingly to be
part of the conversion of a normal cell to cancer.

SUMMARY

The various times during embryonic life when chromosome mutation theoretically can occur and its possible clinical significance are considered.

Although chromosome mutation can be of significance relative to developmental defects, the importance, if any, of environmental agents in producing it has not yet been determined. Furthermore, the proportion of abortions and live births with chromosome imbalance due to breakage and rearrangement is small. Therefore, any additional contribution made by newly introduced environmental mutagens, when considered in relation to the extensive wastage of human conceptuses, could possibly be quite trivial. However, the lack of epidemiological data pertaining to these matters is as impressive as the difficulty inherent in devising and carrying out such studies.

The significance of chromosome breakage in cancer etiology is still poorly understood. Whether mutation is or is not an essential first step in the conversion of a normal cell to a cancerous one, a correlation exists between the chromosome-breaking ability of an environmental agent (and also of certain rare genes) and the incidence of cancer. The implications, then, in relation to cancer seem to be considerably clearer than those related to defective embryonic development.

REFERENCES

Boué, J., Boué, A. and Lazar, P. (1975). *Teratology. 12,* 11.
German, J. (1970). *Clin. Genet. 1,* 15.
German, J. (1972). *In* "Progress in Medical Genetics", Vol. 8. (A.G. Steinberg and A.G. Bearns, eds.), pp. 61–101. Grune and Stratton, New York.
Chaganti, R.S.K., Morillo-Cucci, G., Friis, L., Degnan, M. and German, J. (1976). *Ann. Génét. 19,* 43.
Hertig, A.T. (1967). *In* "Comparative Aspects of Reproductive Failure" (K. Benirschke, ed.), pp. 11–41. Springer-Verlag, New York.
Miller. R.W. (1968). *J. Natl. Cancer Inst. 40,* 1079.
Stern, C. (1973). *In* "Principles of Human Genetics" 3rd ed., pp. 508–571. Freeman and Co., San Francisco.
Stewart, A., Webb, J. and Hewitt, D. (1958). *Brit. Med. J. 1,* 1495.
Witschi, E. (1969). *In* "Proceedings of the Third International Conference on Birth Defects". The Hague, The Netherlands, 13–17 September, 1969. (F.C. Fraser, V.A. McKusick and R. Robinson, eds.), p. 157. Excerpta Medica, New York (1970).

CHROMOSOME DAMAGE IN HUMAN MITOTIC CELLS
IN VITRO EXPOSURE TO MUTAGENS

Anton Brøgger

Genetics Laboratory
Norsk Hydro's Institute for Cancer Research
Oslo
Norway

Chromosome aberrations have been used as a measure of genetic hazards in man for many years. In itself chromosomal changes are mutations. In addition they are indicators - at least one group of them - that the cells have been exposed to agents which also induce point mutation in the same cells and/or in other cells of the organism. They are also warnings that carcinogenic processes may be under way.

The purpose of this communication is to discuss some aspects of the reliability and relevance of the analyses of mitotic chromosomes as a method for the detection of possible genetic damage in man caused by environmental agents.

TYPES OF CHROMOSOME DAMAGE

Mitotic Disturbances (Turbagenic Effects)

Anomalies originating from disturbances of the mitotic process may be observed as:

Scattered Metaphase Chromosomes, which may be due to inhibition of spindle development and function (the so-called colchicine effect).

Chromosome Stickiness, which is now interpreted as due to improper folding of the chromosome fiber into single chroma-

d chromosomes (McGill *et al.*, 1974, Klasterska *et al.*,
. Fibers are intermingled and the chromosomes become
ched to each other, subchromatid bridges may occur and
so anaphase bridges. Incompletely folded chromosomes may
be observed. A border-line type of incomplete folding results
in the chromatid gap (achromatic lesion), where the insufficient folding occurs in a minor chromosome part. This may sometimes be seen as a constriction or attenuation.

Multipolar Spindles have been observed ever since the
first preparations of human chromosomes were made (Arnold,
1879). They may occur when the cycle of the division of the
mitotic spindle organizer is out of phase with the division
of the cell, or after cell fusion.

Endoreduplication is the result of two chromosome division
cycles with no intermediate mitosis and is easily recognized
because of the diplochromosomes. Endoreduplication may lead
to tetraploid cells, but irregularities in subsequent divisions with tri- and tetrapolar spindles and formation of
micronuclei have been described (Schmid, 1966).

Premature Chromosome Condensation. A possible new parameter of chromosome damage is premature chromosome condensation
in micronuclei that stems from lagging chromatin, described by
Kürten and Obe (1975) after irradiation of Chinese Hamster
bone marrow *in vivo*. We have seen similar aberrations in human
lymphocytes irradiated *in vitro* (unpublished data).

Chromosome Breakage (Clastogenic Effects)

The second main type of damage is the chromosome breaks
and aberrations derived from breakage and reunion of the
chromosome fiber, the clastogenic effects, which are now
generally considered to concern the DNA molecule responsible
for the linear continuity of the chromosome. These aberrations
are the result of unfinished repair or misrepair of DNA, the
chromosome being considered as a visualized DNA molecule.

The clastogenic effects are mainly observed on metaphase
chromosomes, but chromatin bridges and fragments may be seen
in anaphase cells, and micronuclei may be observed in interphase. The latter is now a parameter in a specially developed
test system, the micronucleus test (Schmid, 1976).

The Unstable Aberrations, breaks, chromatid exchanges, fragments, rings and dicentric chromosomes are mainly seen in the first mitosis after their induction. Rings and dicentrics may survive for several cell generations.

Damage processed by the cell into break or exchange events before the S phase will appear as *chromosome* type aberrations involving both chromatids in the metaphase chromosome. Damage processed during or after the S phase appears as *chromatid* type aberrations affecting one chromatid only. Exposure to clastogens during G2 may result in damage that is not visible until the cell has passed a second mitosis, the so-called delayed action (Kihlmann, 1966).

The breakage/reunion process may give rise to structurally altered chromosomes observed as *the stable aberrations*, deletion, inversion, duplication or translocation chromosomes.

A very detailed classification system has been given by Savage (1975).

Turbagenic and Clastogenic Effects

Agents which disturb the mitotic process may be called mitoturbagenic (from Greek τυρβαζω, turbázo, to trouble, stir up), or perhaps abbreviated to turbagenic, to coin a term similar to clastogenic, which denotes the chromosome breaking ability. (We have avoided the word turbogenic, which would be a Latin/Greek hybrid involving the Latin turbo, meaning whorl or violent, circular motion.) The turbagenic agents, the turbagens, lead to errors in chromosome segregation. One important class of surviving cells from such errors are aneuploid cells.

Hsu (1976) has recently drawn attention to the importance of screening not only for clastogenic effects but also for segregational errors. Many of the turbagenic agents do not affect the DNA, and they will not be discovered in any of the test systems based upon point mutation, and certainly not in the bacteria with no mitotic apparatus. But they may be observed in the test systems developed for detecting clastogens if colcemid and hypotonic treatment are omitted.

Some substances seem to be only turbagenic, such as colchicine, vincristine, inblastine, mercaptoethanol (Brøgger, 1975), and the inhalation anaestetics halothane, enflurane and methoxyflurane (Kusyk and Hsu, 1976). Other agents have both turbagenic and clastogenic properties, such as ethidium bromide (McGill *et al.*, 1974), actinomycin D, propidium dioxide and possibly all the DNA intercalating substances (Pathak *et al.*, 1975). Some agents such as the base analogues are probably clastogenic only. The alkylating agents are best known for

their clastogenic properties, but since folding defects such
as attenuations and gaps are also induced, other targets than
the DNA must be considered.

METHODOLOGICAL CONSIDERATIONS

Examination of Individuals

Two tissues may be subject to direct examination, *the
bone marrow* and *the testicular tissue* with the spermatogonial
mitoses. Attention has been drawn to the promising possibilities of analyzing germ line cells (Hultén, this Conference).
One advantage of using bone marrow is that we are able to
study mitoses that have developed *in vivo*, where the tissue
is flushed with blood so that substances reaching the blood
will expose the bone marrow cells with no further barrier.
The disadvantages are that sampling is more complicated than
drawing blood, that the number of suitable metaphases is
often low, and that the chromosomes are short, so that gaps
and breaks are not easily scored and may perhaps be overlooked. The material is not suitable for banding studies as are
lymphocyte chromosomes. We have the impression that when bone
marrow and lymphocytes from the same individual are analyzed,
more aberrations are found in the lymphocytes. This may be
due to the technical reason already mentioned, or the sensitivity of the two materials is different. A systematic study
of this has not ben done.

By far the most important technique is *the lymphocyte
culture system*, which involves a short-term *in vitro* culture
of *in vivo* exposed cells. This method is now fairly well
standardized (Buckton and Evans, 1973; Evans and O'Riordan,
1975). But world-wide there are differences among laboratories, which may to some extent hamper the comparison of results.
Undoubtedly, the value of the analyses increases when groups
dealing with the same type of problems agree on a detailed
schedule of the method, such as was done by six laboratories
in the United States (Kilian and Picciano, 1976) and by
European workers involved in vinyl chloride monomer projects
(Meeting arranged by European Council of Chemical Manufacturer's Federations, Milan, February 2, 1976). In Norway, laboratories in Porsgrunn and Oslo have standardized their procedures in minute details (Meeting, September 3, 1976).

There remain, however, several points for worth-while
discussion.

The Gap/Break Scoring Problem

Many authors consider the chromatid gaps (achromatic lesions) not to be true breaks (Revell, 1955, 1963; Evans, 1963; Conger, 1967). From our studies with mercaptoethanol we have concluded that gaps are chromosome fibre folding defects (Brøgger, 1975). In cells prepared by spreading on a water surface the gaps appear to be bridged by chromosome material. In acetic acid/alcohol-fixed material the connecting chromatin is usually ruptured, leaving a gap in the chromatid (Brøgger, 1971a).

This means that with our standard method of preparing chromosomes it is not possible to distinguish between a true break and a folding defect involving a bridge of chromatin. So we have to use an arbitrary definition when we are scoring for the true breaks in which we are interested.

We have been used to score the aberration as a break only when the distal part of the chromatid is dislocated. Other groups score the aberration as a break if the discontinuity is larger than the chromatid's width, irrespective of whether the distal fragment is dislocated or not (Schinzel and Schmid, 1976).

This scoring problem is an important point, since many workers consider the gaps to be innocent with no further consequence as regards chromosome mutation, whereas a break is a serious event leading to irrepairable loss of genetic material. It may, therefore, be asked how reliable our scoring is in this respect. There is some evidence that perhaps we do not make many mistakes. Mercaptoethanol is not considered to damage the DNA. With this substance we find only what we score as folding defects (gaps). Breaks and exchanges are at control level (Brøgger, 1975).

On the other hand, it may be discussed how innocent the gaps really are. It has been suggested that single-strand DNA breaks may result in gaps (Bender *et al*., 1973; Brøgger, 1974). It is possible that such a break will lead to release of supercoiled DNA and a change in the protein/DNA interaction responsible for the perfect folding of the chromatin. If this occurs in a sufficient number of sites it might result in a folding defect detectable at the chromatid level. Other changes in the DNA, such as intercalation of compounds, may also result in folding defects.

It has also been argued that gaps may be true breaks (Comings, 1974). Perhaps we have to consider two types of gaps, morphologically indistinguishable: the clastogenic and the turbagenic type. With the aid of an elegant premature chromosome condensation technique Hittelman and Rao (1974) revealed at least two types of gaps, one probably due to single-strand regions in DNA, the other to altered DNA/protein binding.

At our present stage of knowledge we may conclude that an increased amount of gaps at least is an indicator that turbagenic or clastogenic agents *might* have been at work.

Time of Culture Before Harvest

There is no general agreement between laboratories whether to harvest lymphocyte cultures at 48 hrs or 72 hrs. The irradiation experiments of Heddle *et al.* (1967) show clearly that the aberration yield decreases from 48 to 72 hrs, probably mainly due to the difference in frequency of first and second metaphases. Similar results have been found in cultures from patients treated with x-rays (Buckton and Pike, 1964).

On the other hand, Kilian and Picciano (1976) report that their group has found no essential difference between 48 hrs and 72 hrs, and that this has also been the experience of other investigators.

Most laboratories seem to have the peak of mitotic activity at 72 hrs, so that the yield of analyzable cells is highest at this time of harvest. This is one of the reasons why we have previously chosen the 72 hrs culturing time. Another reason has been that several years ago we found by autoradiography that 1 out of 9 metaphases at 72 hrs were second divisions (Johansen, 1972). Obe *et al.* (1976) observed with BrdU-labelling that at 72 hrs 7% were first, 51.9% were second, and 41.1% were third divisions. Undoubtedly, the dynamics of the lymphocytes depends on both culture conditions and the blood donor. It is possible that the lymphocyte populations from chemically exposed and unexposed individuals behave differently. To be on the safe side, some laboratories harvest both at 48 and 72 hrs, but this doubles the amount of analysis work, which is a bottle-neck already.

Effects of In Vivo Exposure

For how long time after the *in vivo* exposure is it possible to find any damage in the lymphcytes? Is there an accumulation or an exclusion of damage in the course of time? From radiation studies it is known that the frequency of dicentrics remains constant during the first few months after exposure, but may be reduced to one half around three yrs later (Preston *et al.*, 1974). The life span of the small lymphocyte has been estimated to around three yrs (Evans, 1976). But structural chromosome damage of both chromatid and chromosome type has been found more than 20 years after radiotion exposure

(Buckton *et al.*, 1962; Bauchinger, 1968), which suggests the existence of some very long-lived cells among the lymphocytes.

With chemicals we may expect a wide variation in persistence according to the mode of action of the substance. Possibly, non-proliferating cells may retain in their chromatin drug molecules which may interfere with DNA processes after long periods of time (Hsu *et al.*, 1975).

There are many positive findings of chromosome damage in lymphocytes after varying periods of exposure, and varying time intervals between exposure and examination, in individuals exposed to a variety of chemicals: cadmium (Deknudt and Leonard, 1975), benzene (Forni *et al.* 1971; Tough *et al.*, 1970; Fredga *et al.*, this Conference), lead (Bauchinger *et al.*, 1976; Schwanitz *et al.*, 1970), mercury (Skerfving *et al.*, 1971), pesticides (Yoder *et al.*, 1973), LSD (Cohen *et al.*, 1967), vinyl chloride monomer (Ducatman *et al.*, 1975; Hansteen, this Conference), and a number of cytostatics in cancer therapy (Krogh Jensen, 1967; Schmid and Bauchinger, 1973; Stenson and Patel, 1973; Krajincanic *et al.*; Nevstad, this Conference).

However, there are also a number of negative findings in lymphocytes from individuals exposed to agents that are known from *in vitro* studies to be clastogenic. Dobos *et al.* (1974) found a significant increase in structural chromosomal aberrations in lymphocytes from children receiving cyclophophamide as long as the treatment went on. But six months after termination of therapy no anomalies were found.

Schinzel and Schmid (1976) have studied 67 patients under cytostatic therapy. In their material the incidence of chromosome type aberrations was 0-1% in the controls as well as in 19/20 patients who had received anti-metabolites and spindle poisons only and in 22/53 patients who had received therapeutic irradiation and/or well-known clastogenic agents. The incidence of chromatid breaks was 0-2% in the controls and 0-4% in most of the patients. The cultures were harvested at 72 hrs, but is is doubtful whether the possible loss of first divisions would account for all these negative findings. There are other cases with similar negative findings (Sparkes *et al.*, 1968; Dick *et al.*, 1974; Robinson, 1974). We have a similar study under way in collaboration with the Norwegian Radium Hospital, and we have both positive and negative findings in bone marrow and lymphocytes.

It may be argued that on the population level the Schinzel-Schmid study does not give entirely negative results. But this is no escape. It seems that we have to face the fact that our technique has a limited resolving power; some negative findings are false negatives. Still, our technique is very useful, since there are so many positive findings. The

negatives must be considered with caution. They have to be
supplemented by *in vitro* studies and analyses with other test
systems.

Banding Methods

Banding methods are even more time-consuming than aberration scoring and have not been much used in studies of clastogenic effects. But they may be important in two respects:
1) They increase the possibility of discovering stable
structural alterations: deletion, duplication, inversion and
translocation chromosomes. Such cells reveal the existence of
aberrant cell clones in the lymphocyte producing tissue.
In one case of Hodgkin's disease treated with radiation,
we made banded karyotypes of 34 cells 6 yrs after treatment
and found 5 with structural rearrangements and one with an
extra chromosome 21. We must mention that one similar patient
was examined before therapy, and maong 13 karyotypes 2 were
abnormal. Our data are preliminary, but similar results are
mentioned by Evans (1976).
Aberrant cell clones have also been found in bone marrow.
Three subjects among the 20 fishermen exposed to 220-600 rad
during the Bikini test explosion in 1954 proved to have cell
clones with abnormal karyotypes several years after the accident (Ishihara and Kumatori, 1967).
2) Comparative studies of non-random distribution of induced chromosome damage (Brøgger, 1977) raise the question of
agent specificity. X-rays, mercaptoethanol, psoralen/UVA,
quinacrine mustard, mitomycin C and chlorambucil seem to have
different target bands in the chromosomes. In some situations
it could be that this agent specificity might be used as a
parameter in evaluating the damage.

In Vitro Test Systems

The most popular test system is the human lymphocyte system already mentioned. Established lines of diploid human cells
are available or may be developed in the laboratory, either as
fibroblasts or lymphoblastoid cells. Some methods are described by Evans (1976). Any diploid line is preferable to transformed cells with variable karyotypes and possibly different
sensitivities to clastogens compared with diploid cells.
The lymphocyte population is also heterogenous and populations with different sensitivities seem to be present (Beek
and Obe, 1974). There might also be individual variation among
blood donors in the lymphocytes' response to the clastogen,

such as has been found with mitomycin C (Brøgger, 1971b), the alkylating cytostatic Zitostop (Schuler *et al.*, 1972), and azathioprine (Zyl and Wissmüller, 1974).

The *in vitro* systems have all the advantages of the experimental situation and are valuable tools for obtaining dose response information and relations between agent and cell cycle. The real disadvantage is, of course, the lack of the total organism with its metabolism implying excretion, breakdown, activation or inactivation of the agent. One of these obstacles has been overcome by the addition to the culture of a rat microsome fraction capable of biotransformation (Natarajan *et al.*, 1976).

Generally, it must be recommended to perform *in vitro* studies when positive findings in individuals point to a suspected agent. If subsequent *in vitro* experiments produce negative findings, the question may be raised whether the observations in the individuals were due to some other agent than the suspected one.

The problem of false negatives in the *in vivo* studies increases the importance of *in vitro* studies.

CHROMOSOME DAMAGE, WHAT DOES IT MEAN?

We have some information on the fate of chromosome aberrations. A number of them are certainly lethal to the cell, but aberrations may persist for a long time in proliferating systems (Carrano and Heddle, 1973; Sasaki and Norman, 1967). When unstable aberrations are observed, we know that some of them could have resulted in stable aberrations which survived. If this occurred in the germ line, unbalanced gametes might be the result.

Blood cells have been used to reconstitute blood formation after irradiation in mice, rats, guinea pigs and dogs, and human blood also contains hemopoietic cells (Nelson and Andrews, 1976). It is not known if these cells are among those we see as metaphases in our preparations. But it is reasonable to believe that clones may be established also from circulating cells with aberrations. Whether neoplastic disease will develop from such clones is pure speculation.

Quite as important is that the chromosome damage we find in the bone marrow or lymphocyte samples are *indicators* of mutagenic and also carcinogenic actions in these and other tissues.

From the molecular point of view chromosome damage is a very crude way of studying changes in the genome. A change in one single band will be at the limit of the resolving power of our technique. The average amount of DNA in one band is

5.8 pg/2·321= 9.034·10^{-3} pg or approximately 8.15·10^{-6} base
pairs, which is about the equivalent of three coli genomes.
This means, of course, that significant genetic changes will
escape our detection.

We have every reason to believe that the clastogens also
induce point mutation. Concerning the quantitative relation
between chromosome damage and point mutation we have only the
data from the pioneering experiments of Kao and Puck (1969)
with hamster cells. So this is a field for future research.

In this connection may be mentioned that in *Drosophila* a
large number of substances – the indirect carcinogens in
particular – are highly effective in producing mutation but
do not induce chromosome breakage at all, at least not unless
much higher concentrations are used (Sobels and Vogel, 1976).

Quantitative correlations between somatic damage and damage in the germ line cells would allow in individual cases
extrapolation from observations on lymphocytes or bone marrow
to the germ line. So far, human data of this type is totally
lacking, but correlation data from the mouse have been used
in extrapolation to man (Brewen and Preston, 1974).

It is far from clear which role the induction of chromosome damage plays in the initiation and development of carcinogenic processes. But we all agree that when chromosome
damage is observed carcinogenic processes may also be under
way in the organism.

However, very few quantitative data are available on the
correlation between chromosome damage and risk for neoplastic
disease. Ottonen and Ball (1973) found that a dose of 7,12-
dimethylbenz(a)anthracene sufficient to induce thymic lymphomas in 100% of their mice did not produce chromosome damage
in bone marrow or thymus above control level. On the other
hand, Sugiyama (1973) observed a close correlation between
the amount of chromosome damage induced in mouse bone marrow
by various benz(a)anthracene derivatives and the ability of
these substances to cause fibrosarcomas in the animals. Clearly this field of research needs further investigations.

SUMMARY

Standardized methods are now available for both *in vitro*
and *in vivo* studies of chromosome damage in human cells caused
by environmental agents. The value of such studies may be increased when research groups dealing with the same problems
agree on a detailed schedule of the method.

Particularly the methods for detecting clastogenic effects (chromosome breakage) are well developed, whereas methods revealing turbagenic effects (mitotic disturbances) need further development with regard to quantified observations.

The gap/break scoring problem is not solved. In the meantime gaps must be considered as an indicator that turbagenic or clastogenic agents might have been at work.

In some cases there seems to be a lack of correlation between the *in vitro* and *in vivo* findings: some *in vivo* negative findings are false negatives in the respect that the subjects have been exposed to clastogens and still their lymphocyte chromosomes appear undamaged. Negative *in vivo* findings must be supplemented with *in vitro* studies.

Banding methods are time-consuming but will give valuable additional information, particularly a long time after exposure.

We have some – mostly qualitative and theoretical – information on the significance of chromosome damage. More experimental and quantitative data on the correlations between chromosome damage and point mutation, somatic and germ line damage, and chromosome damage and carcinogenic risk are needed and will hopefully come from future research.

REFERENCES

Arnold, J. (1879). *Virchow's Arch. 78*, 279.
Bauchinger, M. (1968). *Strahlentherapie. 135*, 553.
Bauchinger, M., Schmid, E., Einbrodt, J., and Dresp, J. (1976). *Mutation Res. 40*, 57.
Beek, B. and Obe, G. (1974). *Mutation Res. 24*, 395.
Bender, M.A., Griggs, H.G., and Walker, P.L. (1974). *Mutation Res. 20*, 387.
Brewen, J.G. and Preston, R.J. (1974). *Mutation Res. 26*, 297.
Brøgger, A. (1971a). *Hum. Genet. 13*, 1.
Brøgger, A. (1971b). *In* "Fourth Int. Congr. Hum. Genet." *Excerpta Med. 233*, 35.
Brøgger, A. (1974). *Mutation Res. 23*, 353.
Brøgger, A. (1975). *Hereditas. 80*, 131.
Brøgger, A. (1977). *"Helsinki Chromosome Conference."*
Buckton, K.E. and Evans, H.J. (1973). Methods for the Analysis of Human Chromosome Aberrations". *Geneva, World Health Organization,* p. 66.
Buckton, K.E. and Pike, M.C. (1964). *Int. J. Radiat. Biol. 8*, 439.

Buckton, K.E., Jacobs, P.A., Court Brown, W.M., and Doll, R. (1962). *Lancet. ii*, 676.

Carrano, A.V. and Heddle, J.A. (1973). *J. Theoret. Biol. 38*, 289.

Cohen, M.M., Hirschhorn, K., and Frosch, W.A. (1967). *New Engl. J. Med. 277*, 1043.

Comings, D.E. (1974). *In* "Chromosomes and Cancer" (J. German, ed.), p. 95. John Wiley and Sons, New York.

Conger, A.D. (1967). *Mutation Res. 4*, 449.

Deknudt, G. and Leonard, A. (1975). *Environ. Physiol. Biochem. 5*, 319.

Dick, C.E., Schniepp, M.L., Sonders, R.C., and Wiegand, R.G. (1974). *Mutation Res. 26*, 199.

Dobos, M., Schuler, D., and Fekete, G. (1974). *Hum. Genet. 22*, 221.

Ducatman, A., Hirschhorn, K., and Selikoff, I.J. (1975). *Mutation Res. 31*, 163.

Evans, H.J. (1963). *In* "Radiation-Induced Chromosome Aberrations" (S. Wolff, ed.), p. 8. Columbia Press, New York.

Evans, H.J. (1976). *In* "Chemical Mutagens. Principles and Methods for Their Detection", Vol. 4 (A. Hollaender, ed.), p. 1. Plenum Press, New York.

Evans, J.H. and O'Riordan, M. (1975). *Mutation Res. 31*, 135.

Forni, A.M., Cappellini, A., Pacifico, E., and Vigliani, E.C. (1971). *Arch. environm. Hlth. 23*, 385.

Heddle, J.A., Evans, H.J., and Scott, D. (1967). *In* "Human Radiation Cytogenetics" (H.J. Evans, W.M. Court-Brown, and A.S. McLean, eds.), p. 6. North Holland Publ. Co., Amsterdam.

Hittelman, W.N. and Rao, P.N. (1974). *Mutation Res. 23*, 259.

Hsu, T.C. (1976). *Mammalian Chrom. Newsl. 17*, 1.

Hsu, T.C., Pathak, S., and Kusyk, C.J. (1975). *Mutation Res. 33*, 417.

Ishihara, T. and Kumatori, T. (1967). *In* "Human Radiation Cytogenetics" (H.J. Evans, W.M. Court-Brown, and A.S. McLean, eds.), p. 144. North Holland Publ. Co., Amsterdam.

Johansen, J. (1972). *Experientia (Basel). 28*, 416.

Kao, F.-T. and Puck, T.T. (1969). *J. Cell Physiol. 74*, 245.

Kihlman, B.A. (1966). "Actions of Chemicals on Dividing Cells", p. 143. Prentice-Hall Inc., New Jersey.

Kilian, D.J. and Picciano, D. (1976). *In* "Chemical Mutagens. Principles and Methods for Their Detection", Vol. 4 (A. Hollaender, ed.), p. 321. Plenum Press, New York.

Klasterska, I., Natarajan, A.T., and Ramel, C. (1976). *Hereditas. 83*, 153.

Krajincanic, B., Lazarov, A., and Radojicic, B. (1976). *Strahlentherapie. 151*, 459.

Krogh Jensen, M. (1967). *Acta med. Scand. 182*, 445.

Kürten, S., and Obe, G. (1975). *Hum. Genet. 28,* 97.

Kusyk, C.J. and Hsu, T.C. (1976). *Environ. Res. 12,* 366.

McGill, M., Pathak, S., and Hsu, T.C. (1974). *Chromosoma (Berl.). 47,* 157.

Natarajan, A.T., Tates, A.D., van Buul, P.P.W., Meijers, M., and de Vogel, N. (1976). *Mutation Res. 37,* 83.

Nelson, B.M. and Andrews, G.A. (1976). *In* "Adv. Rad. Biol.", Vol. 6 (J.T. Lett and H. Adler, eds.), p. 325.

Obe, G., Brandt, K., and Beek, B. (1976). *Hum. Genet. 33,* 263.

Ottonen, P.O. and Ball, J.K. (1973). *J. Natl. Cancer Inst. 50,* 497.

Pathak, S., McGill, M., and Hsu, T.C. (1975). *Chromosoma (Berl.) 50,* 79.

Preston, R.J., Brewen, J.G., and Gengozian, N. (1974). *Radiat. Res. 60,* 516.

Revell, S.H. (1955). *In* "Proc. Radiobiol. Symp." Liege 1954, p. 243, Butterworth, London.

Revell, S.H. (1963). *In* "Radiation-induced Chromosome Aberrations" (S. Wolff, ed.), p. 41. Columbia Univ. Press, New York.

Robinson, J.T. (1974). *Brit. J. Psychiat. 125,* 238.

Sasaki, M.S. and Norman, A. (1967). *Nature. 214,* 502.

Savage, J.R.K. (1976). *J. med. Genet. 12,* 103.

Schinzel, A. and Schmid, W. (1976). *Mutation Res. 40,* 139.

Schmid, E. and Bauchinger, M. (1973). *Mutation Res. 21,* 271.

Schmid, W. (1966). *Exp. Cell Res. 42,* 201.

Schmid, W. (1976). *In* "Chemical Mutagens. Principles and Methods for Their Detection", Vol. 4 (A. Hollaender, ed.), p. 31. Plenum Press, New York.

Schuler, D., Fekete, G., and Dobos, M. (1972). *Hum. Genet. 16,* 329.

Schwanitz, G., Lehnert, G., and Gebhart, E. (1970). *Germ. med. Mth. 15,* 738.

Skerfving, S., Hansson, K., and Lindsten, J. (1971). *Arch. environm. Hlth. 21,* 133.

Sobels, F.H. and Vogel, E. (1976). *Mutation Res. 41,* 95.

Sparkes, R.S., Melnyk, J., and Bozzetti, L.P. (1968). *Science. 160,* 1343.

Stevenson, A.C. and Patel, C. (1973). *Mutation Res. 18,* 333.

Sugiyama, T. (1973). *Gann. 64,* 637.

Tough, I.M., Smith, P.G., Court-Brown, W.M., and Harnden, D.G. (1970). *Europ. J. Cancer. 6,* 49.

Yoder, J., Watson, M., and Benson, W.W. (1973). *Mutation Res. 21,* 335.

Zyl, J. van and Wissmüller, H.F. (1974). *Hum. Genet. 21,* 153.

MONITORING FOR CHROMOSOMAL
DAMAGE IN EXPOSED INDUSTRIAL
POPULATIONS

D. Jack Kilian
Dante J. Picciano

Occupational Health and
Medical Research
Dow Chemical U.S.A.
Freeport, Texas

The Texas Division of Dow Chemical U.S.A. is located
about 50 miles south of the city of Houston, Texas. While the
area immediately around our 3500 acre site is predominantly
rural and residential, we are part of a large industrial and
petrochemical complex that extends along the northern coast of
the Gulf of Mexico and the southeastern coast of the North
American continent. The Texas Division has some 7000 full-
time employees who work around or use a wide variety of chemicals (Table I). These chemicals have become the subjects of
intense investigation in recent years because of their toxic
potential. For many of these compounds, the potential for
toxicity includes the possibility of genetic alteration in an
individual's somatic or gametic cells.
    The possibility of genetic damage, especially the associ-
ation between an increased rate of chromosomal aberrations and
carcinogenesis, is a matter of serious concern. One of the
participants in this conference, Dr. James German (1972), has
observed that many human cancers arise from a single cell and
that this cell often has one or more chromosomes that differ
in morphology from any in the complement of the noncancerous
cells of the affected person. In addition, chromosomal instability, leading to an increased number of chromosomal abnormalities, mutation, and cancer is found in the several rare
genetic disorders that feature increased chromosomal instability together with a greatly increased expectancy for cancer
to develop. Certain agents - radiation, some chemicals, and
certain viruses - have in common the ability to produce in-

TABLE I. Dow Texas Division, Freeport, Texas.

7000 employees. Manufacturer or user of a variety of chemicals,
including

Benzene	Toluene diisocyanate
Ethylenimine	Toluene diamine
Vinyl chloride	Phosgene
Vinylidene chloride	Carbon tetrachloride
Trichloroethylene	Chloroform
Epichlorohydrin	Hexachlorobenzene
Hexachlorobutadiene	

creased rates of cells with chromosomal abnormalities and to
increase the risk of malignancy.

The comprehensive surveillance program established for
Dow Texas Division workers is a good example of how cytogenetic monitoring can be integrated with a system of industrial
monitoring. The surveillance program (Table II) incorporates
four major areas of investigation: Periodic Evaluation, which
includes a history and physical examination at preset intervals and appropriate laboratory tests; Cytogenetic Evaluation,
done on a routine basis and in the event of accidental overex-
posure; Conventional Epidemiologic Studies, to determine the
mortality and morbidity experience of exposed workers as com-

TABLE II. Vinyl Chloride (VC) Workers

Comprehensive surveillance program

Periodic evaluation
    History
    Physical
    Lab work

Cytogenetic evaluation
    Routine exposures
    Over-exposures

Conventional epidemiology
    Mortality
    Morbidity

Special epidemiology
    Reproductive study
    (VC workers' wives)

pared to a suitable control group; and Special Epidemiology, which, in a study underway now, will compare the reproductive patterns of the wives of vinyl chloride workers to the child-bearing experience of the wives of a matched control group.

The medical management team of Dow's Texas Division first became aware of the need for a program of cytogenetic monitoring of certain workers in the early 1960s when reports of the mutagenicity of ethylenimine (EI) in non-mammalian systems began to appear in the literature. Dow had started research on EI in 1958 and started to develop a commercial process in 1961. Actual production began in 1965. Because of the early reports of possible mutagenicity, cytogenetic monitoring of EI workers was made part of the medical surveillance program already in effect. When the cytogenetic program was started, the list of potential mutagens was short, and many compounds - such as vinyl chloride - were thought to be essentially harmless.

Looking back, we can see that our first efforts at cytogenetic analysis were relatively crude and our findings inconclusive. Continued work on the many technical problems, however, allowed us to become more confident in our results and to expand the cytogenetic surveillance program to a number of other worker groups, including those exposed to benzene and vinyl chloride.

Participation in a collaborative study of rat bone marrow cytogenetics was of particular help to us (Kilian *et al.*, 1977a, Kilian *et al.*, 1977b). Six cytogenetics laboratories, including our own, joined together, first, to devise methods to minimize interlaboratory differences and, second, to measure interlaboratory variation in results following the standardized administration of a known mutagen - triethylenemelamine - to rats from a common source. A preliminary workshop was held to resolve scoring differences, to develop a joint protocol and common glossary, and to reach agreement on uniform reporting methods. This study showed (Fig. 1) a good pattern of agreement among the laboratories, particularly for certain categories of aberration. It also (Table III) allowed us to evaluate the variability of individual scorers in our own laboratory.

As the list of potential chemical mutagens grew longer, the need for well-controlled cytogenetic studies of human effects, using simplified, repeatable methods became more pressing. We have previously published details of our methods (Kilian and Picciano, 1976); I would like to review these and the conditions we find are necessary for meaningful results. The groups being monitored should be of significant size and relatively stable and healthy. Employee turnover rates should be low, there must be adequate technology for accurate monitoring of both chronic and acute exposures, and there has to

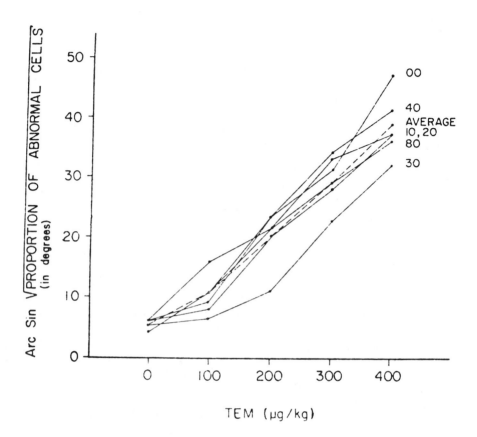

FIGURE 1. Graph showing degree of agreement among six laboratories for average arc sin √proportion of abnormal cells by laboratory and dose level of triethylenemelamine in cooperative study of rat bone marrow cytogenetics.

be medical and genetic expertise for the evaluation of whatever changes might be found in an individual. It is very helpful if exact levels of exposure to various compounds, alone or in combination, are known. The number of individuals available for study should be large enough to permit evaluation of matched control groups. In addition, the availability of findings from periodic medical examination and records from comprehensive health insurance plans - for the workers and their families - allows the development of cooperative study plans involving the industrial physician, epidemiologist, and cytogeneticist to evaluate not only the worker's medical status, but any reproductive consequences as well.

TABLE III. *Proportion abnormal Cells by Scorer and Dose for Laboratory 00*

Scorer number	Dose quantity					All doses
	000	100	200	300	400	
01	.00	.00	.10	.31	.41	.16
02	.00	.03	.19	.185	.66	.21
03	.01	.02	.29	.205	.28	.16
05	.03	.10	.20	.40	.72	.29
07	.01	.02	.08	.21	.53	.17
All scorers	.010	.034	.172	.260	.538	.20

A standardized medical history - including exploration of family and occupational background - is taken prior to pre-employment examination, and the worker's status in regard to those factors - such as exposure to chemicals, drugs, radiation, and virus infections - that are known to affect the chromosomes is recorded. The medical status of workers exposed to toxic chemicals is reassessed at periodic intervals thereafter. Standard culturing and harvesting procedures for peripheral blood are used.

Abnormalities are recorded on raw data sheets by cytogenetics laboratory technicians. The data are transferred to 80-column computer punch cards. Since one of our objectives is to set up longitudinal life-time studies to evaluate the consequences of chromosomal breakage, having the necessary information in computer language is very helpful. Scored cells are identified by vernier setting and slide number. Early in our cytogenetic experience, we scored 20 cells per individual and, later, increased this to 50 cells, whenever possible. At the present time, however, we are analyzing at least 200 cells per culture to ensure the validity of the test system.

As a general rule, results of cytogenetic evaluation of specific worker groups are compared to those found upon evaluation of preemployment examinees, that is, persons who have had chromosome analysis done as part of routine preemployment examination and who were judged, on the basis of the history taken at that time, to have no condition that would predispose them to increased aberration rates. In some cases, it is possible for the worker to serve as his or her own control. During the more than 12 years that the cytogenetics surveillance program has been in effect, more than 5000 cultures on more than 4000 individuals have been available for study.

A description of our cytogenetic study of vinyl chloride-
exposed workers will serve to illustrate the potential usefulness of our surveillance program. Our cytogenetic study group
was composed of 209 currently employed persons who had worked
in the vinyl chloride plant for periods ranging between one
and 332 months - an average of 48 months - at the time of the
study. Findings were compared to cytogenetic data from a group
of 295 preemployment examinees. The records selected for in-
clusion in the control group were matched to those of the
study group, insofar as possible, for sex, number of cells
analyzed, and time period during which the culture was initiated.

A detailed evaluation of the results of our vinyl chloride cytogenetics study is now in press (Picciano *et al.*,
1977). The major finding of the study, however, can be summarized simply: that no cytogenetic differences of significance
were found between the group of vinyl chloride-exposed workers
and the preemployment control group. Table IV shows the comparison of the percentage findings for chromatid and chromosome
breaks; rings, dicentrics, and exchanges; and the proportion
of abnormal cells. Table V displays the results of Chi-square
analysis for differences in the distribution of chromatid
breaks between the two groups. Both groups were divided into
those showing 0-to-5-percent aberrations and those showing
greater-than-5-percent aberrations; the vinyl chloride workers
were separated into those with estimated average exposure
levels of less-than-one part per million (ppm), from 1-to-5
ppm, and greater-than-5 ppm. The same analysis was used to
devaluate for differences in the distribution of chromosome
breaks (Table VI) and the proportion of abnormal cells (Table
VII). We would like to emphasize that our work-force has been
exposed to relatively low levels of vinyl chloride and that
we believe this has a definite bearing on our results. It is
to be hoped that our demonstration of lack of significant

*TABLE IV.   Cytogenetic Study of 209 Workers Exposed to
Vinyl Chloride*

	Workers	Controls
Number of cultures	*209*	*295*
Number of cells	*10,483*	*14,761*
Chromatid breaks	*2.4%*	*3.6%*
Chromosome breaks	*1.0%*	*1.1%*
Rings, dicentrics, and exchanges	*0.4%*	*0.2%*
Abnormal cells	*3.7%*	*4.5%*

TABLE V. Distribution of Chromatid Aberrations Related to Vinyl Chloride (VC) Exposure

	Number in group	Percentage of group with 0-5% aberrations	Percentage of group with >5% aberrations
Exposure to VC[1]	209		
<1 ppm	70	90	10
1-5 ppm	98	77	23
>5 ppm	41	80	20
Controls	295	75	25

$$\chi^2_{(3)} = 7.75 \quad (p \sim 0.06)$$

[1]Exposure levels are estimates based on calculations for specific job classifications; prior exposure for individuals may have been higher or lower; exposure levels for individuals are known to vary within job classification

TABLE VI. Distribution of Chromosome Aberrations Related to Vinyl Chloride (VC) Exposure

	Number in group	Percentage of group with 0-5% aberrations	Percentage of group with >5% aberrations
Exposure to VC	209		
<1 ppm	70	96	4
1-5 ppm	98	94	6
>5 ppm	41	95	5
Controls	295	94	6

$$\chi^2_{(3)} = 0.36 \quad (P \sim 0.95)$$

effect at these levels will be useful in establishing levels of exposure that are genetically safe.

In 1974-75, a group of five reports appeared in the literature concerning chromosome aberrations in vinyl chloride and polyvinyl chloride workers (Table VIIIa). Three of the study teams concluded that an increased rate of chromosomal aberrations was associated with vinyl chloride exposure (Ducatman *et al.*, 1975; Funes-Cravioto *et al.*, 1975; Purchase *et al.*, 1975), while two studies - one of them a preliminary investi-

TABLE VII. *Distribution of Abnormal Cells Related to Vinyl Chloride (VC) Exposure*

	Number in group	Percentage of group with 0-5% aberrations	Percentage of group with >5% aberrations
Exposure to VC	209		
<1 ppm	70	84	16
1-5 ppm	98	71	29
>5 ppm	41	73	27
Controls	295	70	30

$$\chi^2 = 5.97 \quad (P \sim 0.12)$$

TABLE VIIIa. *Published Cytogenetic Studies of Workers Exposed to Vinyl Chloride*

Authors	Cases reported		Exposures		Effect
	Workers	Controls	Type	Magnitude	
Ducatman et al.	11	10	PVC	Estimated to exceed 500 ppm	Positive
Funes-Cravioto et al.	7	3	PVC	Decreased in recent years to 20-30 ppm	Positive
Purchase et al.	56	24	PVC	Not given	Positive
Kilian and Picciano	203	108	VC	1969-74: average 5 ppm; 1974-75: 1-2 ppm	Negative
Fleig and Thiess	10	4	PVC	Until 1974: >100 ppm; 1975: 10-25 ppm	Negative

TABLE VIIIb. *Conditions (Neoplastic or with Neoplastic*
Association) Displaying Chromosome Anomalies[1]

Chronic Myelocytic Leukemia	*Ataxia-Telangiectasia*
Acute Myelocytic Leukemia	*Fanconi's Anemia*
Chronic Lymphatic Leukemia	*Bloom's Syndrome*
Polycythemia Vera	*Kostmann's Infantile*
Myeloid Metaplasia	*Agranulocytosis*
Erythroleukemia	*Glutathione Reductase*
Sideroblastic Anemia	*Deficiency*
Meningioma	*Down's Syndrome*
Burkitt's Lymphoma	*Turner's Syndrome*
	Klinefelter's Syndrome
	Wilms' Tumor-Aniridia Syndrome

Retinoblastoma in D-Deletion Syndrome

[1]*Source: Mulvihill, 1975*

gation from our laboratory - appeared to show no differences
of statistical significance between exposed and control groups
(Fleig and Thiess, 1974; Kilian *et al.*, 1975). These studies
tend to indicate a dose-related response since the exposure
levels of polyvinyl chloride workers to the vinyl chloride
monomer are, generally, much higher than the exposure of vinyl
chloride manufacturing workers (Occupational Safety and Health
Administration, 1974). The indication of a dose-related effect
was reaffirmed by further study of vinyl chloride workers at
Imperial Chemical Industries in which increased aberration
rates were significantly correlated to length and intensity
of exposure (Purchase *et al.*, 1976).

It is known that increased rates of chromosomal breakage,
as have been demonstrated in cases of exposure to high levels
of vinyl chloride, have occurred following x-ray treatment of
ankylosing spondylitis (Buckton *et al.*, 1962), in atomic bomb
survivors (Bloom *et al.*, 1970), radium dial painters (Vaughan,
1962). persons exposed to Thorotrast (Fischer *et al.*, 1966),
and workers exposed to lead (Garza Chapa *et al.*, 1976), benzene (Vigliani and Forni, 1976), uranium dust, and plutonium.
Increased rates of chromosomal breakage are also seen in a
variety of diseases (Table VIIIb) in which an increased risk
of cancer is also evident (Mulvihill, 1975).

Evidence of a relationship between chromosomal abnormalities and increased risk of neoplasia has continued to accumulate, although the exact nature of this relationship is far
from clear. It is because of this evidence that we believe
cytogenetic monitoring can be an extremely valuable tool in
the detection of those environmental situations that may be

associated with an increased risk of cancer and, perhaps,
reproductive problems for the chemically exposed worker as
well.

A cytogenetic program and periodic medical evaluation,
however, are not enough to ensure comprehensive surveillance
of worker groups at risk. Epidemiology programs, both conventional and special, are also essential. Regarding our epidemiology efforts, we are fortunate to be reasonably close to
the University of Texas Medical Branch at Galveston. Under the
direction of University of Texas epidemiologist, Dr. Patricia
Buffler, a unique epidemiology program was started a few years
ago; the heart of this program is the cooperative efforts put
forth by both our own industrial concern and academia. University of Texas employees actually work at the plant site on a
continuous basis to identify and process necessary records for
later analysis at the medical school. Complete confidentiality
is maintained through the use of various coding systems.

Epidemiologic analysis of our vinyl chloride workforce
showed that the employees, past and present, fell into one of
three major assignment areas (Table IX): the vinyl chloride
plant itself, the maintenance units, and the developmental
laboratory. Dozens of specific job classifications were identified, and we were able to arrive at fairly firm estimates of
actual or potential exposure for most of them. The estimates
were based on both personnel and area monitoring; Table X
shows a few of them. It needs to be noted that actual exposures will vary from person to person within the same job classifications and that accidental, short-term exposures of some
workers to concentrations in excess of the mandatory standard
probably occur from time to time. Documentation of such incidents is usually difficult. The current U.S. standard for

*TABLE IX. Vinyl Chloride (VC) Cohort for Historical-Pro-
spective Study of Mortality Experience. Worker Classifications*

VC production plant
        Operators
        Loaders
        Supervisors
        Quality control laboratory

Maintenance workers
        Electricians
        Boiler-makers
        Pipefitters
        Others

Developmental laboratory

TABLE X. *Estimated Exposure to Vinyl Chloride (VC) for Vinyl Chloride-Related Job Classifications*

Job classification	Estimated exposure in parts per million, as time-weighted-average		
	1973-1974	1960-1972	Before 1960
R&D engineer	1.7	4.4	8.2
Production engineer	1.7	4.4	8.2
Quality control laboratory	8.7	11.4	15.2
Instrument man	1.3	4.0	7.8
Painter	0.7	0.7	0.7
Machinist	2.4	5.1	8.9
Material handler	0.5	0.5	0.5
Production foreman	1.3	4.0	7.8
Control A OP. (Vinyl)	2.8	5.5	9.3
Control B OP. (Oxy & Chloride)	0.7	0.7	0.7
Control C OP. (Chlorination)	2.6	5.3	9.1

TABLE XI. *Vinyl Chloride (VC) Workers: U.S. and Dow Standards for Exposure*

1948-1958	Dow	No specified standard; exposure levels unknown
1959	Dow	50 ppm, averaged concentration, established as air quality goal
1959-1967	Dow, Texas Division	Estimated average exposures: below 50 ppm
1968-1973	Dow, Texas Division	Estimated average exposure: below 5 ppm
Before 1974	Other U.S. VC producers	No specified standard; exposure levels unknown; some very high
1974	Other U.S. VC producers	Threshold limit value of 500 ppm recommended by ACGIH[1] as industry guideline; voluntary compliance
1974	Dow, Texas Division	Estimated average exposure; 1-2 ppm
1975	Dow, Texas Division	Estimated average exposure: below 1 ppm

[1]American Conference of Governmental Industrial Hygienists

TABLE XII.   Vinyl Chloride (VC) Cohort for Historical-Prospective Study of Mortality Experience

Cohort membership:	*On Dow payroll*
	Actual exposure, or potential exposure
	Exposed at least 4 consequtive work-weeks
Types of exposure:	*Continuous*
	Intermittent

TABLE XIII.   Vinyl Chloride (VC) Cohort for Historical-Prospective Study of Mortality Experience. Cohort Membership

Currently employed	
Formerly with VC unit	
Presently with VC unit	
Present status determined	*100%*
Formerly employed	
Retired	
Terminated	
Disabled	
Dead	
Present status determined	*100%*

vinyl chloride is 1 ppm as a time-weighted-average over an 8-hour day. Table XI illustrates the changes in the standard for vinyl chloride that have occurred over the years.

In defining our vinyl chloride cohort for epidemiologic study, it was decided to include (Table XII) only those on the Dow payroll with documented exposure, or significant potential for exposure, to vinyl chloride for at least a four-week-consecutive period. We were able to identify a total of 533 persons meeting these criteria; this group is composed of persons who are working now, or who have worked previously, in the vinyl chloride plant and those who worked with vinyl chloride in the past, but are no longer employed by Dow in any department. The present status of our currently employed workers (Table XIII) was relatively easy to ascertain, but it took considerable effort to arrive at the 100% follow-up figure for former employees. It is our belief, however, that close to 100% follow-up is necessary for proper assessment of the risks that may be faced by any group of chemical workers.

The vinyl chloride cytogenetics study is one of several involving Texas Division employees. A special epidemiologic

TABLE XIV. *Reproductive Consequences of Vinyl Chloride Exposure: Survey of Workmen's wives*

	Participation	
	Number of women	Percentage of group
Study group	205	88%
Control group	144	81%

TABLE XV. *Comprehensive Surveillance Program for All Workers Exposed to Toxic Chemicals*

Periodic Evaluations
    History and Physical Examination
    Laboratory Tests
        Computerize Findings

Mutagenic Evaluation
    Cytogenetic Monitoring
        Routine Exposures
        Accidental Over-Exposures
    Other Mutagenicity Tests

Conventional Epidemiology
    Mortality
    Morbidity

Special Epidemiology
    Reproductive Studies
    Hospital Records
    Vital Statistics

study is now underway to investigate the possibility that increased rates of fetal loss or birth defects in offspring are experienced by wives of vinyl chloride workmen. We have had a gratifying response from the women - both in the study group and a control group - who were asked to participate. The study group was composed of women married to currently employed Dow workmen identified as having had at least two consecutive months of exposure to vinyl chloride. The comparison group was made up of wives of currently employed workmen who did not work in production, a laboratory, or the maintenance department. As shown in Table XIV, almost 350 women, 88 per cent of the study group, and 81 per cent of the control group, agreed to discuss these personal matters with the trained, University

of Texas interviewers. The interviewing is complete, the data
are being analyzed, and we hope to have a report of the final
results in the literature soon.

Our major contribution to this conference is to demonstrate the feasibility of studying the whole person and that
person's family through coordinated, long-term investigations
(Table XV). Life-disease processes will be recognized and recorded by medical monitoring and insurance programs, mutagenic
exposures and somatic consequences will be recognized by appropriate test methods, and conventional and special epidemiologic studies will be used to evaluate occupational disease
and germinal mutation problems. It is only through such a coordinated effort that the hazards of chemical exposure to the
human species can be accurately evaluated. Such a program not
only benefits the individual worker and family, but society as
a whole.

REFERENCES

Bloom, A.D., Nakagoma, Y., Awa, A.A., and Neriiski, S. (1970).
    Am. J. Public Health. 60, 641.
Buckton, K.E., Jacobs, P.A., Court Brown, W.M., and Doll, R.
    (1962). *Lancet. 2*, 676.
Ducatman, A., Hirschhorn, K. and Selikoff, I.J. (1975). *Muta-
    tion Res. 31*, 163.
Fischer, P., Golob, E., Kunze-Mühl, E., Haim, A.B., Dudley,
    R.A., Müllner, T., Parr, R.M., and Vetter, H. (1966).
    Radiat. Res. 29, 505.
Fleig, I., and Thiess, A.M. (1974). *Arbeitsmed. Sozialmed.
    Praeventimed. 9*, 280.
Funes-Cravioto, F., Lambert, B., Lindstein J., Ehrenberg, L.,
    Natarajan, A.T., and Golkar, S. (1975). *Lancet. 1*, 459.
Garza Chapa, R., Leal, C.H., Alvarez, M., and Sanchez, F.J.
    (1976). *In* "Abstracts V International Congress of Human
    Genetics" (S. Armendares and R. Lisker, eds), no. 325.
    Excerpta Medica, Amsterdam.
German, J. (1972). *Progr. med. Genet. 8*, 61.
Kilian, D.J., and Picciano, D. (1976). *In* "Chemical Mutagens:
    Principles and Methods for Their Detection", Vol. 4
    (A. Hollaender, ed.), p. 321. Plenum, New York.
Kilian, D.J., Picciano, D.J., and Jacobson, C.B. (1975). *Ann.
    N.Y. Acad. Sci. 269*, 41.
Kilian, D.J. Moreland, F.M., Benge, M.C., Legator, M.S. and
    Whorton, E.B., Jr. (1977a). *Mutation Res.*, in press.
Kilian, D.J., Moreland, F.M., Benge, M.C., Legator, M.S. and
    Whorton, E.B., Jr. (1977b). *In* "Handbook of Mutagen Testing"
    (B.J. Kilbey, ed.), Elsevier/North Holland, Amsterdam.

Mulvihill, J.J. (1975). *In* "Persons at High Risk of Cancer.
    An Approach to Cancer Etiology and Control" (J.F.
    Fraumeni, Jr., ed.), p. 3. Academic Press, New York.
Occupational Safety and Health Administration (1974). *Federal
    Register. 39*, 35889.
Picciano, D.J., Flake, R.E., Gay, P.C., and Kilian, D.J.
    (1977). *J. Occup. Med.*, in press.
Purchase, I.F.H., Richardson, C., and Anderson, D. (1976).
    Proc. Roy. Soc. Med. 69, 32.
Purchase, I.F.H., Richardson, C., and Anderson, D. (1975).
    Lancet. 2, 410.
Vaughan, J. (1962). *Int. Rev. Exp. Path. 1*, 243.
Vigliani, E.C., and Forni, A. (1976). *Environ. Res. 11*, 122.

THE RELATIVE ROLE OF VIRUS INDUCED TRANSFORMATION AND OF CYTO-
GENETIC CHANGES IN THE CAUSATION OF CERTAIN HUMAN AND EXPERI-
MENTAL MALIGNANCIES

George Klein

Department of Tumor Biology
Karolinska Institute
Stockholm
Sweden

EPSTEIN—BARR VIRUS (EBV) SYSTEM

EBV is a lymphotropic herpesvirus, capable of transforming
(immortalizing) human and certain non-human primate B-lymphocytes into established lymphoblastoid cell lines carrying
multiple copies of the viral genome and a virally determined
nuclear antigen, EBNA. It is regularly associated with African
Burkitt's lymphoma (BL), a monoclonal proliferation of EBV-
carrying malignant B-lymphoblasts. It is also known to induce
a self-limiting lymphoproliferative disease, infectious mono-
nucleosis (IM), on primary infection, particularly in teenagers
and young adolescents. In acute IM, there is temporary poly-
clonal proliferation of EBV carrying B-blasts, accompanied by
the rapid appearance of EBV-specific killer T-cells and subsequent complete regression. For some time, it was believed
that the difference between the self-limiting and the malignant
disease was determined by the host response. According to this
idea, the immune surveillance of the BL patient would have
broken down under the impact of some geographically variable
environmental factor, perhaps chronic holoendemic malaria,
leading to the emergence of a progressively growing clone.
While the primary EBV-induced lymphocyte immortalization and
the immune breakdown are still probably part of BL causation,
cytogenetic changes appear to play an essential role, in
addition. Manolov and Manolova (1972) were the first to detect
a characteristic 14q+ marker in 10 of 12 Burkitt lymphoma

biopsies. Subsequently, Jarvis *et al.* (1974) found the same
marker in all 7 BL biopsies they have studied. Our group
found the same marker in 13 of 14 BL biopsies and in 3 of 4
EBV-negative, Swedish lymphomas. The extra band added to the
long arm of chromosome 14 appeared to have been translocated
from chromosome 8. The marker was not found in recently established, EBV-carrying lymphoblastoid cell lines of normal or
monunucleosis origin or in *in vitro* EBV-transformed cord blood
cell lines. EBV-carrying lines of non-malignant (*i.e.* non-BL)
origin were purely diploid in the beginning, as a rule. Later,
they often changed to aneupliody but did not acquire the 14q+
marker in any of the cases studied so far. In contrast, none
of the BL-derived lines were purely diploid. The 14q+ marker
negative minority had other chromosomal anomalies.

It is thus clear that the 14q+ marker is not specifically
associated with either EBV or with Burkitt's lymphoma, but it
appears to be restricted to lymphomas, particularly of the
B-cell type. Also, somatic hybridization studies between EBV-
carrying BL or NPC cells and mouse fibroblasts have shown
that chromosome 14 is *not* required for the maintenance of the
EBV-genome and so is probably not an important integration
site for the virus.

Recently, we have extended this study further, by testing
cytogenetically characterized, EBV-transformed normal and BL-
derived lines for tumorigenicity in nude mice (Nilsson *et al.*,
1977). Diploid lines of non-malignant (*i.e.* non-BL) derivation
failed to grow, whereas chromosomally altered BL and non-BL
lines (including long propagated, normal-derived lymphoblastoid cell lines) grew more or less regularly.

Against this background, it appears probable that Burkitt's
lymphoma develops in three distinct stages. On primary infection, EBV establishes a latent infection in a certain
number of B-lymphocytes, with subsequent persistence of the
virus converted cell. An environmental factor, possibly
chronic holoendemic malaria, leads to a promotion-like process
as the second stage. Conceivably, this could happen through
the stimulation of Bücell proliferation and/or suppression
of the T-cell response. This paves the way for the third step,
the cytogenetic change, required for full autonomy. Specific
chromosomal anomalies, the 14q+ marker in particular, may
play an important, but not exclusive role. The frequent
involvement of a particular chromosome may be related to the
tern of a given target cell, including responsiveness to
growth control. In this connection, it is noteworthy that
McCawe *et al.* found (1975) that chromosome 14 was particularly
prone to chromosomal aberrations in ataxia teleangiectasia, a
condition that predisposes for high leukemia and lymphoma
incidence.

VIRUS INDUCED MOUSE LYMPHOMAS OF T–CELL ORIGIN

Dofoku *et al.* (1975) found that 10 of 11 thymic mouse
lymphomas arising in the AKR strain were trisomic for chromosome 15. The 11th case was trisomic for chromosome 12. Trisomy was not found in normal AKR tissues.

In a recent collaboration with Francis Wiener, Nechama
Haran-Ghera *et al.* (to be published) we have studied the
chromosomal constitution of primary mouse lymphomas induced
by the radiation leukemia virus (RadLV) in C57Bl mice. We
have used both the radiation dependent (D) and the radiation
independent, autonomous (A) variants of the virus. All 10 A-
RadLV induced lymphomas had a chromosome 15 trisomy. Six D-
RadLV leukemias were also trisomic for chromosome 15, 3 appeared to be diploid, and 5 were mixed populations of diploid
and 15-trisomic cells. Since all mice were killed for the
cytogenetic study, it is not yet known whether there was any
biological difference between the diploid and the trisomic
group, *e.g.* with regard to autonomous growth potential. RadLV
and the AKR-associated Gross virus are related but differ with
regard to N- *vs.* B-tropism, host range, and the course of the
pathogenic process. They induce leukemia in the same target
tissue, however, at least in the majority of the cases. Therefore, these findings further support the possibility that neoplastic development in certain tissues is accompanied by
specific chromosomal changes.

The role of cytogenetic changes in tumor progression is
emphasized by the fact that the AKR mouse has been inbred for
high leukemia incidence and is known to carry at least 4
different genetic systems that favor leukemia development in
different ways. They include the germ-line integrated, leuke-
mogenic Gross virus, localized at two specific sites (Akv-1
and 2), the viral amplification system ($Fv-1^n$) and the
MHC-linked $Rgv-1^s$ locus, probably acting as an "immune un-
responsiveness" factor in relation to Gross virus induced,
lymphoma associated membrane antigens. A fourth cytogenetic
system, demonstrated by F_1 transplantation test, conveys a
high susceptibility on the T-lymphocyte itself to undergo a
neoplastic change. In spite of these cooperating systems that
all favor leukemia, development of the disease takes considerable time (3–6 months) in relation to the life span of the
mouse. The regular occurrence of the chromosome 15 trisomy
suggests that cytogenetic changes are important in this process. It is conceivable that chromosome 15 is an integration
site of the proviral genome. If so, trisomy could lead to
viral amplification of the type described by Jaenisch (1976)
in the Moloney system. An alternative is that chromosome 15

does not carry the integrated provirus but is involved in con-
ditioning the responsiveness of the T-lymphocyte to *its* growth
regulation, in analogy with what was postulated above for the
relationship between the 14q+ marker and Burkitt's lymphoma.

OTHER SYSTEMS

Somatic hybridization studies performed by Henry Harris
and our group in collaboration over a number of years have
shown the importance of specific chromosomes for the suppression of malignant behavior in hybrids between malignant and
normal (or low malignant) mouse cells (Bregula *et al.*, 1971;
Klein *et al.*, 1971; Wiener *et al.*, 1973, 1974a, 1974b). It
appeared that the malignant change may have involved some
losses, on the part of the normal genome, that could be
compensated by fusion with a normal cell. This was the case
even with polyoma virus induced tumors, showing that viral
transformation is probably insufficient, by itself, to bring
about full malignancy in the polyoma system; changes in host
control were required as well. Such evidence supports the
early findings of Vogt *et al.*(quoted by Dulbecco, 1976) and
the more recent findings of Renger and Basilico (1972). It is
of the utmost importance now to analyze whether the same or
different normal-cell derived chromosomes are involved in
suppressing the malignant behavior of tumors that arise in
different tissues.

REFERENCES

Bregula, U., Klein, G., and Harris, H. (1971). *J. Cell. Sci.*
    8, 673.
Dofuku, R., Biedler, J.L., Spengler, B.A., and Old, L.J.
    (1975). *Proc. Natl. Acad. Sci. (USA) 72*, 1515.
Dulbecco, R. (1976). *Science 192*, 437.
Jaenisch, R. (1976). *Proc. Natl. Acad. Sci., (USA) 73*, 1260.
Jarvis, J.E., Ball, G., Rickinson, A.B. *et al.* (1974).
    Int. J. Cancer 14, 716.
Klein, G., Bregula, U., Wiener, F., and Harris, H. (1971).
    J. Cell Sci. 8, 659.
Manolov, G. and Manolova, Y. (1972). *Nature 237*, 33.
McCaw, B.K., Hecht, F., Harnden, D. *et al.* (1975). *Proc.*
    Natl. Acad. Sci. (USA) 72, 2071.
Nilsson, K., Giovanella, B.C., Stehlin, J.S., and Klein, G.
    (1977). *Int. J. Cancer 19*, 337.
Renger, H.C. and Basilico, C. (1972). *Proc. Natl. Acad. Sci.*
    (USA) 69, 109.

Wiener, F., Klein, G., and Harris, H. (1971). *J. Cell Sci.* 8, 681.
Wiener, F., Klein, G., and Harris, H. (1973). *J. Cell Sci.* 12, 253.
Wiener, F., Klein, G., and Harris, H. (1974a). *J. Cell Sci.* 15, 177.
Wiener, F., Klein, G., and Harris, H. (1974b). *J. Cell Sci.* 16, 189.
Zech, L., Haglund, U., Nilsson, K., and Klein, G. (1976). *Int. J. Cancer* 17, 47.

THE SYNAPTINEMAL COMPLEX AND NON-DISJUNCTION
IN THE HUMAN

Preben Bach Holm
Søren Wilken Rasmussen

Department of Physiology
Carlsberg Laboratory
Copenhagen, Denmark

The ultrastructure of meiosis in five normal human males
was analyzed using the technique of three-dimensional reconstruction. Twenty-two pachytene nuclei were reconstructed and
the length of the autosomal synaptinemal complexes measured.
The total mean length of these complexes amounted to $230.8 \pm$
16.3 µm. Differences in mean length among individuals or at
different stages of pachytene were not significant. The
centromeric heterochromatin of all autosomal bivalents was
identified, and allowed classification of the bivalents on the
basis of absolute length, relative length and centromere index.
Bivalents 1,9, and 16 display a bipartite organization of the
centromeric heterochromatin. Most bivalents could be properly
classified on the basis of length and centromere index, but
bivalents 6, 7, 8, 10, 11 and 13, 14, 15 could be assigned
only as belonging to group C and D respectively.

The sex chromosomes are condensed into a spherical body
next to the nuclear envelope. In 19 of the 22 nuclei the X
and Y chromosomes were paired with a synaptinemal complex over
a distance of up to 1.0 µm. Each of the five acrocentric bivalents may be associated with a nucleolus, but generally only
one major nucleolus is present and attached to one of the bivalents.

Reconstruction of one diplotene nucleus demonstrated the
degradation of the synaptinemal complex at this stage. The
bivalents could be identified to the same extent as in pachytene from length and centromere index measurements.

Three nuclei at metaphase were reconstructed. The centromeric heterochromatin of the 23 bivalents could be identified
as well as 92 kinetochores. In the chromatin of the bivalents,
30-60 fragments of central region material of the synaptinemal
complex were observed. These may at least in part represent
synaptinemal complex chiasmata.

REFERENCES

Holm, P.B., and Rasmussen, S.W. (1977). *Carlsberg Res. Commun.*
    42. (in preparation).

PRESENT STATUS OF AUTOMATED CHROMOSOME ANALYSIS

Claes Lundsteen[1]
John Philip

Section of Teratology, Department of Obstetrics and
Gynecology and Department of Paediatrics, Rigshospitalet,
University of Copenhagen
Denmark

INTRODUCTION

During the past 20 years the field of cytogenetics has increased rapidly due to the development of new techniques and
to clinical applications of these techniques.

Like in other biological and medical disciplines automation
has been introduced in cytogenetics.

The purpose of this status report is to:

1. Describe the needs for automation of cytogenetics.
2. Describe different approaches to and results of automation
   of cytogenetics, and illustrate the technical problems
   which still remain.
3. Illustrate the present and potential applications of automation of cytogenetics.

APPLICATIONS OF CHROMOSOME ANALYSIS

Chromosome analysis may be used for a number of purposes
as illustrated in Table I. Chromosome analyses are carried
out for profylactic, diagnostic and research purposes and are
used in population studies. Fetal chromosome analysis is

[1]*Supported by the Danish Medical Research Council, the
Danish Council for Technical Research, and the A.P. Moller
and Chastine Mc-Kinney Moller Foundation.*

TABLE I. APPLICATIONS OF CHROMOSOME ANALYSIS

	Profylactic	*Diagnostic*	*Population studies*	*Research*
Fetal chromosome analysis	+		+	+
Suspicion of chromosome diseases		+	+	+
Cancer (leukemia) evaluation of treatment		+		+
Induced chromosome damage	+	+	+	+

indicated in women having increased risk of having children
with chromosome abnormalities, and chromosome analysis is
indicated in individuals suspected of chromosome diseases.
Chromosome analysis is used in cancer diagnostics, especially
in leukemia, and may be used for evaluation of medical treatment of leukemia. Finally the detection of chromosome aberrations and sister chromatid exchanges can be used for measuring damages induced by environmental factors.

THE NEED FOR AUTOMATION IN CYTOGENETICS

    The need for automation in cytogenetics is illustrated
in Table II.

TABLE II. THE NEED FOR AUTOMATION IN CYTOGENETICS

	Clinical Cytogenetics	Environmental Toxicology
Preparation of chromosome specimens	++	+++
Location and selection of metaphases	+ to +++	+++
Karyotyping	++	?
Scoring of chromosome aberr. and SCE [a]	+	+++

[a] *SCE: Sister chromatid exchange*

1. *Preparation of Chromosome Specimens.* Manual preparation of chromosome specimens is not standardized and is a time consuming procedure. Automation is needed when large numbers of samples have to be processed and for standardization of the procedures.

2. *Location and Selection of Metaphases.* Search for and selection of metaphases are time consuming and boring, especially when the slides have a low mitotic index and when large numbers of metaphases have to be examined. Automation of metaphase search is therefore needed and automatic selection of metaphases of high quality will facilitate and speed up most of the work in cytogenetics.

3. *Karyotyping.* The procedure of karyotyping needs to be automated in order to get rid of the photographic work and of the boring and time consuming work of cutting chromosomes out of photographic prints with scissors and in order to standardize and quantitize the procedure.

4. Scoring of Chromosome Aberrations and Sister Chromatid Exchanges. Automation is needed here because large numbers of metaphases may have to be examined and because the scoring is a quantitative analysis which is difficult to standardize manually.

APPROACHES TO AUTOMATION IN CYTOGENETICS.   METHODS, RESULTS
AND TECHNICAL PROBLEMS

1. Automation of Preparation of Chromosome Specimens. A
semi-automated machine for hypotonic treatment and fixation
of cell cultures and a semi-automated aspirator carrying out
the decanting procedures have been constructed by Wulf (1976a).
The machines (Figs. 1 and 2) are commercially available (HETO,
Birkerød, Denmark).

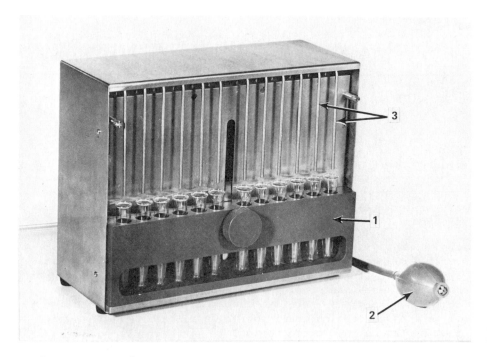

FIGURE 1.   *Aspirator machine.   The machine aspirates the*
supernatant simultaneously from 12 test tubes.
For description see text.   Available from HETO,
Birkerød, Denmark.

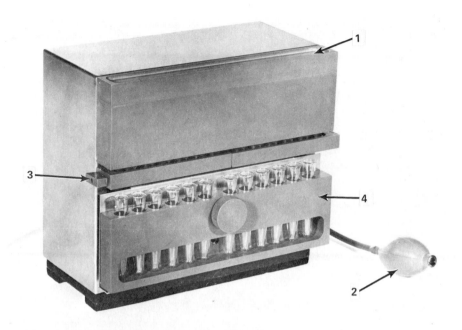

FIGURE 2. Preparation machine. The machine automatically adds hypotonic liquid or fixative to 12 test tubes while these are being shaken. For description see text. Available from HETO, Birkerød, Denmark.

12 samples in test tubes are handled simultaneously. The tubes are placed in a rack (Fig. 1, 1) which can be mounted on both the aspirator and the preparation machine.

After centrifugation the tubes are placed in the rack (Fig. 1, 1), which is mounted on the aspirator (Fig. 1). By raising the rack and pushing a bulb (Fig. 1, 2), 12 needles (Fig. 1, 3) get into contact with the supernatant, and aspiration is carried out simultaneously in all 12 tubes.

The rack is then mounted on the preparation machine (Fig. 2). The hypotonic liquid or fixative is filled into a chamber (Fig. 2, 1). By pushing a bulb (Fig. 2) the liquid starts dripping into each tube, and by pushing a button (Fig. 2, 3) the rack (Fig. 2, 4) starts shaking while the liquid is being added.

Using these machines, one technician is capable of handling approximately 90 samples during an 8-hour workday; this number is about twice as many as can be handled without the

machines. The efficiency of the machines was proved during a
population study of 4,591 tall Danish men (Philip *et al.*,
1976).

Melnyk *et al.*, (1976) have developed an advanced system
for preparation of chromosome specimens. The system is designed around a 12-well culture tray which passes through a number of semi-automated units carrying out the hypotonic treatment, fixation, centrifugation, aspiration, and slide making
procedures in all 12 wells simultaneously.

Test performance has shown an overall reduction of time
by a factor of 8. Using this system, two technicians are
able to process 576 specimens in an 8-hour workday. The
operation of the system is simple. The preparation of the
specimens is standardized and the quality of the slides is
good.

The system is not yet commercially available.

2. Automated Metaphase Finding. The search for and location of
metaphases on chromosome slides is a fundamental procedure in
cytogenetics.

The basic procedure in automatic metaphase finding is registration of small, dark-stained and close-lying objects on
chromosome slides. This registration may be accomplished in
different ways and using different kinds of equipment. One
way is to use an automatic microscope with a TV-camera and
TV-monitor combined with a microprocessor (or minicomputer)
including a mataphase detector and autofocus unit (Fig. 3).

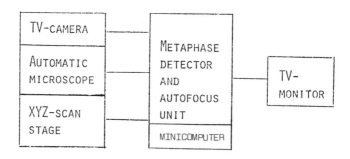

FIGURE 3.   *Basic components of an automatic metaphase finder.*
             For description see text.

The automatic microscope has the facilities of automatic
focusing and automatic moving of the scanning stage which is
necessary for the metaphase finding. The slide is mounted on
the scanning stage which moves stepwize. For each step of
the stage a new area of the slide is viewed by the TV-camera,
and the video signal is passed on to the metaphase detector
unit which registers the presence and position of metaphases.
When enough metaphases have been located and their coordinates
stored in the computer, the metaphases are relocated and displayed on the TV-monitor for evaluation by the operator.

A number of metaphase finders have been constructed and
seem to work well. The fastest show a search rate of about
2 min./slide (Green and Neurath, 1974; Farrow *et al.*, 1976).
Miss rates are about 5-10%, and about 15-20% false positives
are registered (Castleman and Melnyk, 1975). More can be
done to improve the efficiency of the systems. A major step
forward would be a metaphase finder that evaluates the quality
of the metaphase and selects only the good ones (Green and
Neurath, 1974).

Metaphase finders can be used for a number of purposes,
either as a "stand alone machine" where the rest of the cytogenetic work is done manually or in conjunction with different
kinds of automatic chromosome analysis systems as described
below.

3. Semi-automated karyotyping. Because it has been difficult to obtain satisfactorily low error rates in automated
classification of banded chromosomes, and because the human
analyst can classify chromosomes with high accuracy (Lundsteen
et al., 1976). Wulf (1976b) is developing a machine which in
a semi-automated way can produce prints of photographed karyotypes of metaphases, whose individual chromosomes are classified by the operator.

The instrument (Fig. 4) is an ordinary projection microscope modified to project the metaphase image on to a piece
of paper (Fig. 4, 1) in front of the operator. The invididual
chromosomes of the metaphase are framed by using a viewfinder
(Fig. 4, 2) and classified by the operator who presses a button on a keyboard (Fig. 4, 3) corresponding to the chromosome
number. The electronic and optic systems of the instrument
ensure that for each entry on the keyboard the framed chromosome is handled in such a manenr that it is photographed on a
film with the correct orientation and in a position according
to its correct location in a conventional karyotype. When
all chromosomes have been classified and photographed the
final print of the photographed karyotype is obtained.

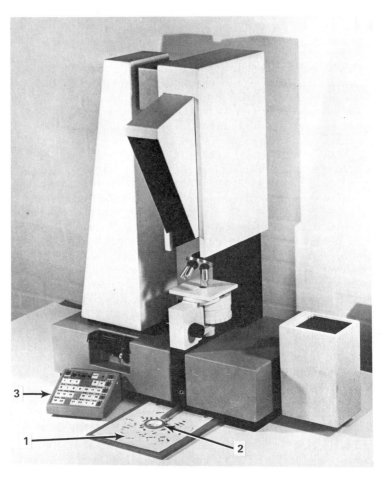

FIGURE 4. Semi-automated machine for production of karyotypes constructed by H.C. Wulf, Copenhagen.
1. Metaphase projected on to a piece of paper
through a modified projection microscope.
2. Viewfinder. 3. Keyboard. For description
see text. (With permission of H.C. Wulf).

A preliminary test of the machine indicates that a karyotype may be produced in 5 - 10 minutes and a simple chromosome count can be carried out in 20 - 40 seconds depending on the quality of the metaphase (Wulf, personnel communication 1977).

Alone or combined with an automatic metaphase finder this instrument may be very useful in clinical cytogenetics.

The most elaborated functional system for production of karyotypes is constructed by Castleman (Castleman and Melnyk, 1975). That system is the first which can produce computer-generated karyotypes with a speed which can compete with the manual methods. Basic components of the system are an automatic microscope with a vidicon TV-camera, a metaphase finder and autofocus unit, an image digitizing system, a PDP 11 minicomputer, a gray level display and curser joystick system for manual interaction with the system, and a hardcopy output device (Fig. 5).

The system operates as follows: A suitable number of metaphases are located by the metaphase finder. Metaphases of acceptable quality are evaluated by the operator and their coordinates are stored in the PDP 11 computer. The metaphases are scanned and digitized and the data are stored in the computer (on disc).

The first step in the analysis of a metaphase is automatic isolation of the individual chromosome in the metaphase using a threshold procedure. At this stage the operator may interact and separate touching chromosomes by means of the display and joystick curser unit. Next step is the measuring (feature extraction) and classification of the individual chromosomes. The constructed karyotype is displayed on the gray level display, and errors can be corrected by the operator by means of the joystick curser unit before a permanent image of the approved karyotype is produced by the hardcopy output device.

With operator interaction, the system can produce 3 - 4 banded karyotypes per hour. This is as fast as it can be done manually when the metaphases are available on photographic prints. Thus the time spent on metaphase finding and obtaining photographic prints is saved. The system can also handle non-banded metaphases. The speed is here about 10 karyotypes per hour which is definitely faster than the manual method.

It is mainly operator interaction for correcting errors in classification of chromosomes made by the system which slows down the speed. Analysis of non-banded metaphases demands less operator interaction than analysis of banded metaphases, and is therefore faster.

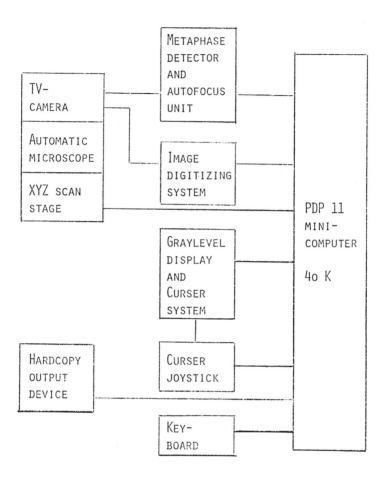

FIGURE 5.   Basic components of a system for computer-assisted karyotyping, (Castleman et al., 1975).   For description, see text.

Some changes in the hardware configuration suggested by Castleman and Melnyk (1975) might speed up the analysis; however, it also seems reasonable to implement more advanced analysing programs to reduce the error rate of the automatic classification of banded chromosomes which is at present about 10 %.

We have done some work to evaluate the possibility for reducing the error rate of automatic chromosome classification. Two studies on the reliability of visual chromosome analysis have been carried out.

In the first study (Lundsteen *et al.*, 1976) we compared the reliability of visual identification of *isolated*, banded photographed chromosomes with the reliability of visual identification of the chromosomes in *intact metaphases*. We name the first process, where isolated chromosomes are identified individually, *Classification*, and the second process, where all chromosomes belonging to a metaphase are available for comparison, *Karyotyping*. In the first (classification) experiment the error rate was 3 % and in the second (karyotyping) experiment the error rate was 0.1 %.

In another study (Lundsteen and Granum, 1977) the classification and karyotyping experiments were carried out on computer generated profiles of the same chromosomes. A profile is a simple data representation of the digitized chromosome image. Here the error rates were 5 % for classification and 0.5 % for karyotyping.

Some of the conclusions of our studies are:
1. Some information which is important for visual classification (and presumably also for computer classification) is lost by the simple intergration by which chromosome profiles are generated from digitized chromosome images. A more sophisticated method for extracting the important chromosome features from the digitized chromosome images is therefore needed.
2. Computer classification of isolated profiles is not optimized. More advanced classification methods are needed.
3. Computer karyotyping should simulate the human way of karyotyping, thus taking advantage of the a priori knowledge of the composition of the metaphase (normally pairs 1 - 22 and two sex chromosomes) and being able to make appropriate comparisons between individual chromosomes within the metaphase.

As to the first point, a method for detection of banding patterns in the digitized chromosome image is being developed (Granum and Lundsteen, 1977). The method is capable of detecting chromosome bands which are not identifiable on the corresponding chromosome profiles.

As to the second point some very preliminary data indicate that the error rate of automatic classification of isolated profiles may be reduced to about 5 %. Following the suggestions of the third point, we expect a final error rate of automatic karyotyping of less than 1 %.

4. (Semi-) Automated Scoring of Chromosome Aberrations and Sister Chromatid Exchanges. Automated analysis of chromosome damage is an underdeveloped field in automated cytogenetics. Only few groups have been working in this area.

One way of speeding up the scoring of aberrations would therefore be to use an automatic metaphase finder to locate the metaphase and to classify and count the aberrations manually.

Wald *et al.*, (1976) have developed such a system. It consists of an automatic microscope with a vidicon TV-camera, an autofocus and metaphase finder unit, a video monitor, a minicomputer, and a hand-controller for fine adjustment of the system and for the manual scoring of aberrations. Initially the microscope automatically locates the metaphases on sets of six slides. The operator sits at the video display, and uses the hand-controller to call up previously located metaphases and to record the number of ring chromosomes, dicentric chromosomes, and fragments in each metaphase. Finally the resultant data of the analysis are printed out.

The system is easy to operate and the rate by which metaphases are scored for aberrations is increased by a factor of 5.

The system has also been used for scoring of sister chromatid exchanges and seems to facilitate the scoring.

In order to increase the speed further and to obtain objectivity in the analysis, some attempts have been made to automate the scoring procedures not only for chromosome aberrations (Aggarwal and Fu, 1974; Wald *et al.*, 1976; Farrow *et al.*, 1976), but also for sister chromatid exchanges (Zack *et al.*, 1976).

In automatic scoring of aberrations the metaphases are scanned and digitized. Using a threshold precedure, the individual chromosomes are separated and by a pattern recognition procedure the individual chromosomes are analysed for the basic features of the aberrations: No centromere (acentric fragment), two centromeres (dicentric chromosome) and ring (ring chromosome). Preliminary data using such methods show an overall recognition accuracy of about 90 % (Aggarwal and Fu, 1974; Wald *et al.*, 1976).

Zack *et al.*, (1976) have recently reported a method for automatic measurement of sister chromatid exchange frequency. Metaphases are digitized by a computer-controlled microscope-television system. The individual chromosomes in the metaphase are identified by a threshold procedure. The principle in detecting sister chromatid exchanges is the recognition of sharp reversals of bright and dull chromatids at the sites of exchanges.

Preliminary data show an error rate of automated scoring of approximately 10% as compared with manual scoring.

5. Chromosome Analysis by Flow Microfluorometry. A completely different approach to automated chromosome analysis is flow microfluorometry (Van Dilla *et al.*, 1976). The principle in flow analysis is that isolated objects (in casu fluorescent stained isolated chromosomes in suspension) successively pass by a fine laser beam which excites the individual chromosomes. The emitted light from each chromosome is measured by a photo-multiplier and the amount of light expressing the DNA content or some other feature of the chromosomes is used for sorting out the chromosomes. Using ethidium bromide as a stain, 7 groups of chromosomes are obtained in analysis of human chromosomes. However, using Hoechst 33258 more than 15 groups can be obtained (Gray *et al.*, 1977).

Obvious advantages of flow analysis is the high speed (100,000 chromosomes/min.) and that slide making is not necessary. Though a number of technical problems remain it is worthwhile mentioning some potential applications of flow analysis of human chromosomes.

Assuming some improvements of the system, *e.g.* the capability of determining the centromeric index most individuals with chromosome aneuploidy and structural rearrangements could be detected and the exact diagnosis confirmed by conventional banding analysis. If 100,000 chromosomes per person were analysed it should be possible to screen about 40 - 50 individuals per day using flow systems (Van Dilla *et al.*, 1976).

Chromosome aberrations like dicentrics and fragments might also be counted by flow analysis. Some preliminary data obtained from flow analysis of irradiated chinese hamster cells support this assumption (Van Dilla *et al.*, 1976).

DISCUSSION AND CONCLUSIONS

Table III is an attempt to illustrate the present and
potential applications of automation of cytogenetics.
1. Some automatic procedures for preparation of chromosome
specimens are already in use and the potential usefulness of
an instrument like the one constructed by Melnyk *et al.*,
(1976) is obvious.
2. Automated location of metaphases is in use to day and will
without any doubt be used in the future, especially where
large numbers of metaphase are needed and when the slides
have a low mitotic index.
    The capability of an automated selection of the good meta-
phases will obviously increase the applicability of metaphase
finders.

TABLE III. PRESENT AND POTENTIAL APPLICATIONS OF
AUTOMATION IN CYTOGENETICS

Methods	Clinical cytogenetics		Environmental toxicology	
	Present	Potential	Present	Potential
Automation of chromosome preparation	+	++	+	+++
Automated methaphase 1) finding	++	++	+++	+++
2) selection	0	++	0	+++
Automated karyotyping 1) banded	+	+++	?	?
2) non-banded	+?	+?	?	?
Automated analysis of chr. damage	+?	+?	+++	+++
Flow analysis of chromosomes	0	+++	0	+++

3. Automated karyotyping using computer assisted classification of banded chromosomes may at present be fast enough to compete with manual methods. However, improved techniques would greatly increase the applicability in clinical cytogenetics. A system like that developed by Wulf (1976b) may also prove useful.

Computer assisted karyotyping of non-banded metaphases is faster than the manual method. However, the accuracy of detection of abnormalities is unknown. If only a few abnormalities are missed using automated karyotyping of non-banded chromosomes, this method might prove useful for screening as well as for diagnostic purposes.

4. Semi-automated analysis of chromosome aberrations is in use and will presumably also be used in the future. It is possible that the methods will be further automated.

Semi-automated analysis of sister chromatid exchanges can be used today and will probably be further automated.

Flow analysis of chromosomes is not in use at present but the potential usefulness is obvious.

Automation is already a useful tool in clinical cytogenetics and in environmental toxicology. Improvements of the automated methods will substantially increase their usefulness and ought to have a high priority in future work.

SUMMARY

Automation of cytogenetic techniques is needed in order to cope with the increasing demands for chromosome analyses.

A semi-automated, commercially available system for handling part of the *preparation procedures* for metaphase chromosomes shows a time saving factor of 2. A more elaborated but not commercially available system shows a time saving factor of 8.

Automatic *metaphase finders* have been constructed and work satisfactorily. Methods for automatic selection of good metaphases are still needed.

A system is being constructed which produces *photographed karyotypes* of metaphases of which the individual chromosomes are classified by the operator. A print of a photographed karyotype may be produced in 5 - 10 minutes.

A system for *computer assisted karyotyping* can, with operator interaction, produce 3 - 4 banded or 10 non-banded karyotypes per hour.

Manual scoring of chromosome aberrations facilitated by using an automatic metaphase finder provides a time saving factor of 5. *Automated scoring* of chromosome aberrations and sister chromatid exchanges is being developed.

Flow analysis of human chromosomes is at the research level; however, potential applications are screening for chromosome abnormalities and scoring of aberrations.

Automation is a useful tool in cytogenetics. Improvements of the automated methods will increase their usefulness and ought to have priority in future work.

REFERENCES

Aggarwal, R.K. and Fu, K.S. (1974). *J. Histochem. Cytochem.* *22*, 561.

Castleman, K.R. and Melnyk, J.H. (1975). "An automated system for chromosome analysis". Final Report, Jet Propulsion Laboratory, Internal Document No. 5040-30.

Farrow, A.S.J., Green, D.K.,and Rutovitz, D. (1976). *In* "Automation of cytogenetics" (M.L. Mendelsohn,ed.), p. 68. Lawrence Livermore Laboratory, Conf. 751158.

Granum, E. and Lundsteen C. (1977). *In* "Poster presentations at the Fifth Engineering Foundation Conference on Automated Cytology" (E.M. Sullivan, ed.), p. 44. Los Alamos scientific laboratory. Conf. LA-6719-C.

Gray, J.W., Carrano, A.V., Langlois, R., and Van Dilla, M.A. (1977). *In* "Poster presentations of the Fifth Engineering Foundation Conference on Automated Cytology" (E.M. Sullivan, ed.), p. 37. Los Alamos scientific laboratory. Conf. LA-6719-C.

Green, D.K. and Neurath, P.W. (1974). *J. Histochem. Cytochem.* *22*, 531.

Lundsteen, C., Lind, A.M., and Granum, E. (1976). *Ann. Hum. Genet. (Lond.)*, *40*, 87.

Lundsteen, C. and Granum, E. (1977). *Ann. Hum. Genet. (Lond.)*, *40*, 431.

Melnyk, J., Persinger, G.W., Mount, B., and Castleman, K.R. (1976). *In* "Automation of cytogenetics" (M.L. Mendelsohn, ed.), p. 51. Lawrence Livermore Laboratory, Conf. 751158.

Philip, J., Lundsteen, C., Owen, D., and Hirschhorn, K. (1976). *Am. J. Hum. Genet. 28*, 404.

Van Dilla, M.A., Carrano, A.V., and Gray, J.W. (1976). *In* "Automation of cytogenetics" (M.L. Mendelsohn, ed.), p. 145. Lawrence Livermore Laboratory, Conf. 751158.

Wald, N., Li, C.C., Herron, J.M., Davis, L., Fatora, S.R., and Preston, K. (1976). *In* "Automation of cytogenetics" (M.L. Mendelsohn, ed.), p. 39. Lawrence Livermore Laboratory, Conf. 751158.

Wulf, H.C. (1976a). *In* "Chromosomes today", Vol. 5. (P.L. Pearson and K.R. Lewis, eds.), p. 439. John Wiley and Sons and Israel Universities Press, New York and Jerusalem.

Wulf, H.C. (1976b). *In* "Excerpta Medica International Congress Series" No. 397, Abstract No. 447.

Zack, G.W., Spriet, J.A., Latt, S.A., Granlund, G.H. and Yong, I.T. (1976). *J. Histochem. Cytochem. 24*, 168.

ON THE POSSIBILITIES OF DETECTING
CHROMOSOME ABNORMALITIES INDUCED
BY ENVIRONMENTAL AGENTS IN HUMAN
GERM LINE CELLS

Maj Hultén

Regional Cytogenetics Laboratory
East Birmingham Hospital, Bordersley Green East
Birmingham, England

J.M. Luciani

Laboratoire d'Histologie et Embryologie II
Faculté de Medicine
Marseille, France

Cytogenetic information about the effects of potentially
hazardous environmental agents on *human germ line cells* is indispensible. Recent experiments in mammals have clearly shown
that there is no simple way in which the rapidly increasing
knowledge about the effects on *somatic mitotic cells* (as reported by other colleagues at this Conference) may be extra-
polated to the *germ line cells* (review in UNSCEAR, 1977; van
Buul and Roos, 1977). Still, only the amount of a mutagen that
penetrates the germ line is of heritable significance.

A large body of literature exists concerning mutagenic
effects on the gametogenic tissues of various species. But,
again, extrapolation to man is not easy even from primates
(Jagiello and Lin, 1974; Brewen, 1976; van Buul, 1976; UNSCEAR,
1977). However, interspecific cytogenetic studies provide us
with insight into the kind of problems that we are likely to
encounter at the human level, so that the right questions can
be asked, and adequate techniques applied (Sankararanarayanan,
1976).

To date, little is known about human meiosis (reviews in
Hultén and Lindsten, 1973; Chandley, 1975; Ferguson-Smith,

1976; Jagiello *et al.*, 1976b), and almost nothing about the
effects of environmental mutagenic agents on this process. One
of the main reasons for this lack of knowledge is obviously
the difficulty involved in obtaining human gonadal biopsies
for investigation. There also seems to be a general reluctance
among human cytogeneticists to touch this type of material,
which is supposed to be very difficult to process, analyse and
interpret.

This presentation does not primarily attempt to summarize
and evaluate the present cytogenetic knowledge on human germ
line cells. Rather, a number of points will be discussed, with
special reference to the possibilities of detecting damage induced by environmental agents at the chromosome or DNA levels.
The different human germ cell stages will be searched for
practical diagnostic potential in the light of techniques used
within animal experimentation. In addition, some current hypotheses on the meiotic origin of human mitotic chromosome abnormalities will be discussed. It should be added that no new
data will be presented.

Clearly, meiosis is a much more complex process than mitosis. In order to facilitate the presentation, we have, as an
introduction, allowed ourselves to repeat some exceedingly
basic information on normal meiosis. Some of the apparently
complicated first meiotic pairing configurations may be quite
easily understood, by using any simple string model, as we have
found from personal experience.

A. *What Are the Differences between Human Male and Female Meiosis?*

The most obvious difference between male and female meiosis is the time of commencement and duration of events
(Sybenga, 1975; John, 1976). Human male meiosis is a post
pubertal event, continuing throughout the lifespan. In the
female, by comparison, the complete supply of oocytes is formed before birth and consequently they age with the individual.

Of the two meiotic divisions, first prophase, which is
genetically the most important one, is greatly prolonged and
specialized, and has been subdivided into five stages -
leptotene when the 46 chromosomal filaments individualize;
zygotene when the maternal and paternal partner chromosomes,
the homologues, start to pair to form bivalents; pachytene
when the maternal and paternal chromosomes are intimately
paired along their entire length, and non-sister chromatid
exchange, *i.e.* crossing-over occur; diplotene when the half
bivalents start to separate, held together only at the exchange
points, *i.e.* the chiasmata; and diakinesis when the bivalents

are more condensed, A further condensation occurs up to first
metaphase.

The segregation of the maternal and paternal centromeres
and the recombined chromosomes at first anaphase results in
two daughter cells with 23 chromosomes. The segregation of
their centromeres is delayed until second anaphase. Normally,
in the male one first spermatocyte results in four spermatozoa, while in the female one first oocyte produces one functional ovum only, due to the elimination of one of the daughter
cells at both first and second divisions as non-functional
polar bodies.

In the human male, the primitive germ line cells undergo
a series of changes during childhood leading to the development of early forms of spermatogonial stem cells (Heller and
Clermont, 1964; Oakberg, 1968; Monesi, 1972; Skakkebaek, 1976).
It is not until puberty that these cells proliferate by mitosis, reaching maturity as primary spermatocytes after approximately one month. First meiotic division then occurs with
a preleptotene stage of one day's duration. Leptotene and zygotene are brief, occurring within one week. Pachytene lasts
approximately two weeks, the final phases of the first and
all of the second meiotic divisions thereafter being completed within one day. Throughout a final three week period, the
four resultant spermatids go through spermiogenesis, a complex transformation without cell division. The spermatozoa so
produced undergo a final maturation phase acquiring motility
and fertilizing capacity, the entire process having taken
approximately 64 days.

It is important to note that in addition to dividing to
produce primary spermatocytes, the spermatogonia are also
replenished by cell division, so that the potential number of
spermatozoa which may be produced is vast.

In sharp contrast to the situation in the male, the female
germ line cells rapidly mature into oogonia during intrauterine life. Following a finite number of mitotic divisions, the
oogonia are transformed into oocytes. The population of germ
cells in the human female therefore has a fixed upper limit,
which is depleted with increasing age by atresia and ovulation.

Meiosis is not synchronized in the human female. Oocytes
enter first meiotic prophase at around the ninth week of
fetal gestational age (reviewed by Baker and Wai Sum O, 1976;
Luciani *et al.*, 1977a). Leptotene and zygotene are thought to
be brief compared to pachytene, which occupies about two weeks,
and is in evidence from the eleventh week followed by diplotene
from the thirteenth week of gestational age (Luciani, unpublished observations). At birth all the oocytes are in the diplotene stage of first prophase (Franci, 1977) in which they
remain until ovulation. Thus, the first and second meiotic

divisions do not occur until ovulation and fertilization.

B. *Which Sources of Material Are Available for Meiotic Invest-*
   igation in Man?

The materials traditionally used for investigations are
gonadal biopsies obtained from adults, usually men rather than
women.

BI. *Material From Male Gonads.*
   1. *Testicular biopsies from adult men.* Generally,
three categories of men have been investigated. Firstly, men
attending infertility clinics, when in addition to a testicular biopsy for histologically diagnostic purposes, a tiny
extra fragment of testicular parenchyma is taken for meiotic
chromosome analysis. Secondly, a small group of men, with
mitotic chromosome abnormalities, as well as a few healthy
men, with normal mitotic chromosomes, have volunteered to do-
nate a testicular fragment for research purposes. In addition,
testicular material obtained in connection with therapeutic
testicular surgery, for vasectomy, inguinal hernia or prosta-
tic cancer has been utilized to get some idea of "normal"
meiosis. In testicular biopsies from these types of source,
usually all the stages of spermatogenesis can be analysed.
   A great advantage with meiotic investigations from testicular biopsy material in the estimation of risks from *in vivo*
exposure to a potential mutagenic agent is that only a few
subjects have to be examined. Conclusive evidence as regards
induction of specific meiotic irregularities may readily be
obtained. On the other hand, the sensitivity of the meiotic
investigations is counteracted by the improbability that testicular biopsies will ever be obtained, on a large scale, from
mutagenically exposed living adult men. From a practical point
of view, we should therefore also look for other possibilities
as regards mutagenicity testing *in vivo*. There are two types
of material which are potentially useful, namely testicular
material obtained postmortem from postpubertal men, and semen
specimens.
   Autolysis of the spermatogenetic epithelium occurs very
rapidly postmortem, and most of the stages usually used for
meiotic analyses have degenerated in testicular parenchyma
obtained within as short a time as one hour after death
(Hultén, personal experience). However, as has been pointed
out by Edwards and Guli (1963), some first meiotic prophase
chromosomes may retain their morphology for several days and
may allow the detection of even small induced chromosome

abnormalities, as will be described in some detail below
(section *FA*).

   2. *Semen samples.* To reveal improper exposure to a
mutagen hazard and estimate risks for induction of heritable
damage, particular attention should be given to the potential
advantage of using semen samples. This is the only source of
human germ line cells obtainable without any inconvenience. In
men, analyses of spermatozoa may be needed to give full insight
into heritable mutations. Screening of large numbers of men is
possible. Semen samples may be banked when deep frozen (Richardson, 1975). Smears may be analysed many years afterwards if
kept in a dust-free environment (Eliasson, 1975). It is well
known that under stress, sloughing of various cell stages from
the germinal epithelium occurs. The effect of mutagenic agents
might thus be detected not only in spermatozoa and spermatids.
Also some dividing spermatocytes may be cytogenetically analysed (Sperling and Kadern, 1971). Although Jacobson (personal
communication) has found this latter approach of little help,
van der Hagen (personal communication) has confirmed that a
few cells of good quality at diakinesis-first metaphase may be
obtained even from normal semen samples.
   There is considerable disagreement as to whether the sensitivity of the germinal epithelium to various types of stress,
with subsequent variation of the semen parameters commonly used
in the diagnosis and treatment of infertility, is likely to
create too much background noise to allow early detection of
exposure to a suspected genetic hazard. Some andrologists argue
that with adequate initial pre-exposure values some such parameters should be useful with serial semen samples (Eliasson,
1975, 1976a,b). Large scale investigations would be of particular interest to evaluate the effects of exposure to radiation and to chemical compounds in cell killing of the spermatogenetic epithelium - and subsequent sperm depletion. As regards exposure to various chemical compounds, there are some
indications that sperm head morphology might be more affected
than density or motility, as also demonstrated in mouse experiments (Lancranjan *et al.*, 1975; Wyrobek and Bruce, 1975).
Even high doses of colchicine have been reported to have no
effect on sperm density (Bremner and Paulsen, 1976).
   The crucial question is, however, which techniques may be
the most useful in detection of gross chromosome and DNA damage when applied to ejaculated human semen samples. Some techniques already utilized will be summarized, and some specula-
tions on new approaches will be presented below (section *FE*).

BII. Material from Female Gonads.
        1. Ovarian biopsies from adult women. Ovarian biopsi-
es for meiotic studies have usually been obtained at hyster-
ectomy or sterilization in adult women. However, change in
surgical routine in many countries has limited the possibilities of obtaining ovarian biopsies from these operations. Aspiration of oocytes at laparoscopy has been suggested (Steptoe
and Edwards, 1970), but has not been commonly utilized. Investigations have so far only been performed with ovarian
biopsies from women supposed to have a normal mitotic karyotype.

        2. Ovarian material from miscarriages. The stage
which is optimal for the detection of small structural chromosome abnormalities is pachytene. In the human female fetus
this stage begins at about 11 weeks, increases in frequency,
and is prominent from 12 to 28 weeks gestational age (Luciani
et al., 1977a; Luciani, unpublished observations).

        BIII. Complementary Data. To adequately evaluate the
implications of any chromosome abnormality in meiotic preparations by light microscopy, the following complementary data
are essential:

        1. Clinical reproductive data.

        2. Gonadal histology.

        3. Mitotic data. Comparisons of aberrations detected
in somatic mitosis and in germ line cells will give us insight
into the complex question as to the extrapolation from the
first to the second.

        4. Electron microscopy data. The material must be
absolutely fresh - every second counts (for details see Baker
and Franchi, 1967; Hultén *et al.*, 1974; Holm and Rasmussen,
1977; Holm, this Conference; Moses *et al.*, 1975; Moses and
Solari, 1967).

*C. What Are the Techniques for Obtaining Meiotic Chromosome
    Preparations?*

        CI. Direct Preparations. It is technically easy to obtain
direct meiotic preparations from adult testes and fetal ovaries
for light microscopic analyses. The air-drying technique is the
most commonly used. The cells are fixed and dropped in suspension on to ordinary (thoroughly cleaned) glass slides.

1. Direct preparations from adult testis material. In order to obtain material suitable for analysis of all meiotic stages, the testicular biopsy has to be dealt with immediately, because the spermatogenetic epithelium degenerates rapidly. A time delay of a few minutes in the operating theatre reduces the chances of obtaining good slides. Most laboratories use variations of the air-drying technique as described by Evans *et al.*, 1964 (Luciani *et al.*, 1972a; Hultén and Lindsten, 1973; Leonard, 1973; Chandley, 1975). Light squashing of individual short pieces of testicular tubules could be useful as the organization of the spermatogenetic epithelium is preserved.

The screening of the slides to find the appropriate, spontaneous, cell divisions may be a real difficulty in the investigation of human male meiotic chromosomes. It is much more time consuming than screening meiotic preparations from male mice or a human lymphocyte culture. In air-dried preparations from a testicular biopsy of ordinary size, there are usually numerous cells at pachytene, 10-100 cells at diakinesis – first metaphase (MI) and at second metaphase (MII), respectively, but only few premeiotic spermatogonial metaphases. The air-drying technique of Meredith (1969) provides more gonial mitoses for analysis than other methods.

Some stages are very rarely observed, presumably because they are passed rapidly, *e.g.* diplotene, as well as all the anaphase stages.

2. Direct preparations from fetal ovaries. Optimal slides for light microscopy meiotic analyses may be obtained from fetal ovaries even up to four days after intrauterine death. The ovary may then be left in fixative (3:1 methanol/ glacial acetic acid) for 12-14 hours, and a further delay of up to three days is not a disadvantage. The fetal ovary cells may be dispersed on ordinary glass slides using acetic acid (Luciani *et al.*, 1974).

CII. Preparations from In Vitro *Cultures.*
1. Short term in vitro *cultures of oocytes and spermatocytes. In vitro* short term cultures of oocytes, from adult ovarian biopsies, which is necessary for obtaining first and second metaphases, is a more exacting technique (Tarkowski, 1966). One advantage is that the oocyte may be exposed to mutagenic agents during *in vitro* maturation (Jagiello and Lin, 1974). The value of this approach is evident from mouse experiments (Jagiello *et al.*, 1968; Masui and Pedersen, 1975; Brewen and Payne, 1976). Short term *in vitro* culture of human testicular cells in suspension, primarily used to increase the number of cells at diakinesis-MI (Dutrillaux, 1971) might also be of help in this respect.

2. Long-term in vitro *culture of oogonia and sperma-togonia.* In vitro organ cultures of human gonadal material, oogonia and oocytes from fetal ovaries (Baker and Neal, 1969; Blandau, 1969; review in Baker and Wai Sum O, 1976) and spermatogonia from adult testicular biopsies (Matte and Sasaki, 1971; Steinberger and Steinberger, 1975; Fraccaro, personal communication), might also eventually become helpful tools for human mutagenicity testing, where experimental *in vivo* exposure is impossible. But there is still one hurdle to overcome. It has so far not been possible to obtain a high frequency of later stages of mammalian meiosis, using these *in vitro* culture techniques.

A wide range of experimental manipulations are obviously available. For example DNA replication/DNA repair processes (Lambert, this volume; Hotta and Stern, 1976; Söderström and Parvinen, 1976) and SCE (Wolff, this volume; Nevstad, this volume; and Allen and Latt, 1976) may be investigated; and non-SCEs may pinpoint crossing-over positions.

3. Transplantation into nude mice. The difficulties involved with long term *in vitro* cultures have also been encountered when transplanting adult human testicular tissue into nude mice (Müller, 1976).

D. Which Cytochemical Techniques May Be Used with Human Meiotic Slides?

DI: Conventional Staining Techniques. The most commonly used conventional staining techniques such as aceto-orcein (Fig. 1), Carbol Fuchsin, Feulgen, Giemsa, Toluidine blue, have all been used for the analysis of human meiotic chromosomes.

DII. Banding Techniques.
1. Banding techniques - constitutive heterochromatine. The centromeric heterochromatin (Fig. 1) can be visualized by the C-staining technique (Paris Conference, 1971).

So-called polymorphic or heteromorphic variants, constituting extreme variations in morphology of heterochromatic chromosome segments (*e.g.* involving the secondary constriction of chromosomes nos. 1, 9, and 16, the pericentromeric region of no. 3; the short arm, the secondary constrictions, the satellite stalks and the satellites of the acrocentric chromosomes, *i.e.* nos. 13, 14, 15, 21, and 22; and the distal part of the long arm of the Y chromosome) are revealed by the C-, Q- (Figs. 1 and 2), N-/Ag-R, and modified G-staining techniques (Paris Conference, 1971; Bobrow *et al.*, 1972; Geraedts

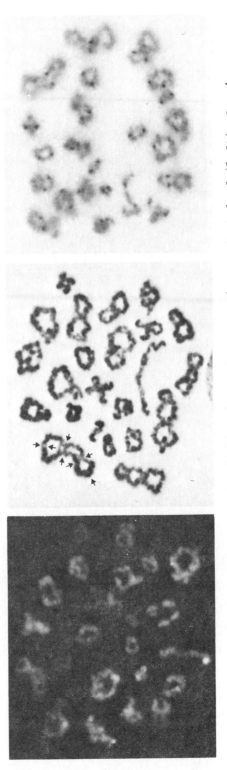

Fig. 1. Three spermatocytes at diakinesis-MI from normal men, Q-stained (left), Orcein stained (middle) and C-stained (right). The chiasmata (arrows) may easiest be counted in the Orcein stained cell. Total number of autosomal chiasmata 54–56.

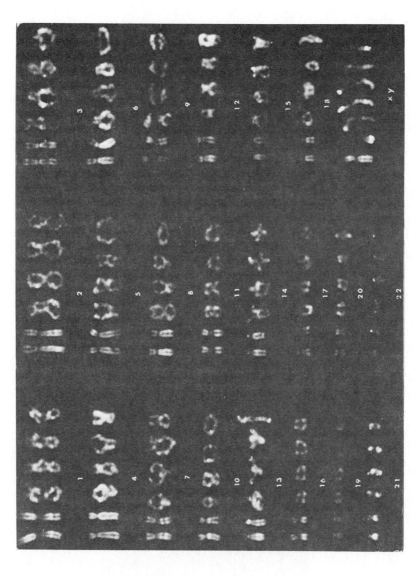

Figure 2. Q-banding pattern of mitotic chromosomes and meiotic bivalents at diakinesis–MI from a normal man (Caspersson et al., 1971a).

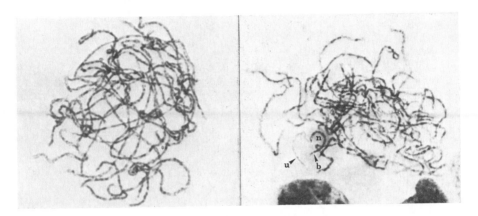

Fig. 3. Pachytene chromosomes stained by conventional techniques. Left: Oocyte from a normal fetus. Right: Oocyte from a +21 Down's syndrome fetus. Note bivalent 21 (b) and univalent 21 (u) associated to the nucleolus (n) (cont. next page).

and Pearson, 1973; Matsui and Sasaki, 1974; Funaki *et al.*, 1975; Goodpasture and Bloom, 1975; Pearson, 1975; Evans, 1976; Hayata *et al.*, 1977, Verma and Lubs, 1975). A combination of the techniques may be applied to any particular preparation.

Many opinions as to the function of the heteromorphic variants, their specific role in meiotic chromosome pairing, their exchange pattern and their influence on chromosome segregation, have been put forward (Hsu, 1975; John, 1976; Craig-Holmes, 1977; Yamamoto and Miklos, 1977). Analysis of their behaviour in the human in both sexes, during the different mitotic stages, in comparison with mitosis, and of their condensation, their specific staining pattern, and their interaction, would be useful in order to gain some further insight into their normal biological significance and their possible

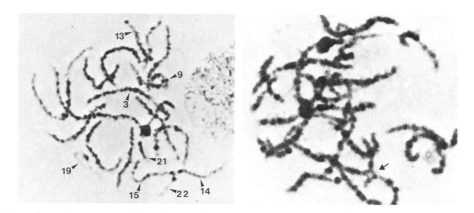

Fig. 3 (cont.). Pachytene chromosomes stained by convention-al techniques. Left: Spermatocyte from a normal man. Some bi-valents, identifiable according to their chromomere pattern are indicated by arrows. Right: Spermatocyte from a carrier of a re-ciprocal translocation involving chromosomes 10 and 11 with breakpoints at 10q 23 and 11q 24. Note the pathytene cross figure (arrow) (Ferguson-Smith and Page, 1973).

contribution to meiotic errors. By such analysis, we may gain some further insight into the reason for *e.g.* the high number of miscarriages trisomic for the acrocentric chromosomes and for chromosome nr. 16 (Carr and Gedeon, 1977; Jacobs, 1977a); the high proportion in the liveborn population of structural abnormalities - centric fusions/translocations and inversions - engaging the acrocentric chromosomes and chromosome no. 9 (Jacobs, 1977b; Hook and Hamerton, 1977; Patil *et al.*, 1977; Walzer and Gerald, 1977; see also de la Chapelle *et al.*, 1974; Boué *et al.*, 1975; Lindenbaum and Bobrow, 1975; Lurie *et al.*, 1976; Hecht, 1977); the increased frequency of positive 21 Q-satellites in liveborn trisomy 21 reported by Robinson and Newton (1977); "jumping" markers under particular circumstances (Gimelli *et al.*, 1976); the segregation distortion of some

particular variants (Fitzgerald, 1973; Das and Winter, 1969;
Palmer and Schroeder, 1971; review in Baker *et al.*, 1976), and
the low reproductive fitness of variant carriers (Morton *et
al.*, 1975).

It may be added that the frequency of Q- and C- variants
vary between races (Lubs *et al.*, 1977; Buckton *et al.*, 1976),
and that although the majority of variants investigated in
this respect seem to show simple Mendelian inheritance as dis-
crete units (Robinson *et al.*, 1976; Verma and Lubs, 1976;
Barker *et al.*, 1977), exceptions have been observed (Ferguson-
Smith, 1975; Baker *et al.*, 1976; Gimelli *et al.*, 1976; Robinson
et al., 1976, Craig-Holmes, 1977; Magenis *et al.*, 1977;
Nagakome *et al.*, 1977).

2. Banding techniques – other segments. As regards
the banding patterns of other segments along the chromosome
arms, there is no obvious difference between mitosis and meiosis (Figs. 2 and 3). The Q-banding pattern is very similar, as
shown for diakinesis in the human male (Caspersson *et al.*,
1971a). The Q- and G-banding patterns are apparently identical
at mitosis, and correspond to the chromomere pattern at pachytene (Luciani *et al.*, 1975, 1977a). Analysis of the detailed
banding pattern is more difficult with the increasing compaction of the chromosomes as first meiosis progresses. Nevertheless, by using a combination of Q- and C-banding the chromosomes may be individually identified at diakinesis-MI (Hultén,
1974a). T- or R-banding may also be applied (Dutrillaux, personal communication). To our knowledge G-banding has not so
far been informative. When performing successive staining
techniques on the same slides it is necessary to start with
Q-banding and finish with C-banding, as with mitotic cells.
Meiotic cells are more sensitive to harsh treatment.

3. Other techniques. Any of the more sophisticated
techniques applicable to mitotic slides are worth considering
for meiotic studies, as may be exemplified by reports using
in situ hybridisation of DNA satellite III in the human male
at pachytene (Seuanez *et al.*, 1976) and rRNA in the human female at diplotene (Wohlgemuth-Jarashow *et al.*, 1977).

E. When and how Do Mutagenic Agents Affect Meiosis?

The damage induced by mutagenic agents in germ line cells
is, in principle, similar to that produced in somatic mitotic
cells (Brøgger, this Conference).

The types of abnormality which are seen at the different
meiotic divisions depend on the stage at which they were in-

duced; specific damage may occur during homologous chromosome pairing, at first meiotic prophase. The susceptibility of the different meiotic stages varies with species, with sex and with agent. The present knowledge on radiation effects are summarized in Searle (1975); Baker (1977), Franchi (1977); UNSCEAR (1977). For instructive reviews including chemical mutagens, readers are referred to reviews by Leonard (1973); Russell (1976); and the recent report by Generoso *et al.* (1977).

In the present context one of the most important facts to stress is that many aspects of the normal process of gameto-genesis will have to be more thoroughly understood before the sensitivity to mutagenic agents of experimental animals and man can be adequately compared (Vogel and Rathenberg, 1975). For example, high sensitivity to cell-killing, characteristic of the dictyate nuclei of mouse makes this species a poor model for generalizations on ovarian response in other mammals, including man, where the long "resting stage" is of amphibian lamp-brush rather than dictyate type (Oakberg, 1968; Franchi, 1977). The heritable genetic damage, which may be induced in human oocytes is likely to be higher than in the mouse oocytes, considering the experimental evidence for a much higher radio-sensitivity to cell killing of human oocytes.

Two further important points should be appreciated. First-ly, meiotic DNA metabolism is a complex process (Hotta and Stern, 1976), only recently investigated in any detail in any mammal *i.e.* mouse (Chandley *et al.*, 1977; Hotta *et al.*, 1977). It is affected differently by different mutagenic agents. Secondly, some chemical compounds have been demonstrated to differ drastically from radiation in yield of scorable meiotic abnormalities (see section FB) in spite of producing the same degree of genetic damage in the offspring (Generoso *et al.*, 1977).

F. *Which Abnormalities Are Most Easily Detected and When?*

FA: *Early First Prophase?*
FAI 1. Identification of minor structural abnormalities - animal experiments. Using standard cytogenetic techniques, the early first prophase stages, leptotene and zygotene, are much less informative than the following pachytene stage. This latter stage is the one of choice for the detection of minor structural abnormalities, such as deletions/duplications. The pachytene chromosomes are decondensed, showing a natural G-banding, the chromomeres (Fig. 3). The great advantage with the pachytene chromosomes in comparison with similarly decon-densed late prophase mitotic chromosomes is the intimate pair-

ing of the homologues. Thus, any structural difference between
them may be detected. This has recently been demonstrated in
mice by Jagiello *et al.* (1976a), using conventionally stained
air-dried meiotic preparations, analysed by light microscopy.
Small loops were seen at pachytene, in both males and females
heterozygous for seven different *c* albino locus alleles, indicating that these point mutations comprise deletions.

2. *Possible implications in man.* There is no evident
reason to believe that the pachytene analysis should be more
difficult in the human than in the mouse. The individual human
chromosomes may be identified by the polymorphic variant marker bands, in combination with C-staining demonstrating the
centromeres, and the G-banding/chromomere pattern.

About five per cent of miscarriages at 12-20 weeks gestation age have chromosome abnormalities in the mitotic karyotype (Carr and Gedeon, 1977). Only a minority of these are
structural, and more may be identified by pachytene analysis.
Both these types of cytogenetic investigations should provide
valuable complements for the epidemiological studies on miscarriages as described by other colleagues at this Conference
(Beckman *et al.*, Fredga *et al.*, Hansteen, Infante, Jacobson,
Meretoja and Vainio), some of which should correspond to the
dominant lethal tests in animal experiments (Brinkert and
Schmid, 1977; Hitotsumachi and Kijuchi, 1977; UNSCEAR, 1977).

Naturally, the identification of a structural abnormality,
expected to occur in only a proportion of cells, is a much
more difficult task. Anyhow, the pachytene test in adult men
postmortem seems worth trying.

3. *Present light microscopical experience in identification of structural aberrations at pachytene in man.* So far
no deletion/duplication loop has been reported at pachytene
either in meiotic preparations from adult human males or from
fetal females.

No detailed pachytene screening was performed in the few
human male carriers of inversions or the single man with an
insertion, so far investigated (McIlree *et al.*, 1966;
Hauksdottir *et al.*, 1972; Therkelsen *et al.*, 1973; de la
Chapelle *et al.*, 1974; van der Linden *et al.*, 1975; Ferguson-
Smith, 1976). However, the pachytene cross figure formed by
the two reciprocal translocation chromosomes when pairing with
the homologous segment of their normal partner chromosomes has
been clearly demonstrated by Ferguson-Smith and Page (1973),
in a human male carrier of a balanced translocation 10/11, as
identified in the mitotic karyotype. The breakpoints may be
exactly identified by a detailed examination of the pachytene
cross (Fig. 3).

The three chromosomes no. 21, forming a trivalent (Fig. 3), has been demonstrated both in males (Hungerford *et al.*, 1970) and in females (Luciani *et al.*, 1976) and two female triploid fetus have been analysed (Gosden *et al.*, 1976; Luciani *et al.*, 1977).

FAII. Disturbance in homologous chromosome pairing – electron microscopy investigations. Synapsis of the homologous chromosomes and chiasma formation may be upset by environmental agents, but as yet there is only little direct cytogenetic information in mammals (Vig, 1977). Experience in the human is so far restricted to disturbances observed among childless men. Some of these may be due to meiotic mutations (Hultén and Lindsten, 1973; Baker, 1976; Ferguson-Smith, 1976; Hendry *et al.*, 1976). By light microscopy the effects are more easily demonstrated at diakinesis-MI than at pachytene (see FBIV).

Two human males have been described with a gross disturbance in homologous chromosome pairing, demonstrating abnormalities in the formation of the Synaptonemal Complex, as seen by EM analysis (Hultén *et al.*, 1974; Ferguson-Smith, 1976). Examination of the Synaptonemal Complex at pachytene (Baker and Franchi, 1967; Moses *et al.*, 1975; Moses and Solari, 1976; Holm, this Conference, Holm and Rasmussen, 1977) provides a complement to elucidate all the problems mentioned above in section FAI.

FB: Diakinesis – First Metaphase?
FBI: Translocations
    1. Identification. Analysis of the pairing configurations at diakinesis-first metaphase (MI) has become the classical cytogenetic test for identification of (non-homologous reciprocal) translocations, and the most commonly used technique to quantify such translocations induced by mutagenic agents in animals, usually male mice (Leonard, 1973, 1976; UNSCEAR, 1977; Generoso *et al.*, 1977). Spermatogonia have been considered to be the most important cells as regards induced heritable damage. The argument for this is that spermatocytes and spermatids have a relatively much shorter duration in comparison with reproductive life-span.

Translocations which have been induced in surviving spermatogonia may be easy to detect at diakinesis – easier than in the spermatogonia themselves – as large ring or chain quadrivalents, usually readily distinguishable from the normal bivalents. Depending on the formation or non-formation of chiasmata in the four segments of the pachytene cross different configurations will appear at diakinesis-MI. Chiasmata on all four arms will lead to a ring. Lack of chiasma

formation on one or more of the arms will make the pairing
arms open to a chain-of-four, a chain-of-three plus a univalent, two bivalents, one bivalent plus two univalents, or
four univalents.

Experience gained from investigations on human male constitutional carriers of translocations is of relevance as to
the problems involved in the identification at diakinesis-MI
of translocations induced at an earlier stage of meiosis. To-
date meiotic investigations have been published on about 20
human male subjects with balanced reciprocal translocations
and 15 with Robertsonian centric fusion/translocations. In
addition, unbalanced carrier patients (two Robertsonian and
four non-Robertsonian) have been examined (Hultén and Lindsten, 1973; Chandley, 1975; Chandley *et al.*, 1976a;
Dutrillaux and Geuguen, 1975; Laurent and Dutrillaux, 1976;
Curtis, 1977).

Quadrivalents are regularly seen in carriers of balanced
reciprocal translocations. A chain-of-three plus a univalent
is a very rare exception, and the other possible configurations have not been observed. The multivalent appearance
varies much with different types of human translocations, involving chromosomes with large differences in size and centromere index, and with a comparatively high chiasma frequency
(Figs. 4, 5, 6). Some translocation multivalents are easily
identified by conventional staining techniques. However, even
when one pairing figure less than normal is consistently observed in all cells precisely countable, the presumptive
quadrivalent might sometimes be difficult or impossible to
identify as such in any given cell without karyotyping, using
banded preparations. Especially with insufficient chromosome
spreading the translocation scoring may thus be quantitatively unreliable when only conventional staining techniques are
applied.

To our knowledge no translocation multivalents have been
observed in any control subject.

*2. Induction of translocations by environmental
agents.* Two negative reports attempting to assess the effect
of environmental agents on human meiotic chromosomes *in vivo*,
one dealing with the possible effect of LSD in a human male
(Hultén *et al.*, 1968), and one concerning contraceptives in
human females (Jagiello and Lin, 1974) demonstrate some other
abnormalities which might be detected (primarily breakage),
and the caution needed in the cytogenetic interpretations.

The only report so far published on meiotic investigations in human subjects exposed to a known mutagenic agent
concerns six male volunteers, receiving an acute dose of 78,
200 and 600 rad X-radiation, with testicular biopsies obtain-

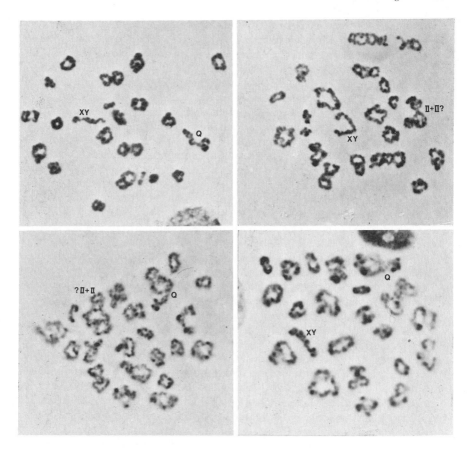

Fig. 4. Spermatocytes at diakinesis-MI from carriers of balanced reciprocal translocations, t16;18 (cells a and b, top) and t1;21 (cells c and d, bottom). The counts of autosomal associations were in both patients consistently either definitely 21 (as in cells a and d) or uncertain but less than 21 (as in cells b and c). The quadrivalents (Q) are easily identified in cells a and c, but might escape detection in cells b and d. Overlapping bivalents which may be misinterpreted as quadrivalents are indicated in cells b and c (II + II?) (Partly from Hultén and Lindsten, 1973).

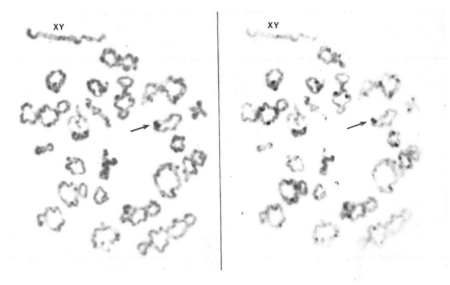

Fig. 5.   *Robertsonian 14;14 translocation difficult to identify at diakinesis-MI. Dicentric 14;14 translocation univalent arrowed. From Hultén and Lindsten (1973).*

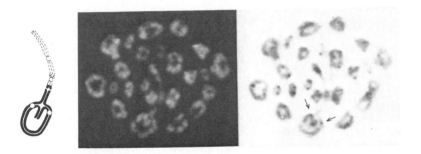

Fig. 6.   *Example of translocation identification problem at diakinesis-MI. Constitutional carrier of an unbalanced translocation 2p:Yq. From Caspersson et al. (1971b).*

ed after various time intervals (Brewen and Preston, 1975).
Four, seven and six per cent, respectively, of the cells at
diakinesis-MI contained translocation multivalents, as judged
by conventional staining techniques. No chromosome counts were
given, the translocation figures were not precisely defined,
no other abnormalities were recorded, and no other stages were
analysed.

   *3. Implications – calculation of transmissable re-
ciprocal translocations.* The inference that man is twice as
sensitive as mouse to radiation damage in spermatogonia, and
the critical question, *i.e.* the risk of inducing transmissable
reciprocal translocations (calculated to 0.96 x 10^{-4} - 1.93 x
10^{-4} translocations per gamete per rad) remains questionable
for multivalent identification uncertainty, but also for other
reasons beyond the scope of this presentation (Searle, 1975;
Sankaranarayanan, 1976). Transmission to offspring was calculated from mouse data. The notion that radiation might not induce the same degree of Robertsonian rearrangements was not
considered, as has been pointed out by Ford (personal communication).
   On the other hand, relevant information in man on reproductive fitness of spontaneous non-Robertsonian reciprocal
translocation carriers is still scarce in comparison with
Robertsonian or Robertsonian-like centric fusion/translocati-
ons (Lindenbaum and Bobrow, 1975; Lurie *et al.*, 1976; Jacobs,
1977b).
   Although the frequency of balanced translocations in
childless men is not significantly increased (Chandley *et al.*,
1975), some specific types of balanced sex chromosome-autosome
and some autosome-autosome translocations may *per se* induce
maturation arrest and male infertility, analogous to the
situation in the mouse (Chandley *et al.*, 1976a; Turleau *et al.*,
1976, 1977; Evans, 1976). Balanced autosomal translocations
with particular breakpoints, when forming chains at diakinesis-MI, may interfere with later progress of spermatogenesis
(Chandley *et al.*, 1976a). If induced in spermatogonia in men
such translocations may be harmless to future generations.
Two points should, however, be stressed. Firstly, the same
types of translocations would not be eliminated if induced
post meiosis, in spermatids and spermatozoa (see FE). Secondly,
no elimination is expected in human females. The lack of
chiasma formation on one arm may lead to a fairly frequent
production of viable tertiary trisomics, as a result of non-
disjunction (Searle, 1975; Lindenbaum and Bobrow, 1975; Lurie
et al., 1976). In short, this type of translocation presents a
special risk, and seems more likely to occur after exposure of
females or postmeiotic germ cells in males.

   4. Heritability estimations by chiasma data. An
exacting analysis of the chiasma distribution on the translocation pairing configurations at first meiosis is of theoretical and practical importance, since this determines the
number and type of balanced and unbalanced chromosomes in relation to the normal - and thus the genotype of the offspring
(Sybenga, 1975).
   The chiasma distribution on the translocation multivalent
at diakinesis-MI has been examined in detail by the aid of C-
staining in 5 subjects, constitutional carriers of balanced
reciprocal translocations with breakpoints known from banded
mitotic preparations (Chandley *et al.*, 1976a). The chiasma
distributions predicted for the human situation by Ford and
Clegg (1969), and Hamerton (1971), was largely confirmed, but
with one exception, *i.e.* that chiasmata should be less likely
to form in very short segments near breakpoints. On the contrary, some very short translocated segments had a much higher
chiasma frequency than expected on the basis of a random distribution. This indicates that chiasmata may have a natural
tendency to be formed much more often in some segments than
in others, corresponding to the pattern seen in the normal
situation (Hultén, 1974a). A remarkably high efficiency in
homologous chromosome pairing with localized chiasma formation
in short segments is seen also in some other subjects with
balanced translocations as well as in one patient with an unbalanced translocation (Ferguson-Smith and Page, 1973; Hultén,
1974b).
   Any variation in frequency and localization of chiasma
formation, with genetic background, with sex and age, between
individuals - and with environmental influence - is important
for the risk estimations with different types of structural
rearrangements, *e.g.* translocations. A further complication
is that the rearrangement in itself may change the chiasma
pattern.
   As yet we know nothing about the environmental influence
on the chiasma formation in man. The normal mean total number
of chiasmata per cell is estimated to be somewhat higher in
men (see Fig. 3) than in women (Hultén and Lindsten, 1973;
Chandley, 1975; Ferguson-Smith, 1976; Jagiello *et al.*, 1976b).
There is an indication but no definite evidence for increase
with age in men (Lange *et al.*, 1975). No corresponding data
are available in women (Jagiello *et al.*, 1976b, Polani and
Jagiello, 1976).
   The chiasma distribution along the individual chromosomes
seems to be quite different in men (Hultén, 1974a) and women
(Jagiello *et al.*, 1976b). However, all these comparisons are
complicated due to the fact that the analyses of the chromosomes of oocytes cultured *in vitro* from adult ovaries is a

very difficult task. A change in chiasma pattern might also
occur during culture of the oocyte *in vitro*.

    FBII: Other structural aberrations. Other structural
aberrations have generally received much less attention in
estimating risks for induction of heritable damage (Searle,
1975; UNSCEAR, 1977). We know even less about the meiotic
behaviour in man of other structural aberrations than translocations. To-date investigations have been published on six
human male carriers of inversions, one of an insertion, two of
rings and one of a dicentric chromosome (Hultén and Lindsten,
1973; de la Chapelle. *et al.*, 1974; Chandley, 1975; van der
Linden *et al.*, 1975; Ferguson-Smith, 1976). Some of these
aberrations, *e.g.* small rings, may be readily detected at diakinesis-MI even with conventional staining techniques. Ascertainment bias for subfertility makes it difficult to judge the
risk for transmission if induced in spermatogonial stem cells.
The chromosome abnormality may cause spermatogenetic depletion,
in which case the risk of transmission would be low.

    Other structural aberrations, such as insertions and in-
versions may remain undetected (Fig. 7) without a detailed
analysis using for instance C-banding. The risk for formation
of unbalanced spermatozoa in inversions has been estimated
from the chiasma distribution of the inversion pairing configuration at diakinesis-MI as analysed by C-banding (van der
Linden *et al.*, 1975). Chiasma formation in the inverted chromosome segment was found to be the same as that in the normal
situation.

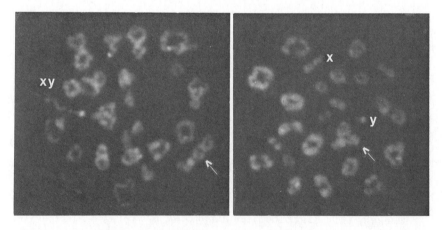

    *Fig. 7. Structural abnormality difficult to identify at
diakinesis-MI. From human male carrier of a direct insertion
(2) (q34p13p24). The odd looking number 2 bivalent is arrowed.
From Therkelsen* et al. *(1973).*

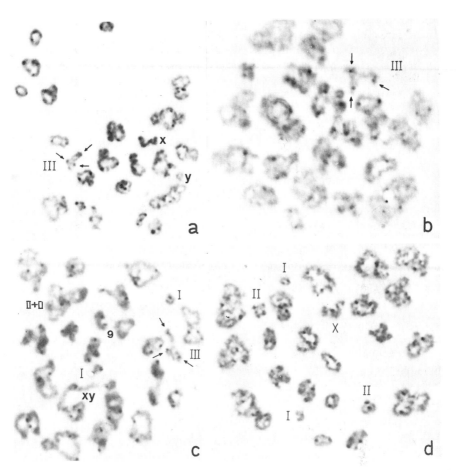

Fig. 8.   Examples of problems in identification of numer-ical and structural abnormalities at diakinesis-MI. From a patient with 21+ Down's syndrome.
a. *Twenty-two autosomal associations, one of which is the assymetrical trivalent 21 (III).*
b. *The number of associations is difficult to count, but the trivalent 21 (III) is clear.*
c. *Twenty-four autosomal structures. The three chromosome 21 are paired into a trivalent (III), and one bivalent 22 is separated into two univalents (I). Overlapping of XY and one autosome, as well as of two autosomes (II + II).*
d. *Twenty-five structures. The X and Y chromosomes are separa-ted. Presumably one of the small univalents (I) is the Y chromosome, the other the extra chromosome 21 which has not paired. Normal symmetrical bivalents 21 and 22. From Hultén and Lindsten (1973).*

FBIII: Numerical aberrations.
    1. General comments. Numerical aberrations are an
important genetic load in man. Over 20 per cent of all karyotyped spontaneous abortions comprise autosomal trisomies or
monosomy X. They represent together over 70 per cent of the
chromosome abnormalities found among miscarriages. About 0.4%
of liveborn children carry sex chromosomal or autosomal numerical abnormalities (Carr and Gedeon, 1977; Jacobs, 1977a;
UNSCEAR, 1977). With the exception of maternal age, the roles
of factors affecting this, including the effects of radiation
(Cohen *et al.*, 1977; Uchida, 1977) are not unequivocally
established.
    Quite much is known about the induction and meiotic beha-
viour of numerical chromosome aberrations in lower organisms,
above all in plants (Khush, 1973). In comparison only little
information is available on mammals (Russell, 1976; UNSCEAR,
1977).
    Many hypotheses concerning the meiotic origin of human
numerical aberrations have been put forward. None have been
confirmed by cytogenetic meiotic observations.

    2. Identification of extra chromosomes - loss dur-
ing spermatogonial mitosis - transmission. Some relevant information has been obtained from meiotic analyses of subjects
with extra chromosomes. Human male carriers of extra small
metacentric markers, XYY men and male patients with trisomy 21
Down's syndrome have been investigated (Hultén and Lindsten,
1973; Chandley, 1975; Chandley *et al.*, 1976b; Faed *et al.*,
1976).
    A small extra metacentric marker chromosome seems to pass
through meiosis without difficulty. This chromosome of unknown
nature, does not seem to pair with any other chromosome. It is
easily identified in the majority of cells at all stages of
meiosis. An increased frequency is found among childless men
(Chandley *et al.*, 1976a) but it occurs also sporadically, and
in families, in phenotypically normal as well as abnormal
patients. Testicular histology shows severe impairment in most
cases (Skakkebaek *et al.*, 1973b), irrespective of ascertainment. Thus, these extra small chromosomes are unlikely to be
transferred to progeny from male carriers.
    The extra chromosone no. 21, on the other hand, seems to
be eliminated in some cells. In the majority of cells it is
retained and is often seen paired with the two other chromosomes no. 21. Whether occurring unpaired as an univalent or
paired as a trivalent, the extra chromosome number 21 should
be easily detectable even in conventionally stained preparations (Fig. 8) although Q staining is certainly helpful (Fig.
9). Testicular histology varies much, from almost normal to

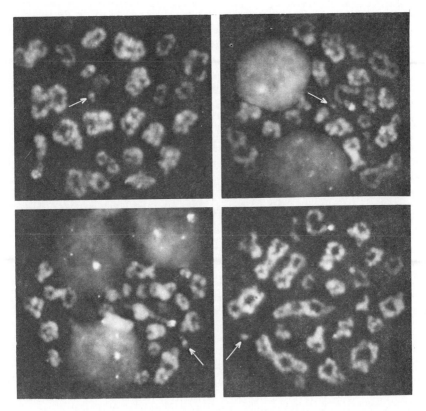

Fig. 9. Examples of advantages in Q-staining even when not of optimal quality. Spermatocytes from a patient with 21+ Down's syndrome. The univalent 21 (arrow) is identifiable in all the cells.

gross depletion of the spermatogenetic epithelium (Skakkebaek *et al.*, 1973b). Apparently morphologically normal spermatozoa have been observed in meiotic preparations (Hultén, unpublished observations). No semen samples have been obtained. Transmission to offspring has not been reported in males. To what extent this is dependent on psychological factors is unknown.

The extra Y chromosome finally, is apparently lost early during spermatogenesis in the majority of cases. Mosaicism with a normal cell line has been demonstrated in spermatogonial metaphases, and most XYY men have shown only a single Y chromosome at diakinesis-MI. In the few cases where the Y chromosome is retained it pairs almost exclusively with the other Y chromosome. It might easily escape detection unless banded preparations, preferably Q-stained, are utilized (Hultén, 1970;

and Hultén and Pearson, 1971). The mechanism for the elimination of the extra Y chromosome is not known. Testicular histology shows much variation from normal to Sertoli-cell-only
(Skakkebaek *et al.*, 1973a). The majority of XYY men investigated have not been ascertained through infertility, and it is
thus likely that their risk to have impaired spermatogenesis
is generally increased. Only two cases are reported to have
XYY sons, one of which had a mosaicism with several cell
lines.

　　　　　　*3. Spermatogonial non-disjunction - gonadal mosaic-
ism.* Somatic and gonadal mosaicism, with a cell line with an
extra chromosome transmitted to offspring is a commonly suggested reason for trisomies.

　　An extra chromosome, which also may be caused by mitotic
non-disjunction in gonial divisions, would not present any
difficulties in identification at diakinesis-MI. The detection of an extra Y chromosome would require Q-staining, but an
additional X or an autosome should be readily demonstrated
even by conventional staining techniques. To our knowledge
none have been observed in any control subject. No cells with
either a trivalent, a bivalent and a univalent, or three univalents have been observed. Thus, there is a slender indication that spontaneous gonial non-disjunction may be rare.

　　Technical artefacts complicate an adequate evaluation of
hypoploidy, and a low grade mosaicism (Jones *et al.*, 1976) is
obviously a very difficult diagnostic problem.

　　　　　　*4. Inferences on meiotic non-disjunction from
somatic mitotic investigations.* Since the introduction of the
banding techniques in 1970, much mitotic work has been devoted
to tracing the parental meiotic origin of the extra chromosome
in trisomies, in particular trisomy 21. This has mainly been
performed by a comparison of Q-fluorescing satellites on chromosome 21, between children with trisomy 21 and their parents.
These studies infer that both maternal and paternal first and
second meiotic non-disjunction are to be blamed (Langenbeck
et al., 1976; Jacobs, 1977a). This may well be true, but our
limited knowledge about the behaviour of these marker segments at meiosis imposes a diagnostic uncertainty. If a marker
switch between the two homologous chromosomes would occur at
first meiotic pairing it becomes impossible to differentiate
between first and second meiotic non-disjunctional events,
by a comparison between mitotic markers of parents and offspring. A Q-marker switch (see Fig. 2, chromosome 22) could
occur by crossing-over in the short arm of the chromosome, or
between the satellite stalks or the Q-satellites themselves.

Regular crossing-over does not seem to be common in the short arm of chromosome 21 in men (Hultén, 1974a), but the behaviour of the satellite stalks and the satellites has not been investigated.

Further insight into these problems may be achieved in some favourable individuals heterozygous for Q-markers. Entangling of the satellite stalks or the Q-satellites, should they stick together, might be difficult to distinguish from a chiasma on the short arm. Both these events would, however, lead to a misclassification with regular segregation. Segment stickiness may predispose to non-disjunction at first anaphase. The increased frequency of positive Q-markers on chromosome 21 in children with trisomy 21, observed by Robinson and Newton (1977) is interesting.

5. *Background of first meiotic non-disjunction.* The most commonly discussed reason for first meiotic non-disjunction is random segregation of univalents at first anaphase. Unpaired univalents could be caused either by failure of chromosome pairing and lack of chiasma formation at first meiotic chromosome pairing, or alternatively by precocious separation of the two homologous chromosomes. The latter has been supposed to more often occur in chromosome pairs forming a single chiasma, and in particular if this chiasma is distally localized.

At diakinesis-MI the X and Y chromosomes are separated in a proportion of cells (see Fig. 7) which varies considerably between individual human males (Hultén and Lindsten, 1973; Chandley, 1975; Ferguson-Smith, 1976). In the normal situation at least one chiasma is regularly seen on all autosomal bivalents. A single chiasma is generally observed in the G-group, sometimes in the D-group and occasionally on chromosome no. 18. Distal single chiasmata are almost exclusively restricted to the G-group chromosomes. This chiasma is more often distally localized in chromosome 21 than in 22 (Hultén, 1974a and Hultén, unpublished observations). Either of these chromosomes may appear non-associated as univalents, but this is very rarely seen in normochiasmatic men (Hultén, 1974b). Larger univalents have not been observed.

The chiasma pattern at diakinesis-MI in women as judged from oocytes cultured *in vitro* (Jagiello *et al.*, 1976b) differs from that in men. Single chiasma bivalents are much more common, occurring even in the B-group but more often in C-, D-, E-, F- and G-groups. This single chiasma is often distally localized. Unpaired G-group univalents have been found in a proportionately much higher frequency in women than in men (Edwards, 1970; Jagiello *et al.*, 1976b). It has as yet not been possible to draw any conclusions as to age

differences, corresponding to the situation in the mouse
(Polani and Jagiello, 1976).

The nature of the univalents cannot be established by
analysis of diakinesis-MI. Technical artefacts may play a
role. Therefore, it is difficult to speculate about the im-
lications of the above sex differences in relation to first
meiotic non-disjunction, as discussed in some detail by
Jagiello *et al.* (1976b) and Polani and Jagiello (1976). Comparisons with the pairing pattern by EM analyses of the Synaptonemal Complex at pachytene, in direct preparations in
both sexes, would be valuable. Examination of diplotene bivalents by light microscopy in direct preparations from fetal
ovaries, and adult testes might also be rewarding. Preliminary
observations indicate that chiasmata have a general tendency
to be more distally localized in males than in females, in bivalents of similar contraction (Hultén and Luciani, unpublished).

FBIV. Distance in chromosome pairing and chiasma formation.
The mechanism of chromosome pairing and chiasma formation may
be upset by environmental agents (though there are only few
direct cytological studies in mammals (Vig, 1977)).

The information in human is limited to infertile men.
Striking disturbances in chiasma frequency (Fig. 10) and
chiasma distribution over all the chromosomes have been found
in a proportion of childless men with severe depletion of
spermatozoa in ejaculated semen specimens (Hultén and Lindsten, 1973; Chandley, 1975; Ferguson-Smith, 1976; Hendry *et
al.*, 1976). There is some indication that this may be associated with segregating recessive genes, affecting homologous
chromosome pairing and chiasma formation (Baker *et al.*, 1976).

Defective pairing affecting some chromosomes only may
provide one of the explanations for unrelated chromosome abnormalities in some families, *e.g.* the tendency for couples
with repeated miscarriages to have another trisomic one following a first (Baker, 1976; Teese and Jones, 1976; Carr and
Gedeon, 1977; Jacobs, 1977a; Hecht, 1977). Alternatively,
meiotic non-disjunction may be the result of translocations,
which might remain undetected in the mitotic karyotype. A
prerequisite for so-called distributive pairing (Grell and
Valencia, 1964; Grell, 1971) is the occurrence of non-homologous univalents of similar size, joining the distributive
pool. Unrelated G-univalents have been observed in the occasional cell, in a few instances, *i.e.* in one XYY man (Hsu *et
al.*, 1970) and in one male patient with trisomy 21, Down's
syndrome (Hultén and Lindsten, 1973), but as far as we are
aware, not in any defined translocation carriers (Hultén,
1974b).

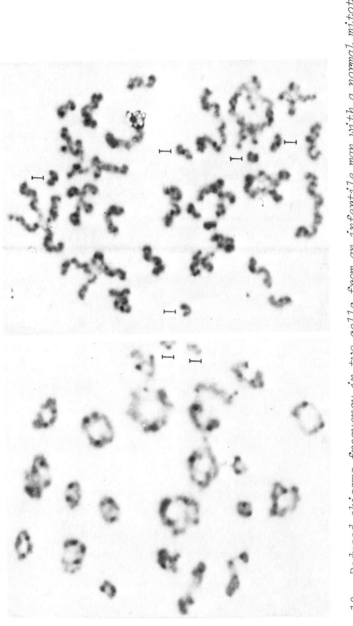

Fig. 10. Reduced chiasma frequency in two cells from an infertile man with a normal mitotic karyotype. The number of structures are difficult to count exactly. Univalents are indicated by symbols I. The chiasma frequency in the cell to the left is about 30, and in the one to the right about 20. From Hultén and Lindsten (1973).

In summary, light microscopy investigations at diakinesis-MI indicate that the mechanism of chromosome pairing
and chiasma formation normally shows a high degree of efficiency in man. Deviations from the normal situation are readily detectable, although the underlying mechanisms and the
effect as to the production of chromosomally abnormal offspring may sometimes be difficult to evaluate at this stage.
We have stressed the caution needed in the interpretation of
the data to avoid underestimation, particularly of structural
chromosome abnormalities. In this connection, it should also
be pointed out that the yield of abnormalities seen at diakinesis-MI induced by chemical compounds is expected to be
much lower than that by radiation even with the same amount
of heritable damage (Generoso *et al.*, 1977). These warnings
should not discourage the analysis, since some of the abnormalities expected are not at all difficult to identify, and
seem to be very rare in control subjects.

FBV: Meiotic polyploidy. Some infertile oligochiasmatic
men have shown an increased frequency of polyploid cells, and
this has also been observed as an isolated phenomenon. Multivalents at pachytene or diakinesis-MI would provide unambiguous evidence for "true" polyploidy. This has not been observed in any men, either infertile or controls, but apparently polyploid cells are quite frequently observed at both pre-
meiotic spermatogonial divisions and at first and second meiotic metaphase. It is not known to what extent they constitute
adjacent cells which are dividing simultaneously, but it seems
possible that they may finally be included in the same nucleus. Polar body suppression, which may be a likely mechanism for polyploidy in the human female has been observed with
in vitro cultured oocytes from higher mammals (Jagiello and
Lin, 1974).

Mitotic investigations by Q-fluorescing markers to trace
the origin of triploidy in man has demonstrated dispermy to be
a frequent cause (Kajii and Niikawa, 1977). Unequivocal evidence for first maternal non-disjunction has also been provided.

There is some indication that polyploidy may be induced
by irradiation of early first meiotic prophase stages in male
mice (Szemere and Chandley, 1975). High doses of cholchicine
do not induce a drop in sperm count (Bremner and Paulsen,
1976).

FC. First Anaphase. First meiotic mal-segregation can
be investigated by scoring first anaphase for chromosome
lagging. However, this is virtually impossible in the human
male, since this stage is extremely rarely seen at least in

air-dried meiotic preparations. Light squashing and analyses of testicular histology might be helpful to detect paracentric inversions (Roderick and Haws, 1974) or ring chromosomes where anaphase bridges are expected.

Scoring of first anaphase for chromosome lagging might be a more realistic approach in the human female, viz. in cultured oocytes (Jagiello and Lin, 1974).

FD: Second Metaphase. First meiotic mal-segregation, anaphase lag as well as non-disjunction, is most adequately screened by counting the number of chromosomes at second metaphase (MII).

1. *Animal experiments.* MI aneuploidy in male mice is normally rare, but may be induced by radiation of spermato-cytes during first meiotic prophase (Szemere and Chandley, 1975a). Its spontaneous frequency is higher in aged mice in comparison to the young (Martin *et al.*, 1976). Low-dose radi-ation effects are enhanced in aged females (Uchida and Freedman, 1977). However, the implications of these findings as to the production of trisomic offspring are still not clear (Szemere and Chandley, 1975a; Eich, 1977a,b; Max, 1977).

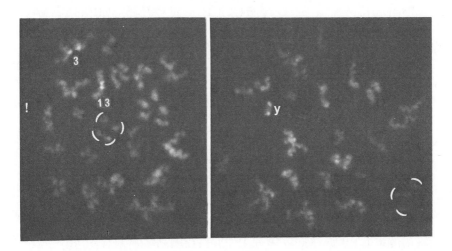

Fig. 11. Q-stained second metaphases from a normal man. There are 23 chromosomes in each cell. Chromosome 20 in the cell to the left is barely visible (!). In both cells chromo-some 9 (brackets) may easily be misinterpreted as a broken chromosome or three small chromosomes. From Hultén and Lindsten (1973).

2. Implications in man . Analogous human data would
be of interest with respect to a possible radiation effect
(Cohen *et al.*, 1977) and the maternal age dependent increased
risk for begetting some types of trisomic offspring, *i.e.* an
extra X, no. 18, or any acrocentric chromosome. No effect of
maternal age has been found for an extra Y or no. 16, and not
for monosomy X (Jacobs, 1977a).

3. Present cytogenetic experience in man . Prior to
the introduction of modern banding techniques analysis of MII
was very difficult in human males, and there are still no
counts to illustrate the frequency of spontaneous first meiotic mal-segregation. Examples of normal Q-stained cells at MII
are given in Fig. 11.

Some extra chromosomes with particular morphology, for
instance small extra centric fragments, are easy to identify
even without banding techniques (Hultén and Lindsten, 1973;
Chandley, 1975). Other numerical and structural abnormalities
have to our knowledge been identified in only four instances,
all in translocation heterozygotes. This concerns a few mono-
somic and disomic MII cells in two carriers of Robertsonian
translocations, one homologous (Hultén and Lindsten, 1970,
Fig. 12) and one non-homologous (Luciani *et al.*, 1972b); an
extra chromosome no. 9 identified by C-staining in a 9;22
translocation carrier (Chandley *et al.*, 1976); and asymmetric-
al chromatids in a translocation carrier, implying chiasma
formation in the segment between the centromere and the breakpoints (Laurent, personal communication).

In contrast, Jagiello *et al.* (1976b) have been able to
perform reliable MII counts in 411 *in vitro* cultured MII
oocytes from adult women of various ages. In four instances,
one cell was observed with 24 bodies from women aged 26, 42
46 and 50 years, and two were seen from a subject who was 49.
No conclusions could be drawn other than that 6/411 MII cells
demonstrated hyperploidy.

With the aid of a combination of staining techniques,
e.g. C- and G-banding (Szemere and Chandley, 1976) reliable
MII counts should now be obtainable in both men and women. The
combination of all current staining techniques should also
give insight into other questions, for instance the segregation pattern of numerical and structural chromosome aberrations, as well as heteromorphic variants.

In subjects heteromorphic for a long secondary constriction of chromosome no.9, Page (1973) observed three classes of
MII cells, indicating crossing over between the centromere
and the proximal end of the paracentric region. In our own
experience, there is much variation in length of the secondary constriction between MII cells in a given individual. In

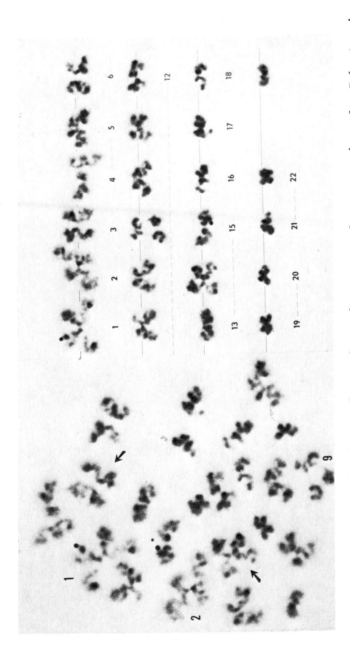

Fig. 12a. Orcein stained second metaphase chromosomes from a carrier of a Robertsonian 14;14 translocation. The translocation chromosome is marked with an arrow. From Hultén and Lindsten (1973).

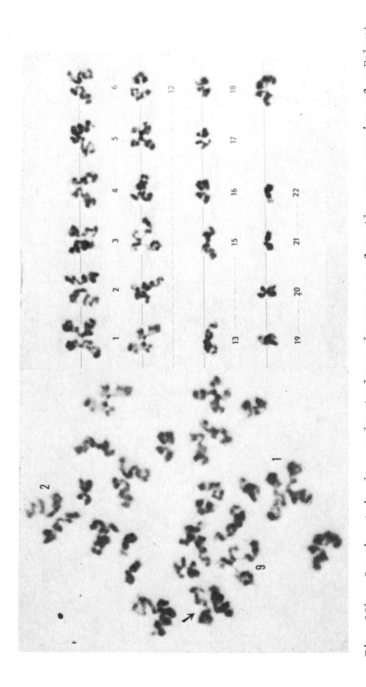

Fig. 12b. Orcein stained second metaphase chromosomes from the same carrier of a Robertsonian 14;14 translocation. The translocation chromosome is missing in this metaphase. From Hultén and Lindsten (1973).

some cells this segment may be extremely elongated (Hultén and Lindsten, 1973). We have therefore so far found it difficult to quantitatively estimate the frequency of crossing-over, equal or unequal.

FE: Semen Analysis - Spermatozoa? There are strong reasons to consider not only damage induced during premeiotic divisions and during meiosis, but also post meiosis - during the maturation stages of spermatids to spermatozoa. It is understood that the fertilization capacity of spermatozoa is generally not affected even with gross damage (UNSCEAR, 1977; Salisbury *et al.*, 1977).

FEI 1: Animal experiments. Exposure to mutagenic agents of spermatids/spermatozoa *in vivo* and spermatozoa *in vitro*, which are commonly used experimental techniques with various species, have shown that a heavy genetic load may be imposed in these cells. Much of this damage may be eliminated before implantation, as zygotes with gross structural chromosome abnormalities, acting as dominant lethals, will not survive further development (Brinkert and Schmid, 1977; Hitotsumachi and Kikuchi, 1977). However, post-zygotic sex chromosome loss (Russell, 1976) and heritable translocations (Land and Adler, 1977) may also be induced by exposure of spermatids and spermatozoa to mutagenic agents. In *Drosophila*, radiation of mature spermatozoa has been found to cause not only sex chromosome loss in the zygote but also structurally abnormal sex chromosomes (Maddern and Leigh, 1976).

2: Possible implications in man. Pre-meiotic spermatogonia may not be the most important male germ line cells as regards induction of constitutional chromosome abnormalities in the next generation (Brewen, 1976). In man, monosomy X is the most common of all karyotype abnormalities (Carr and Gedeon, 1977; Jacobs, 1977a), and liveborn show a predominance of paternal fault (Race and Sanger, 1975). Gametic induction of different types of structurally abnormal X and Y chromosomes, with subsequent variations in malsegregation and selection patterns may, bearing in mind the high tolerance for sex chromosome aneuploidy, provide one of the explanations for the high frequency of this karyotype in spontaneous abortions (Carr and Gedeon, 1977; UNSCEAR, 1977; Jacobs, 1977a). Mosaicism is more often found in early miscarriages than in later. The earlier the chromosome loss during fetal life, the lower the frequency of mosaicism is expected to be in any individual liveborn patient. Obviously some constitutional structural abnormalities still escape detection, as may be exemplified by our past experience in unexpectedly finding

an unbalanced 2p/Yq translocation in a 45,X/46,XY/47,XYY
mosaic patient, by meiotic analyses (Caspersson *et al.*, 1971b).
    It is not excluded that minor structural autosomal damage
induced post meiosis may predispose for malsegregation in
the zygote. And, it is interesting to note that the X and no.
21 chromosomes may *in vitro* be particularly susceptible to somatic non-disjunction with low doses of radiation (Uchida,
1977). This brings us back to the second diagnostic problem
with mitotic marker comparisons between parents and offspring
to trace the origin of an extra chromosome, *e.g.* no. 21 (see
section *FBIII 4*). The problem is that second meiotic non-disjunction cannot be distinguished from post zygotic mitotic
non-disjunction, using these techniques. In other words, although it might be inferred whether the extra chromosome is
the maternal or the parental one, it is not possible to tell
if the failure of segregation occurred at meiosis or in the
zygote, in a proportion of cases - and this is generally neglected (Langenbeck *et al.*, 1976; Cohen *et al.*, 1977; Jacobs,
1977a; Robinson and Newton, 1977; Uchida, 1977, Kajii *et al.*,
1977). Some of the patients with Down's syndrome may then
have been misdiagnosed as caused by a meiotic fault, although
the damage may in fact have been induced post meiosis, during
gametic maturation or at early cleavage divisions in the zygote. Chromosomes damaged in spermatozoa may not be replicat-
ed and breaks not repaired until after sperm penetration of
the ovum at fertilization (Maddern and Leigh, 1976). Maternal
factors, for instance genotype and oocyte age may apparently
play a vital role here (Würgler and Maier, 1972; Bürki, 1974).
There are thus reasons to focus further attention on exposure
to mutagenic agents in men during the maturation period of
the spermatozoa, 22-23 days before conception, and in women
during the period immediately thereafter. In this context it
is of interest to mention that the only maternal factor except
for age, related to trisomy 21 found in the second Baltimore
study (Cohen *et al.*, 1977) was therapeutic radiation years
before conception - and in particular that this effect was
enhanced when including exposure during the first month of
pregnancy. As regards paternal factors the only question mark
was attendance to Military Service. Exposure to any agent
during the critical period in question, just before conception
does not seem to have been specifically covered.

    *FEII: Identification of chromosome damage in ejaculated
semen specimens.* What facilities are available to demonstrate
chromosome/DNA damage in spermatozoa and other cells in ejaculated human semen samples?
    The possibilities of detecting chromosome damage induced
in spermatogonia or during first meiotic prophase, by direct

examination of first spermatocytes in semen samples described
by Sperling and Kaden (1971) need to be confirmed.

The compaction of the chromosomes in the spermhead pre-
vents direct cytogenetic investigation with standard techniques. The polymorphic variants have been used as spermatozoan
markers to detect numerical aberrations of the Y chromosomes,
and chromosomes nos. 1 and 9 (Pearson *et al.*, 1973). There is
one report to show a time related increase in YY bodies (suggested to correspond to second meiotic non-disjunction) in
serial semen samples from seminoma patients undergoing radiation treatment and Hodgkin's patients on chemotherapy (Jacobson,
this Conference). However, there is much concern about the reliability of the quantitative technique (Beatty, 1977). Thus
again, the practical value of this type of gametic chromosome
markers in genetic hazard evaluation must be further assessed
by comprehensive studies using conventional spermiograms
(Eliasson, 1975, 1976a,b) as well as parameters to quantify
cell volume and DNA content (Beatty, 1977).

To date no spermatozoan markers have been utilized to detect structural chromosome abnormalities. The individual chromosomes of (bull) spermatozoa have been visualised by *in
vitro* hybridization with somatic cells (Johnson *et al.*, 1970).
This approach is technically cumbersome (Ringertz and Savage,
1976). Karyotyping by swelling of spermheads with detergents
combined with chromosome banding (Jacobson, personal communication) might eventually become a more practical technique
to reveal both structural and numerical aberrations.

Even with karyotyping some damage will remain undetected
due to lack of DNA replication and DNA repair. However, premutational unrepaired DNA single-strand breaks (Legator, this
Conference) might be quantified in spermatozoa from semen
samples as performed with *in vitro* techniques on testicular
cells in suspension from mice (Ono and Okada, 1977). Other
possibilities which may become helpful are quantification of
alkylation of DNA, or any other nucleophile (Aaron, 1976).
These approaches are attractive, since such techniques are
quite simple and sensitive, and thus potentially useful in
screening programs and for spot checking.

Experiments on mice demonstrate DNA repair capacity to
successively decrease from early spermatids to nil in spermatozoa (Sega *et al.*, 1976). Differential DNA repair of matur-
ing spermatids, which may be more numerously shed into the
semen in stress situations, might thus be measurable in human
ejaculates. It is tempting to speculate about the potential
of such techniques to detect exposure of a mutagen by induced
DNA spermatid repair (Sega, 1976), and even individual variations in this capacity. On the basis of administered dose, the
sensitivity of detection of DNA repair appeared to be 5-10

times greater than the dominant lethal test or translocation
studies in mice.

SUMMARY

    Reliable estimates of induced heritable chromosome damage
in the human population are difficult to obtain other than
from human germ line cells. Current techniques of handling
gonadal biopsies and present relevant information on human
meiosis are reviewed. The most commonly used parameter in
mammalian experimentation, the spermatocyte first metaphase
translocation test, may underestimate heritable damage, particularly when using conventional staining techniques to investigate the effect of chemical compounds. Additional new information might be obtained from gonadal material from female miscarriages, and from adult men postmortem, whereby minor structural abnormalities corresponding to dominant lethals might
be detected, as demonstrated in mouse experiments. Sensitive,
and inexpensive, *in vitro* techniques exist for evaluation of
genetic damage induced in spermatids/spermatozoa, but they
have hitherto only been applied to mammals other than man.

ACKNOWLEDGMENTS

    We want to thank C.E. Ford, E.A. Evans, D.G. Harnden and
B. Lambert for valuable discussions, J. Lindsten for critical
comments on the manuscript, and D. Laurie and V. Kirton for
assistance in its preparation. A research grant to M. Hultén
from the West Midlands Health Authority (T) and to J.M.
Luciani from the French Medical Research Institute (INSERM,
AT no. 257548) are gratefully acknowledged.

REFERENCES

Aaron, C.S. (1976). *Mutation Res. 38*, 303.
Allen, J.W. and Latt, S.A. (1976). *Chromosoma (Berl.). 58*, 325.
Baker, B.S., Carpenter, A.T.C., Esposito, M.S., Esposito, R.E.
      and Sandler, L. (1976). *Ann. Rev. Genet. 10*, 53.
Baker, T.G. and Franchi, L.L. (1967). *Chromosoma (Berl.). 22*,
      358.
Baker, T.G. and Neal, P. (1969). *Biophysik. 6*, 39.
Barker, P.E., Mohandas, T. and Kàback, M.M. (1977). *Clin.
   Genet. 11*, 243.

Beatty, R.A. (1977). *Cytogenet. Cell Genet. 18*, 33.

Blandau, R.J. (1969). *Am. J. Obstet. Gynaecol. 104*, 310.

Bobrow, M., Madan, K., and Pearson, P.L. (1972). *Nature New Biol. 238*, 122.

Brewen, J.G. (1976). *Mutation Res. 41*, 15.

Brewen, J.G. and Preston, R.J. (1975). *Nature. 253*, 468.

Brewen, J.G. and Payne, H.S. (1976). *Mutation Res. 37*, 77.

Brinkert, F. and Schmid, W. (1977). *Mutation Res. 46*, 77.

van Buul, P.P.W. (1976). *Mutation Res. 36*, 223.

van Buul, P.P.W. and Roos, R.A. (1977). *Mutation Res. 42*, 1.

Carr, D.H. and Gedeon, M. (1977). *In* "Population Cytogenetics" (E.B. Hook and I.H. Porter, eds.), p.1. Academic Press, New York.

Caspersson, T., Hultén, M., and Zech, L. (1971a). *Hereditas. 67*, 147.

Caspersson, T., Hultén, M., Jonasson, J., Lindsten, J., Therkelsen, A., and Zech, L. (1971b). *Hereditas. 68*, 317.

Chandley, A.C. (1975). *In* "Modern Trends in Human Genetics" (A.E.H. Emery, ed.), p. 31. Butterworth, London and Boston.

Chandley, A.C., Edmond, P., Christie, S., Gowans, L., Fletcher, J., Frackiewicz, A., and Newton, M. (1975). *Ann. Hum. Genet. 39*, 231.

Chandley, A.C., Seuanez, H., and Fletcher, J.M. (1976a). *Cytogenet. Cell Genet. 17*, 98.

Chandley, A.C., Fletcher, J., and Robinson, J.A. (1976b). *Hum. Genet. 33*, 231.

Chandley, A.C., Hotta, Y., and Stern, H. (1977). *Chromosoma (Berl.). 62*, 243.

Chapelle, A. de la, Schroder, J., Stenstrand, K., Feliman, J., Herva, R., Saarni, M., Anttolainen, I., Tallila, I., Tervila, L., Husa, L., Tallqvist, G., Robson, E.B., Cook, P.J.L., and Sanger, R. (1974). *Amer J. Hum. Genet. 26*, 746.

Cohen, B.H., Lilienfeld, A.M., Kramer, S., and Human, L.C. (1977). *In* "Population Cytogenetics" (E.B. Hook and I.H. Porter, eds.), p. 301. Academic Press, New York.

Craig-Holmes, A.P. (1977). *In* "Population Cytogenetics" (E.B. Hook and I.H. Porter, eds.), p. 161. Academic Press, New York.

Curtis, D.J. (1977). *Hum. Genet. 37*, 249.

Das, H., and Winter, S.T. (1969). *J. Med. Genet. 6*, 298.

Dutrillaux, B. (1971). *Ann. Génét. 14*, 147.

Dutrillaux, B., and Gueguen, J. (1975). *Humangenetik. 27*, 241.

Edwards, J.H., and Guli, E. (1963). *Nature. 199*, 1114.

Edwards, R.G. (1970). *In* "Human Population Cytogenetics" (P. Jacobs, W.H. Price and P. Law, eds.) vol. 5, p. 9, Edinburgh University Press, Edinburgh.

Eiche, A. (1977a). *Hereditas. 85*, 63.

Eiche, A. (1977b). *Hereditas. 85*, 254.

Eliasson, R. (1975). *In* "Progress in Infertility" (Behrman and Kistner, eds.) 2nd ed. Little, Brown & Co.

Eliasson, R. (1976a). *In* "Progress in Sperm Action, Reproductive Biology" (P.O. Hubiant, ed.), vol. 1, p. 284. S. Karger, Basel.

Eliasson, R. (1976b). *J. Reprod. Fert. Suppl. 24*, 163.

Evans, E.P., Breckon, G., and Ford, C.E. (1964). *Cytogenetics. 3*, 289.

Evans, E.P. (1976). *In* "Chromosomes Today" (P.L. Pearson and K.R. Lewis, eds.),vol. 5, o. 75. John Wiley & Sons Inc., New York.

Evans, H.J. (1976). *In* "Chemical Mutagens" (A. Hollaender, ed.), vol. 4, p. 1. Plenum Press, New York, London.

Faed, M., Robertson, J., MacIntosh, W.G., and Grieve, J. (1976). *Hum. Genet. 33*, 341.

Ferguson-Smith, M.A. (1976). *In* "Chromosomes Today" (P.L. Pearson and K.R. Lewis, eds.), vol. 5, p. 33. John Wiley & Sons, Inc., New York.

Ferguson-Smith, M.A., and Page, B.M. (1973). *J. Med. Genet. 10*, 282.

Ferguson-Smith, M.A. (1975). *In* "Medical Genetics Today" (D. Bergsma, ed.), p. 19. Birth Defects, Orig. Art. Ser., Vol. 10, No. 10. National Foundation, New York.

Ford, C.E., and Clegg, H.M. (1969). *Brit. Med. Bull. 25*, 110.

Franchi, L.L. (1977). *In* "Scientific Foundations of Obstetrics and Gynaecology" (E.E. Philipp, J. Barnes and M. Newton, eds.). Heinemann and Co.

Funaki, K., Matsui, S., and Sasaki, M. (1975). *Chromosoma (Berl.). 49*, 357.

Generoso, W.M., Krishna, M., Sotomayor, R.E., and Cacheiro, N.L.A. (1977). *Genetics. 85*, 65.

Geraedts, J.P.M., and Pearson, P.L. (1973). *Humangenetik. 20*, 171.

Gimelli, G., Porro, E., Santi, F., Scappaticci, S., and Zuffardi, O. (1976). *Hum. Genet. 34*, 315.

Goodpasture, C., and Bloom, A.D. (1975). *Chromosoma (Berl.). 53*, 37.

Gosden, C.M., Wright, M.O., Paterson, E.G., and Corant, K.A. (1976). *J. Med. Genet. 13*, 371.

Grell, R.F., and Valencia, J.I. (1964). *Science. 145*, 66.

Grell, R.F. (1971). *Ann. Génét. 14*, 165.

Hamerton, J.L. (1971). "Human Cytogenetics", vol. 1. Academic Press, New York.

Hauksdottir, H., Halldorsson, S., Jensson, O., Mikkelsen, M., and McDermott, A. (1972). *J. Med. Genet. 9*, 413.

Hayata, I., Oshimura, M., and Sandberg, A.A. (1977). *Hum. Genet. 36*, 55.

Hecht, F. (1977). *In* "Population Cytogenetics" (E.B. Hook and I.H. Porter, eds.), p. 237. Academic Press, New York.

Hendry, W.F., Polani, P.E., Pugh, R.C.B., Sommerville, I.F., and Wallace, D.M. (1976). *Brit. J. Urol. 47*, 899.

Hitotsumachi, S., and Kikuchi, Y. (1977). *Mutation Res. 42*, 117.

Holm, P.B., and Rasmussen, S.W. (1977). Human meiosis I. In press.

Hotta, Y., and Stern, H. (1976). *Chromosoma (Berl.). 55*, 171.

Hotta, Y., Chandley, A.C., and Stern, H. (1977). *Chromosoma (Berl.). 62*, 255.

Hultén, M., Lindsten, J., Lidberg, L., and Ekelund, H. (1968). *Ann. Génét. 11*, 4.

Hultén, M. (1970). *Lancet. i*, 717.

Hultén, M., and Lindsten, J. (1970). *In* "Human Population Cytogenetics" (P. Jacobs, W.H. Price and P. Law, eds.), vol. 5, p. 24. Edinburgh University Press, Edinburgh.

Hultén, M., and Pearson, P.L. (1971). *Ann. Hum. Genet. 34*, 273.

Hultén, M., and Lindsten, J. (1973). *In* "Advances in Human Genetics" (H. Harris and K. Hirschhorn, eds.), vol. 4, p. 327. Plenum Press, New York and London.

Hultén, M. (1974a). *Hereditas. 76*, 55.

Hultén, M. (1974b). "Cytogenetic Aspects of Human Male Meiosis". Thesis. Civiltryck AB, Stockholm.

Hultén, M., Solari, A.J., and Skakkebaek, N.E. (1974). *Hereditas. 78*, 105.

Hungerford, D.A., Mellman, W.J., Balaban, G.U., LaBadie, G.U., Messalzzia, L.R., and Haller, G. (1970). *PNAS. 67*, 221.

Jacobs, P.A. (1977a). *Am. J. Epidemiol. 105*, 180.

Jacobs, P.A. (1977b). *In* "Population Cytogenetics" (E.B. Hook and I.H. Porter, eds.), p. 81. Academic Press, New York.

Jagiello, G., Karnicki, J., and Ryan, R.J. (1968). *Lancet. i*, 178.

Jagiello, G., and Lin, J.A. (1974). *Am. J. Obst. Gynecol. 120*, 390.

Jagiello, G., Fang, J.S., Turchin, H.A., Lewis, S.E., and Gluecksohn-Waelsch, S. (1976a). *Chromosoma (Berl.). 58*, 377.

Jagiello, G., Ducayen, M., Fang, J.-S., and Graffeo, J. (1976b). *In* "Chromosomes Today" (P.L. Pearson and K.R. Lewis, eds.), vol. 5, p. 43. John Wiley & Sons, Inc., New York.

Johnson, R.T., Rao, P.N., and Hughes, S.D. (1970). *J. Cell Physiol. 76*, 151.

Jones, T.M., Amarose, A.P., and Lebowitz, M. (1976). *J. Clin. Endocrinol. Metab. 42*, 888.

Khush, G.S. (1973). "Cytogenetics of Aneuploids". Academic Press, New York.

Lancranjan, I., Popescu, H.I., Gavanescu, O., Klepsch, I.,
    and Serbanescu, M. (1975). *Arch. Environ. Health. 30*,
    396.

Lang, R., and Adler, J.D. (1977). *Mutation Res. 48*, 75.

Lange, K., Page, B.M., and Robert, C.E. (1975). *Am. J. Hum.
    Genet. 27*, 410.

Langenbeck, U., Hansmann, I., Huiney, B., and Honig, V. (1976).
    Hum. Genet. 33, 89.

Laurent, C., and Dutrillaux, B. (1976). *Ann. Génét. 19*, 207.

Leonard, A. (1973). *In* "Chemical Mutagens" (A. Hollaender,
    ed.), vol. 13, p. 21. Plenum Press, New York and London.

Leonard, A. (1976). *Radiat. Environ. Biophys. 13*, 1.

van der Linden, A.G.J.M., Pearson, P.L., and van de Kamp,
    J.J.P. (1975). *Cytogenet. Cell Genet. 14*, 126.

Lindenbaum, R.H., and Bobrow, M. (1975). *J. Med. Genet. 12*,
    29.

Luciani, J.M., Capodano-Vagner, A.M., Devictor-Vuillet, M.
    (1972a). "Techniques D'Analyse de la Meiose chez
    L'Homme. Biologie et Génétique". L'Expansion ed., Paris.

Luciani, J.M., Mattei, A., and Stahl, A. (1972b). *In* "Fécon-
    dité et Stérilité du Male" (C. Thibault, ed.).
    Acquisitions récentes Colloque de la Societé Nationale
    pour l'etude de la Stérilité et de la Fécondité, Paris.

Luciani, J.M., Devictor-Vuillet, M., Gagné, R., and Stahl, A.
    (1974). *J. Reprod. Fertil. 36*, 409.

Luciani, J.M., Morazzani, M.R., and Stahl, A. (1975). *Chromo-
    soma (Berl.). 52*, 275.

Luciani, J.M., Devictor-Vuillet, M., Morazzani, M.R., and
    Stahl, A. (1976). *Chromosoma (Berl.). 57*, 155.

Luciani, J.M., Devictor-Vuillet, M., Morazzani, M.R., and
    Stahl, A. (1977a). *Hum. Genet. 36*, 197.

Luciani, J.M., Devictor-Vuillet, M., Boue, J., Freund, M.,
    Stahl, A. (1977b). *Cytogenet. Cell Genet.*, in press.

Lurie, I.W., Lazjuk, G.I., Gurevich, D.B., and Usoev, S.S.
    (1976). *Hum. Genet. 32*, 23.

Maddern, R.H., and Leigh, B. (1976). *Mutation Res. 41*, 255.

Magenis, R.E., Palmer, C.G., Wang, L., Brown, M., Chamberlin,
    J., Parks, M., Merritt, A.D., Rivas, M., and Yu, P.L.
    (1977). *In* "Population Cytogenetics" (E.B. Hook and
    I.H. Porter, eds.), p. 179. Academic Press, New York.

Martin, R.H., Dill, F.J., and Miller, J.R. (1976). *Cytogenet.
    Cell Genet. 17*, 150.

Masui, Y., and Pedersen, R.A. (1975). *Nature. 257*, 705.

Matsui, S., and Sasaki, M. (1974). *Nature. 246*, 148.

Matte, R., and Sasaki, M. (1971). *Cytologia. 36*, 298.

Max, C. (1977). *Hereditas. 85*, 199.

McDermott, A. (1974). *Fertil. Steril. 25*, 79.

McIlree, M.E., Tulloch, W.S., and Newsom, J.E. (1966). *Lancet.
    i*, 679.

Meredith, R. (1969). *Chromosoma (Berl.). 26*, 254.

Moses, M.J., Counce, S.J., and Paulson, D.F. (1975). *Science. 187*, 363.

Moses, M.J., and Solari, A.J. (1976). *J. Ultrastruct. Res. 54*, 109.

Müller, H. (1976). *Stain Technol. 51*, 287.

Nakagome, Y., Kitagawa, T., Iinuma, K., Matsunaga, E., Shinoda, T., and Ando, T. (1977). *Hum. Genet. 37*, 255.

Oakberg, E.F. (1968). *In* "Effects of Radiation on Meiotic Systems". Int. Atomic Energy Agency, Vienna.

Ono, T., and Okada, S. (1977). *Mutation Res. 43*, 25.

Page, B.M. (1973). *Cytogenet. Cell Genet. 12*, 254.

Paris Conference (1971). "Standardization in Human Cytogenetics". Birth Defects, Original Article Series Vol. 7, 1972. The National Foundation, New York.

Pearson, O.L., Geraedts, J.P.M., and Pawlowitzki, I.H. (1973). *In* "Les Accidents Chromosomiques de la Réproduction" (A. Boue and C. Thibault, eds.). Centre International de l'Enfance, Paris.

Pearson, P.L. (1972). *J. Med. Genet. 9*, 264.

Polani, P.E., and Jagiello, G.M. (1976). *Cytogenet. Cell Genet. 16*, 50.

Richardson, D.W. (1975). *In* "Modern Trends in Human Genetics" (A.E.H. Emery, ed.), vol. 2, p. 404. Butterworth, London.

Ringertz, R.N., and Savage, R.E. (1976). *In* "Cell Hybrids", chapt. VI-VII. Academic Press, New York.

Robinson, D., Rock, J., and Menkin, M.F. (1968). *J. Amer. Med. Assoc. 204*, 290.

Robinson, J.A., and Newton, M. (1977). *J. Med. Genet. 14*, 40.

Roderick, T.H., and Haws, O.N.C. (1974). *Genetics. 76*, 109.

Russell, L.B. (1976). *In* "Chemical Mutagens" (A. Hollaender, ed.), vol. 4, p. 55. Plenum Press, New York.

Sankararanarayanan, K. (1976). *Mutation Res. 35*, 371.

Salisbury, G.W., Hart, R.G., and Lodge, J.R. (1977). *Amer. J. Obstet. Gynecol. 28*, 342.

Searle, A.G. (1975). *In* "Modern Trends in Human Genetics" (A.E.H. Emery, ed.), vol. 2, p. 83. Butterworths, London.

Sega, G.A. (1976). *Mutation Res. 38*, 317.

Sega, G.A., Owens, J.G., and Cumming, R.B. (1976). *Mutation Res. 35*, 193.

Seuanez, H., Mitchell, A.R., and Gosden, J.R. (1976). *Cytobios. 15*, 79.

Skakkebaek, N.E., Hultén, M., Jacobsen, P., and Mikkelsen, M. (1973a). *J. Reprod. Fertil. 32*, 391.

Skakkebaek, N.E., Hultén, M., and Philip, J. (1973b). *Acta path. microbiol. scand. 81*, 112.

Skakkebaek, N.E. (1976). *In* "Pathology of the Testis" (R.C. B. Pugh, ed.). Blackwell Sci. Publ.

Sperling, K., and Kaden, R. (1971). *Nature.* *232*, 481.

Steinberger, E., and Steinberger, A. (1975). *In* "Handbook of Physiology" (D.W. Hamilton and R.O. Creep, eds.), section 7, vol. 5, p. 1. Am. Physiological Society, Washington.

Steptoe, P.C. and Edwards, R.G. (1970). *Lancet.* *i*, 683.

Sybenga, J. (1975). *In* "Monographs on Theoretical and Applied Genetics" (R. Frankel, M. Grossman, H.F. Liskens, and D. de Zeeuw, eds.), vol. 1, p. 25. Springer Verlag, Berlin.

Szemere, G., and Chandley, A.C. (1975). *Mutation Res.* *33*, 229.

Szemere, G., and Chandley, A.C. (1976). *Stain Technol.* *51*, 64.

Söderström, K.O., and Parvinen, M. (1976). *Hereditas.* *82*, 25.

Tarkowski, A.K. (1966). *Cytogenet.* *5*, 394.

Tease, C., and Jones, G.H. (1976). *Chromosoma (Berl.).* *57*, 33.

Therkelsen, A.J., Hultén, M., Jonasson, J., Lindsten, J., Christensen, N.C., and Iversen, T. (1973). *Ann. Hum. Genet.* *36*, 367.

Turleau, C., Croquette, M.F., Fourlinnie, J.-C., Desmons, F., and de Grouchy, J. (1976). *Ann. Génét.* *19*, 210.

Turleau, C., Chavin-Colin, F., de Grouchy, J., Repesse, G., and Beauvais, P. (1977). *Hum. Genet.* *37*, 97.

Uchida, I.A. (1977). *In* "Population Cytogenetics" (E.B. Hook and I.H. Porter, eds.), p. 285. Academic Press, New York.

Uchida, J.A., and Freeman, C.P.V. (1977). *Nature.* *265*, 186.

United Nations (1977). *In* "Report of the United Nations Scientific Committee on the Effects of Atomic Radiation". Annex H.

Verma, R.S., and Lubs, H.A. (1975). *Hum. Genet.* *30*, 226.

Verma, R.S., and Lubs, H.A. (1976). *Hum. Hered.* *26*, 315.

Vig, B.K. (1977). *Mutation Res.* *49*, 189.

Vogel, F., and Rathenberg, R. (1975). *In* "Advances in Human Genetics" (H. Harris and K. Hirschhorn, eds.), vol. 5, p. 223. Plenum Press, New York.

Wolgemuth-Jarashow, D., Jagiello, G.M., and Henderson, A.S. (1977). *Hum. Genet.* *36*, 63.

Wyrobek, A.J., and Bruce, W.R. (1975). *PNAS.* *72*, 4425.

Yamamoto, M., and Miklos, G.L.G. (1977). *Chromosoma.* *60*, 283.

CHROMOSOME STUDIES IN WORKERS
EXPOSED TO BENZENE

Karl Fredga
Juhan Reitalu
Maths Berlin

Institute of Genetics and
Department of Environmental Health
University of Lund
Sweden

INTRODUCTION

The consequences to human health of an increase in the
aromatic content of motor fuels have been the object of a
comprehensive literature survey by Berlin *et al.* (1974). It
was concluded that the greatest occupational risk to health,
associated with the aromatic content of fuels, originated from
exposure to benzene. (The benzene content of the motor fuels
handled, determined through the cooperation of the Swedish
Petroleum Institute, varied between 1 and 5%).

We decided to investigate the extent of benzene exposure
among workers who handle motor fuels, to develop methods for
the routine control of the extent of exposure, and to ascertain whether any detectable effects on health attributable to
benzene exposure were present in any of the occupational
groups. Since chromosome aberrations are an early effect of
exposure to benzene, which can persist for a long period
(Tough and Court Brown, 1965; Forni *et al.*, 1971 a, b), we
decided to investigate the incidence of chromosome aberrations.

The results of the chromosome studies on 65 persons are
reported in the present communication together with a summary
of the data on their exposure to benzene. The methods developed for analytical determination of the extent of benzene
exposure and the magnitude of the dose absorbed by the

187

workers, will be reported in detail elsewhere (Berlin *et al.*,
to be published).

MATERIALS AND METHODS

After having analyzed a number of work situations involving the handling of oil and motor fuels, we decided in 1975
to investigate three groups of workers (the number in each
group is indicated in brackets):

(1) Road tank drivers, who delivered petrol from a depot
in Stockholm to consumers in the area (11),
(2) Crew members of coastal tankers carrying petrol from
a refinery in Nynäshamn to destinations on the Baltic Sea
coast and Lake Mälar (9), and
(3) Employees at petrol filling stations, which had not
yet been converted to customer self-service (9). The men were
partly engaged in filling motor-car tanks with petrol and
partly in service and repair work.

In 1976 the investigation was extended with the following
groups:

(4) A new group of road tanker drivers who delivered petrol (12),
(5) Road tanker drivers who delivered milk in the same
area (12), and
(6) Workers at industrial gasworks who were exposed to
benzene in their jobs (12).

Only males were included in the study and their age varied from 17 to 64 years. The age and time in job for the individual subjects are given in Tables I-VI. The mean age of the
subjects in the six groups were 45, 31, 43, 37, 37, and 42
years, respectively (Table VII). The last two groups of road
tanker drivers (4 and 5) were carefully selected and matched
in respect to age and time in job. In connection with the
blood sampling the medical history of each subject was recorded in a special questionnaire (*e.g.* concerning X-ray investigations, stay in hospital, medicines taken for longer periods,
handling of organic solvents, pesticides or other chemicals,
infections with fever within the last months, smoking habits
etc.).
Blood samples were taken in 10 ml vacutainers and sent to
Lund. Blood cultures (micro method) were initiated from all
subjects within 24 hr after sampling according to routine
methods (medium: 80% Parker 199 (series 1) or McCoy's

5a (series 2); 20% fetal calf serum). The cultures were harvested after 48 or 68-72 hr. Ten to twenty chromosome preparations were made from each individual by the air-dry method; in 1976 the Heto Chromofix System was used. The preparations were stained with orcein and/or Giemsa, and some were pretreated and stained for G- and C-banding (GTG and CBG techniques).

The slides were coded, mixed and analyzed blindly, *i.e.* the investigators were unaware of the exposure, occupation, medical history *etc.* of the subjects. One person performed all the analyses in 1975, another in 1976, and all aberrant, as well as doubtful, cells were further checked by a third cytologist, the same in both years.

From each subject 100 cells were analyzed (usually from the 68-72 hr cultures) and each cell was registered in a standardized examination record. In the microscopical examination, special attention was directed to the A-, D- and G-group chromosomes in all cells, in order to disclose any possible rearrangement of a stable nature. Up to 20 cells were photographed and karyotyped from each subject.

The recommendations elaborated by Buckton and Evans (1973) were followed for the classification of aberrant cells. In order to facilitate comparison with earlier investigations, the aberrant cells were registered in Tables I-VI as proposed by Buckton *et al.* (1962). B cells: Those with chromatid-type aberrations. A distinction was made between gaps and true breaks. C cells: Those with chromosome-type aberrations. A distinction was made between unstable (u) and stable (s) types of aberrations.

The figures in the tables give the number of independent aberrations, not the frequency of aberrant cells. These two parameters usually coincide, however, since very few cells had more than one aberration. The frequencies of diploid and tetraploid cells, and also of aneuploid cells (with 45 or 47 chromosomes) are given in the first main column.

The statistical methods employed were the Kruskal-Wallis one-way analysis of variance, the Mann-Whitney U test (first series), and the Wilcoxon matched-pairs signed-ranks test (second series) (Siegel, 1956).

RESULTS

Chromosome Studies

The results of the first series (1975) are compiled in Tables I-III and of the second series (1976) in Tables IV-VI. The materials from both series are grouped together in Table VII.

TABLE I.   Chromosome Aberrations in Road

Subject No.	Age yr	Time in job yr	Analysed cells No.	Dipl. %	Aneupl. %	Sum tetrapl.+ aneupl.%	No. in B Gaps %	
							Chromatid	Isochromatid
1	58	10	100	97	7	10	5	12
2	62	21	100	98	2	4	4	1
3	47	5	100	98	15	17	1	4
4	28	7	100	99	2	3	6	7
5	27	6	100	99	7	8	1	-
6	53	19	58[x]	100	10.3	10.3	6.9	3.4
7	28	4	100	100	2	2	3	1
8	27	4	100	99	4	5	4	1
9	64	24	100	98	4	6	6	4
10	52	22	100	100	2	2	6	3
11	50	25	100	98	5	7	3	3

[x] Only 58 cells suitable for analysis

TABLE II.   Chromosome Aberrations in Ship

Subject No.	Age yr	Time in job yr	Analysed cells No.	Dipl. %	Aneupl. %	Sum tetrapl.+ aneupl.%	No. in B Gaps %	
							Chromatid	Isochromatid
4	43	8.0	100	99	4	5	2	1
5	51	12.0	100	100	4	4	1	-
6	18	0.1	100	99	3	4	4	1
7	62	-	100	99	4	5	-	5
8	25	10.0	100	99	1	2	1	2
9	19	3.5	100	98	0	2	-	2
10	17	0.5	100	99	0	1	2	5
11	17	0.9	100	100	1	1	4	3
12	23	0.3	100	100	1	1	1	3

Tanker Drivers (Petrol), 1975

Cells		No. in C Cells					Sum
Breaks (B') %		C_u %				C_s %	B'+
Frag-ment	Inter-change	Acentr. Fragm.	Dicentr. Chrom. (+ Fragm.)	Ring	Chrom.	Peri-centr. Invers.	C
				Acentr.	Centr. + Fragm.		%
3	–	3	1	–	–	–	7
–	1	2	2	–	–	–	5
1	–	1	–	–	–	–	2
5	–	4	3	–	–	–	12
1	–	1	–	–	–	–	2
–	1.7	1.7	–	–	–	–	3.4
1	–	–	–	–	–	–	1
–	–	1	2	–	–	–	3
1	–	1	–	–	–	–	2
–	–	1	1	–	–	–	2
2	–	–	–	–	–	–	2

Tanker Crews, 1975

Cells		No. in C Cells					Sum
Breaks (B') %		C_u %				C_s %	B'+
Frag-ment	Inter-change	Acentr. Fragm.	Dicentr. Chrom. (+ Fragm.)	Ring	Chrom.	Peri-centr. Invers.	C
				Acentr.	Centr. + Fragm.		%
–	–	–	–	–	–	–	0
–	–	3	–	–	–	–	3
1	–	1	1	–	–	–	3
–	–	1	–	–	–	–	1
1	–	1	1	–	–	–	3
–	–	1	–	–	–	–	1
–	–	–	–	–	–	–	0
–	–	–	–	–	–	–	0
–	–	–	–	–	–	–	0

The first series showed that the road tanker drivers were exceptional, having the highest values for aneuploid cells, gaps, chromatid breaks and chromosome breaks. The sum of chromatid and chromosome breaks was 3.76, 1.22 and 0.66 per hundred cells in the road tanker drivers, ship tanker crews, and petrol station staff, respectively. A statistical treatment of the results was performed by B.O. Bengtsson. The percentage of breaks in the different individuals in the three occupational groups was first analyzed by the Kruskal-Wallis one-way analysis of variance, which showed that it was unlikely that the distributions of chromosome breaks per individual were the same for the three groups (P less than 0.01 for the null hypothesis that the distributions were identical). The occupational groups were then compared in pairs by the Mann-Whitney U test. It emerged that a highly significant difference existed between the road tanker drivers and the petrol station attendants (P less than 0.002), and a significant difference between the drivers and the ship's crew (P less than 0.05). No difference of significance was found between the petrol station attendants and the ship's crews. All tests were "two-tailed", that is, they did not include any previous knowledge of expected tendencies in the data.

The second series of investigation was undertaken in 1976 in order to elucidate whether the increased frequency of chromosome breaks in the road tanker drivers was caused by benzene or not. A new group of drivers who delivered petrol, and a carefully selected control group of road tanker drivers who delivered milk, were investigated. In addition, a group of gas plant workers, exposed to benzene in their occupations but not exposed to any other known potentially chromosome-breaking factor, was studied. The frequency of chromatid + chromosome breaks was 3.41, 2.75, and 3.08 in the three groups, respectively, which was nearly as high as in the drivers previously studied. The Kruskal-Wallis test showed that there was no reason to believe that the samples originated from groups having different frequencies of chromosome aberrations. A special analysis of the two groups of road tanker drivers by the Wilcoxon matched-pairs test showed that the difference found between the groups could well be due to chance.

Benzene Exposure

The employees at petrol stations (10, each on two successive days), were found to be exposed to a daily time-weighted average benzene concentration of 0.084 ppm (range 0.31 to 0.02 ppm); the exposure bore no relation to the amount of petrol handled. Road tanker drivers who delivered petrol (20) were exposed to an average concentration of 4.0 ppm (range

45.5 to 0.03 ppm) during loading, 0.33 ppm (range 2.87 to
0.01 ppm) during unloading, and 0.026 ppm (range 0.11 to less
than 0.005 ppm) while driving the tanker. Their mean time-
weighted average daily concentration was 0.4 ppm (range 4.13
to 0.01 ppm). Higher benzene exposure was encountered in the
crew members loading ship tankers (10); the mean time-weighted
average concentration was 6.56 ppm (range 22.8 to 0.33 ppm).
However, the crew members were less frequently exposed than
the road tanker drivers, who were exposed daily.

No determination of benzene exposure was made in road
tanker drivers delivering milk.

The gasworks workers (9) were occasionally exposed for
short periods to high levels of benzene. The estimated time-
weighted average, estimated from exhaled amount of benzene
before start of work shift, was between 5 and 10 ppm.

DISCUSSION

Benzene is known to be a mutagenic agent inducing chromosome aberrations (Tough and Court Brown 1965; Forni *et al.*,
1971 a, b; Hartwich and Schwanitz 1972), and this may eventually lead to the development of abnormal cell clones capable of
malignant growth. Benzene is also known to cause leukemia, and
the number of known cases of leukemia attributed to benzene
has been estimated to at least 150 (Forni and Vigliani 1974).
A recent survey of the relationship between benzene exposure
and different forms of anemia and leukemia is given by Vigliani and Forni (1976). According to Infante *et al.*, (1977) there
is a significant excess of leukemia in benzene workers in
spite of the fact that the benzene levels in the environment
of the workers were generally below the recommended guidelines.

The aim of the present investigation was to disclose
whether workers, occupationally handling motor fuels, are subjected to any detectable health risks attributable to benzene
exposure.

Our first series of investigation showed that road tanker
drivers, who delivered petrol, had a frequency of chromosome
+ chromatid breaks significantly increased as compared to the
other two groups studied. Determinations of the benzene exposure indicated that the highest long-term integrated exposure
was most likely to be found among the road tanker drivers.
However, it was not proved that the chromosome aberrations
were caused by benzene, and the second series of investigation
was undertaken in order to obtain more evidence as to whether
benzene was the aetiological factor. As a result, no significant difference was found among the three groups studied,
and the frequency of breaks was of the same magnitude as in

TABLE III.   Chromosome Aberrations in

Sub-ject No.	Age yr	Time in job yr	Ana-lysed cells No.	Dipl. %	An-eupl. %	Sum te-trapl.+ Aneupl.%	No. in B Gaps %	
							Chro-matid	Isochro-matid
1	38	4	100	99	2	3	1	–
3	26	7	100	99	1	2	2	–
4	42	19	100	100	6	6	1	5
5	42	23	100	100	2	2	6	2
6	65	43	100	97	5	8	6	2
7	36	12	100	100	2	2	3	5
8	37	19	100	99	0	1	4	2
9	59	25	100	98	5	7	4	2
10	38	20	100	100	2	2	–	5

TABLE IV.   Chromosome Aberrations in

Sub-ject No.	Age yr	Time in job yr	Ana-lysed cells No.	Dipl. %	An-eupl. %	Sum te-trapl.+ Aneupl. %	No. in B Gaps %	
							Chro-matid	Isochro-matid
1	33	9	100	100	6	6	13	1
3	38	5–11	100	99	7	8	9	3
5	32	9	100	100	1	1	4	1
7	29	9	100	100	3	3	14	4
9	33	12	100	100	3	3	4	3
11	51	8–9	100	100	7	7	4	6
13	37	8–14	100	98	2	4	1	–
15	48	14–16	100	100	3	3	2	5
17	33	13–14	100	100	3	3	4	1
19	39	17–18	100	100	4	4	3	1
21	35	11	100	100	0	0	2	1
23	32	4–10	100	100	5	5	–	–

Petrol Station Staff, 1975

Cells		No. in C Cells					Sum
Breaks (B') %		C_u %				C_s %	B'+
Frag-ment	Inter-change	Acentr. Fragm.	Dicentr. Chrom. (+Fragm.)	Ring Acentr.	Chrom. Centr. + Fragm.	Peri-centr. Invers.	C %
–	–	–	–	–	–	–	0
1	–	–	1	–	–	–	2
–	–	–	–	–	–	–	0
–	–	1	–	–	–	–	1
–	–	1	–	–	–	–	1
–	–	–	–	–	–	–	0
–	–	–	–	–	–	–	0
–	–	1	–	–	–	–	1
1	–	–	–	–	–	–	1

Road Tanker Drivers (Petrol), 1976

Cells		No. in C Cells					Sum
Breaks (B') %		C_u %				C_s %	B'+
Frag-ment	Inter-change	Acentr. Fragm.	Dicentr. Chrom. (+Fragm.)	Ring Acentr.	Chrom. Centr. + Fragm.	Peri-centr. Invers.	C %
1	–	–	–	–	–	–	1
4	–	2	3	–	–	–	9
3	–	–	–	–	–	–	3
2	–	–	–	–	–	–	2
1	–	–	–	–	–	–	1
2	–	–	2	–	–	2	6
2	–	–	–	–	–	–	2
2	–	–	1	–	–	1	4
–	–	1	–	–	–	–	1
1	–	4	–	–	–	–	5
–	–	1	–	–	–	–	1
1	–	3	2	–	–	–	6

the drivers previously investigated. It was surprising that
our "control" group, the milk tanker drivers, did not differ
significantly from the petrol tanker drivers.

The question was then raised about possible technical
errors and how to select the control group. The investigations
were carried out with the maximum practical standardization of
conditions. The time between sampling and inoculation of the
cultures was constant as was also the culture period (with
few exceptions, as when the 48-hr slides had to be analyzed
instead of the 72-hr slides due to few mitosis in the latter).
The age distribution among the groups was not uniform, but the
mean age differences that existed, 31-45 yr, were hardly of
any significant importance. The average time in job for the
road tanker drivers of the first series was 13.4 yr, and of
the second series 9.9 - 11.8 yr (petrol) and 9.1 - 14.5 yr
(milk). The first figures give the time for driving a road
tanker loaded with petrol or milk, respectively, and the
second figures give the total time as a lorry and road tanker
driver in general (Tables IV and V).

The analysis of the medical records did not show any
clear difference between the groups or reveal any factor which
could be correlated to the chromosome aberrations. There was
no evident relation between the chromosome damage of an individual person and his age, time in job, exposure to benzene,
etc. The material did not permit any conclusions as to differ-
rences between smokers and non-smokers.

In Table VII and Fig. 1, a control group has been included, which Dr. Felix Mitelman, Department of Clinical Genetics,
Lund University Hospital, kindly put at our disposal. These
persons were analyzed as coded controls in one of his investigations (Mitelman *et al.*, 1976; extended, in preparation 1977).
The reason why these subjects could be used as controls also
in the present study is that the blood was sampled, handled,
and cultured in the same way as in the present study, and also
that the chromosome preparations were made in exactly the same
way. In addition, the slides were analyzed by the same person
who analyzed our first series. (However, both men and women
were included among these controls, and the mean age (27) was
somewhat lower than in the present study). The frequency of
chromosome + chromatid breaks was definitely lower among these
controls than in our groups of road tanker drivers and gasworks
workers, but of the same magnitude as that of the ship tanker
crews and the employees at petrol stations. The figure was
0.83 as compared to 1.22 and 0.66, respectively.

Figures for comparison from three other investigations
are collected together in Table VIII. It is evident that the
chromosome damage observed in the road tanker drivers and gasworks workers is of the same size order as demonstrated earlier
in individuals heavily exposed to benzene, while the other

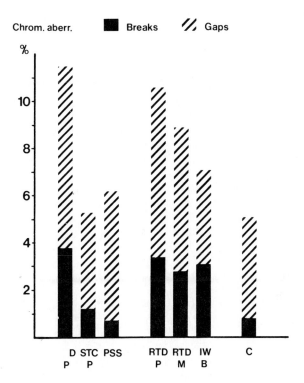

FIGURE 1.   *Diagrammatic representation of the results of the investigation (cf. Table VII). Chromosome and chromatid breaks black, gaps striped. RTD = Road tanker drivers, STC = Ship tanker crews, PSS = Petrol station staff, IW = Industrial workers, C = Controls, P = Petrol, M = Milk, B = Benzene*

groups do not markedly differ from the control values of the investigations quoted. It should be pointed out, however, that the frequency of "stable" chromosome aberrations in our material is low (0.00 - 0.25%) as compared to the other investigations referred to in Table VIII. The difference in frequency of C_s cells between our investigation and the other three investigations may be attributed to the fact that the workers in the latter ones were truly highly exposed. The relatively low frequency of C_u cells in these workers may be due to the relatively long period between benzene exposure and chromosome investigation.

Another investigation of relevance to the present discussion was performed by Hartwich and Schwanitz (1972), who studied the chromosomes of 9 refinery workers, exposed to benzene for 3 to 7 years. The frequency of aberrant cells was

TABLE V.   Chromosome Aberrations in Road

Sub- ject No.	Age yr	Time in job yr	Ana- lysed cells No.	Dipl. %	An- eupl. %	Sum te- trapl.+ Aneupl. %	No. in B Gaps %	
							Chro- matid	Isochro- matid
2	33	2-12	100	100	2	2	6	–
4	38	1-17	100	100	6	6	1	3
6	32	9-13	100	100	3	3	11	4
8	27	6-9	100	100	4	4	11	2
10	36	12	100	100	0	0	8	4
12	56	14-31	100	100	0	0	4	1
14	36	13	100	99	2	3	3	3
16	49	10-20	100	100	4	4	–	1
18	33	15	100	100	1	1	–	4
20	40	11-13	100	100	2	2	3	1
22	33	11-14	100	100	0	0	3	–
24	26	5	100	100	0	0	1	–

TABLE VI.   Chromosome Aberrations in

Sub- ject No.	Age yr	Time in job yr	Ana- lysed cells No.	Dipl. %	An- eupl. %	Sum te- trapl.+ Aneupl. %	No. in B Gaps %	
							Chro- matid	Isochro- matid
25	29	6	100	99	0	1	3	1
26	63	7	100	98	1	3	3	5
27	34	10	100	100	1	1	8	1
28	40	3	100	98	1	3	–	–
29	64	10	100	100	1	1	6	–
30	40	13	100	100	3	3	3	2
31	52	–	100	100	0	0	–	1
32	41	13	100	100	0	0	3	2
33	29	3	100	100	5	5	1	–
34	52	14	100	100	2	2	3	1
35	22	2	100	100	4	4	1	–
36	32	14	100	100	0	0	3	1

Tanker Drivers (Milk), 1976

Cells		No. in C Cells					Sum
Breaks (B') %		C_u %				C_s %	B' +
Frag-ment	Inter-change	Acentr. Fragm.	Dicentr. Chrom. (+ Fragm.)	Ring Chrom. Acentr.	Ring Chrom. Centr. + Fragm.	Peri-centr. Invers.	C %
2	–	–	–	–	–	–	2
1	–	2	3	–	–	–	6
–	–	1	–	–	–	–	1
2	–	1	–	–	–	–	3
2	–	–	3	–	–	–	5
1	–	2	1	–	–	–	4
1	–	2	2	–	–	–	5
–	–	–	–	–	–	–	0
2	–	–	–	–	–	–	2
–	–	–	–	–	–	–	0
4	–	1	–	–	–	–	5
–	–	–	–	–	–	–	0

Industrial Workers Exposed to Benzene, 1976

Cells		No. in C Cells					Sum
Breaks (B') %		C_u %				C_s %	B' +
Frag-ment	Inter-change	Acentr. Fragm.	Dicentr. Chrom. (+ Fragm.)	Ring Chrom. Acentr.	Ring Chrom. Centr. + Fragm.	Peri-centr. Invers.	C %
3	–	–	–	–	–	–	3
–	–	–	–	–	–	1	1
1	–	–	–	–	–	–	1
–	–	1	–	–	1	1	3
2	–	1	1	–	–	–	4
2	–	2	2	–	–	–	6
–	–	1	–	–	–	–	1
–	–	–	–	–	–	–	0
4	–	1	–	1	1	–	7
2	–	1	4	–	–	–	7
–	–	–	–	–	–	–	0
2	–	2	–	–	–	–	4

TABLE VII. Chromosome Studies in 1975 and 1976, Summary of Aberrations

Group	No. in Group	Mean Age Yr	Total No. of Cells	Gaps %	Chromatid Breaks, %	Chromosome Breaks, %	Sum Breaks, %	Sum Aberr. %
1975								
Road Tanker Drivers (Petrol)	11	45	1.058	7.76	1.52	2.24	3.76	11.52
Ship Tanker Crews (Petrol)	9	31	900	4.11	0.22	1.00	1.22	5.33
Petrol Station Staff	9	43	900	5.56	0.22	0.44	0.66	6.22
1976								
Road Tanker Drivers (Petrol)	12	37	1.200	7.17	1.58	1.83	3.41	10.58
Road Tanker Drivers (Milk)	12	37	1.200	6.17	1.25	1.50	2.75	8.92
Industrial Workers (Benzene)	12	42	1.200	4.00	1.33	1.75	3.08	7.08
Controls (Mitelman et al.)	15	27	3.000	4.27			0.83	5.10

TABLE VIII. Frequency of Chromosome Breaks, Comparison with 3 other Investigations

Group	No. in Group	Total No. of Cells	C_u Cells %	C_s Cells %	Sum C Cells %	Reference
Workers examined 2 yr after exposure to benzene for 1–20 yr	20	2000	1.4	1.1	2.5	Tough and Court Brown, 1965
On-site controls	5	500	0.6	0.4	1.0	
Random controls age 15–64 yr	38	1060	0.6	0.8	1.4	
Workers examined 14–15 yr after exposure to benzene for 1–22 yr	10	964	1.66	0.62	2.28	Forni et al., 1971a
Controls matched for age and sex	34	3262	0.61	0.09	0.70	
Workers studied 1–18 yr after recovery from benzene poisoning	25	2380	1.89	1.22	3.11	Forni et al., 1971b
Controls matched for age and sex	25	2450	0.49	0.04	0.53	
Road tanker drivers (petrol)	11	1058	2.24	0.00	2.24	Present investigation (1975)
Ship tanker crews (petrol)	9	900	1.00	0.00	1.00	
Petrol station staff	9	900	0.44	0.00	0.44	
Road tanker drivers (petrol)	12	1200	1.58	0.25	1.83	Present investigation (1976)
Road tanker drivers (milk)	12	1200	1.50	0.00	1.50	
Industrial workers, exposed to benzene for 2–14 yr	12	1200	1.58	0.17	1.75	

on an average 10.4% (range 8 - 12%), which was significantly
higher than in their control group, in which the average frequency of aberrations was 5.1%. In this German investigation,
cells with gaps were included in the frequency figures, which
should therefore be compared with our values in the last column of Table VII, *viz.* 11.52, 5.33, 6.22, 10.58, 8.92, and
7.08. The number of C_S-type cells was also very low in the
German investigation (2 out of 900 cells were recorded as
"Umbaufiguren"). The authors claimed that the extent and duration of exposure of the workers in their investigation was
relatively low. This comparison also supports the conclusion
that road tanker drivers and gasworks workers are groups at
risk while the ship tanker crews and petrol station staff
can not be regarded as such.

In Conclusion

An increased frequency of chromosome aberrations was
found in road tanker drivers, and industrial gasworks workers
exposed to benzene, but not in ship tanker crews and petrol
station staff. The industrial workers were exposed to benzene
at relatively low levels (time-weighted average of 5-10 ppm),
but since no other aetiological factor was considered to be
involved, the increased frequency of chromosome aberrations
in this group was regarded as due to the benzene exposure.
On the other hand, it is unlikely that benzene is the main
factor responsible for the chromosome aberrations in the petrol tanker drivers, since the milk drivers also had an increased frequency of chromosome breaks. It may be concluded
that there is some other factor in the working milieu of road
tanker drivers (and perhaps all lorry drivers) which causes
chromosome damage. It is certainly highly important to find
this factor, even though it is impossible at present to
assess the health risks associated with moderate amounts of
chromosome changes.

SUMMARY

Are workers, occupationally handling motor fuels, subject
to any detectable health risks attributable to exposure to
benzene? In order to elucidate this question, we studied the
chromosomes in cultured lymphocytes from 65 persons.
A moderate, but statistically significant, increase in
frequency of chromosome aberrations was found in road tanker
drivers and industrial workers exposed to benzene, but not in
ship tanker crews and petrol station staff. It is unlikely that

exposure to benzene is the (main) reason for the chromosome aberrations in the road tanker drivers, since both those delivering petrol and those delivering milk had the same incidence of chromosome aberrations.

ACKNOWLEDGMENTS

We wish to thank Miss Birgitta Håkansson, Mrs. Margareta Sunner, and the staff at the Department of Clinical Genetics, Lund University Hospital, for skilful technical assistance. We are indebted to Dr. B.O. Bengtsson for the statistical analysis of the results, and to Dr. Felix Mitelman for permission to quote unpublished data. Financial support was obtained from the Swedish National Board of Occupational Safety and Health and the Swedish Petroleum Institute.

REFERENCES

Berlin, M., Gage, J., and Johnsson, E. (1974). *Work-Environmental-Health 11*, 1.
Buckton, K.E., and Evans, H.J. (1973). "Methods for the analysis of human chromosome aberrations". World Health Organization, Geneva.
Buckton, K.E., Jacobs, P.A., Court Brown, W.M., and Doll, R. (1962). *Lancet ii*, 676.
Forni, A., and Vigliani, E.C. (1974). *Ser. Hematol. VII*, 211.
Forni, A., Pacifico, E., and Limonta, A. (1971a). *Arch. Environ. Health. 22*, 373.
Forni, A., Capellini, A., Pacifico, E., and Vigliani, E.C. (1971b). *Arch. Environ. Health 23*, 385.
Hartwich, G., and Schwanitz, G. (1972). *Dtsch. med. Wschr. 97*, 45.
Infante, P.F., Rinsky, R.A., Wagoner, J.K., and Young, R.J. (1977). *Lancet ii*, 76.
Mitelman, F., Hartley-Asp, B., and Ursing, B. (1976). *Lancet ii*, 802.
Siegel, S. "Nonparametric Statistics for the Behavioral Sciences". McGraw-Hill (1956).
Tough, I.M., and Court Brown, W.M. (1965). *Lancet i*, 684.
Vigliani, E.C., and Forni, A. (1976). *Environ. Res. 11*, 122.

CHROMOSOMAL ABERRATIONS IN WORKERS
EXPOSED TO ARSENIC

Gunhild Beckman[1]
Lars Beckman
Ingrid Nordenson
Stefan Nordström

Department of Medical Genetics
University of Umeå
Umeå
Sweden

INTRODUCTION

Arsenic has been found to have carcinogenic effects in clinical and epidemiological investigations (Brown, 1958; Fiertz, 1965) but not in animal experiments (Hueper and Payne, 1962). Petres *et al.* (1970) found an increased frequency of chromosome aberrations in short term cultured lymphocytes from wine growers exposed to arsenic as pesticide, and from psoriatic patients treated with arsenic. Rossman *et al.* (1975) showed in experimental studies in *Escherichia coli* that arsenic (sodium arsenite) inhibits DNA repair. They suggested that, if arsenic can be shown to inhibit DNA repair also in eukaryotic cells, it may act as a cocarcinogen. This hypothesis would explain the apparent contradiction between epidemiological data and animal experiments. Emissions from smelting operations and coal-fired power plants have been found to give rise to increased concentrations of arsenic in air, soil and water. The present state of knowledge on environmental arsenic has been summarized in a recent international conference (Fowler, 1977).

[1]*Supported by the Swedish Work Environmental Fund and the Swedish Medical Research Council*

The emissions of arsenic from the Rönnskär smelter in
northern Sweden present serious occupational and environmental
problems. Since the start of the smelting operations in 1930
the emissions are known to have decreased considerably, but
exact measurements are not available for different periods of
time.

In this report we present the results of a study of
chromosome aberrations in a series of arsenic exposed workers
at the Rönnskär smelter.

MATERIAL AND METHODS

Blood samples were collected from a series of workers at
the Rönnskär smelter and sent coded to the Clinical Genetics
Laboratory in Umeå for examination. Lymphocyte cultures were
prepared according to Hungerford (1966) with slight modifica-
tions. Culture medium was Parker 199 (Flow Lab.) supplemented
with fetal calf serum (Flow Lab.), streptomycin, penicillin
and phytohemagglutinin. The incubation time was 72 h, and
90 min before harvest Velbe (Lilly) was added (0.05 mg/ml
medium). As hypotonic solution 0.75 M KCl was used, and the
cells were fixed in methanol: acetic acid 3:1. Air dried
slides were stained with 10% Giemsa (Merck) and screened for
suitable metaphases under low magnification. Final analyses
of chromosome aberrations were made under high magnification,
and the aberrations were recorded according to the WHO recommendations (Buckton and Evans, 1973). In most cases 100
cells from each person were analyzed. In the tables the results
are presented as gaps, chromatid aberrations and chromosome
aberrations.

It was agreed that the Health Center at Rönnskär should
collect blood samples from workers with different degrees of
arsenic exposure and also from some newly employed persons.
Data on age, time of employment at Rönnskär, arsenic in urine,
smoking habits and other types of exposure (*e.g.* lead) were
sent to us in return for the completed results of the cytogenetic analysis. The Health Center also provided a scoring
of the degree of arsenic exposure into high, medium and low
(Table I). This scoring was based on known facts about the
working situation and the worker such as type and duration
of work with arsenic compounds, septum perforations and
"arsenic burns". Data on arsenic in urine were available for
all individuals except the newly employed. A number of determinations over a period of time were available for each individual. The mean of these determinations was calculated.
Determinations of arsenic in urine have only been made during

TABLE I: Age, Time of Employment at Rönnskär, Urinary
Arsenic and Degree of Arsenic Exposure in 39
Employees. Standard Deviations in Brackets

Exposure	Age	Years at Rönnskär	Urinary arsenic [a]	No. examined
High	54.05 (7.74)	22.05 (10.67)	290.55 (107.40)	18
Medium	32.90 (13.64)	6.54 (6.89)	386.36 (261.92)	11
Low	31.75 (10.68)	4.00 (1.41)	167.50 (48.56)	4
New employees	32.50 (14.29)	–	–	6

[a] µg/l

the last few years. They are therefore not expected to correlate well with the total arsenic exposure.

The groups of newly employed cannot be considered as a
proper control group of people not exposed to arsenic, since
most of them had lived close to Rönnskär before seeking
employment there. Four out of 6 were smokers, one had been
a painter for 23 years, and another a lorry driver for 17
years. The control series consist of apparently healthy males
from Umeå examined by the same standard method as the workers
from Rönnskär.

RESULTS

When the code was broken it was found that the distribution of high, medium and low exposure was rather uneven
(Table I). Thus there were only 4 workers with low exposure.
It appears that age and years at Rönnskär correlates with
high arsenic exposure. The highest level of urinary arsenic
is, however, found in the group of workers with medium exposure. This is mainly due to extreme values in one person.

Table II shows the frequencies of gaps, chromatid aberrations and chromosome aberrations in workers at Rönnskär
and a control group from Umeå. The frequency of all aber-

TABLE II.   *Chromosome Aberrations in Arsenic Exposed*
Workers from Rönnskär, Sweden and Controls

	Workers with arsenic exposure			Newly employed	Controls
	High	Medium	Low		
No. of cells	2,006	1,100	400	600	1,312
Gaps					
No.	112	64	14	31	15
per cell	0.056	0.058	0.035	0.052	0.011
Chromatid aberrations					
No.	21	23	2	9	5
per cell	0.011	0.021	0.005	0.015	0.004
Chromosome aberrations					
No.	32	14	3	3	1
per cell	0.016	0.013	0.008	0.005	0.001
All aberrations					
No.	165	101	19	43	21
per cell	0.082	0.092	0.048	0.072	0.016

rations together was significantly ($P < 0.001$) higher in all
groups of workers compared to the controls. All three types
of aberrations: gaps, chromatid aberrations and chromosome
aberrations were significantly ($P < 0.001$) increased compared
to the controls. The correlation between the frequency of
aberrations and arsenic exposure was rather poor. Thus the
newly employed workers had a higher, though not significantly
higher frequency than the group with low exposure. The
frequency of aberrations in the low-exposure group was, however, significantly lower than in the medium-exposure
($P < 0.01$) and high-exposure ($P < 0.05$) groups. The frequency
of all aberrations together was significantly ($P < 0.05$)
higher in the "high" and "medium" exposure groups compared
to the "low" exposure and "newly employed" groups. Only the
chromosome aberrations showed a good correlation with ex-

posure (see Table II). The frequency of chromosome aberrations was significantly (P < 0.05) higher in the "high" and "medium" groups compared to the "low" and "newly employed" groups.

The individual variations were very large. Thus in the high-exposure group significant variations between individuals were found (X^2 = 64.1, 17 d.f., P < 0.001).

Table III shows the combined effects of arsenic exposure and smoking on the frequency of chromosome aberrations. The trends for both chromosome aberrations and all aberrations together suggest, however, that there may be an interaction between smoking and arsenic exposure, which should be re-examined in a larger material.

There was no apparent relationship between frequency of aberrations on the one hand and age and employment time at Rönnskär on the other. The individual with the highest frequency of aberrations had been exposed to arsenic, lead and selenium. One worker with low arsenic exposure had in addition an elevated level of lead. He had 4 aberrations in 100 cells.

DISCUSSION

The present results show that persons exposed to arsenic have an increased frequency of chromosome breaks and aberrations. This is in agreement with the results by Petres *et al.* (1970). The mechanism is not known, but it seems likely that arsenic acts by inhibiting DNA repair and not by causing

TABLE III. Combined Effects of Smoking and Arsenic Exposure on Chromosome Aberrations

Arsenic[a] exposure	Smoking	Chromosome aberrations		All aberrations		No. of cells
		n	%	n	%	
+	+	28	1.84	142	9.35	1,519
+	−	18	1.13	124	7.81	1,587
−	+	6	0.75	55	6.88	800
−	−	0	0.00	7	3.50	200

[a] *+ denotes "high" or "medium" exposure (cf. Table I)*

breaks (cf. Rossman *et al.*, 1975). The rather poor correlation
between arsenic exposure and frequency of aberrations found in
this investigation is in accord with this hypothesis. The
chromosome aberrations may thus have been caused by other
agents in the occupational environment and/or smoking, while
the arsenic has prevented their repair.

In a recent study of 6 patients with skin cancer previously treated with arsenic, Burgdorf *et al.* (1977) found a normal
frequency of chromosome breaks, but a significantly increased
frequency of sister chromatid exchanges. In most of these
patients the arsenic treatment had been terminated many years
ago and therefore a certain number of aberrations may have
been lost with time. On the other hand Petres *et al.* (1970)
reported chromosome aberrations in patients where the last
treatment with arsenic had been given more than 20 years ago.
We have made repeated studies of two arsenic exposed individuals with very high frequencies of chromosome aberrations
and found statistically significant variations in time, which
concerned chromosome aberrations, but not gaps. Both individuals had, however, on all occasions a higher than normal
frequency of breaks.

It is a frequent finding in studies of exposed groups that
the individual frequencies of aberrations may show large
variations (from 0 to 25 aberrations per 100 cells) even if
the exposure is very similar. This is true also for our unpublished data on lead exposed workers. Different studies are
simply too similar considering the different exposures. It
is tempting to speculate that this similarity in response is
caused by genetic disposition. Thus in a group of individuals
exposed to a variety of genotoxic agents some individuals may
have a genetic constitution which allows them to cope with the
toxic insults, e.g. via repair systems, while others may accumulate lesions.

The biological significance of gaps may be questioned. In
our laboratory, gaps have been recorded rigorously with the
strict definition that the width of the gap should exceed that
of the chromosome.

In the controls, most of the aberrations were gaps, and in
the arsenic exposed workers relatively more aberrations were
of the chromosome-type. Gaps may originate as the result of
in vitro cultivation, *e.g.* be due to an increased production of
radicals, but nevertheless they may reflect individual variations with respect to preexisting lesions or damage caused by
toxic agents. In the series of arsenic exposed workers we
found a correlation between (chromosome and chromatid) breaks
and gaps. Thus individuals with 5 or more breaks per 100 cells
had an average of 0.081 gaps per ⌐ll, while individuals with
less than 5 breaks per 100 cell⌐ ⌐d an average of 0.044 gaps

per cell. This difference was highly significant ($\chi^2 = 22.03$, 1 d.f., P < 0.001). These results suggest that gaps at least in part should be looked upon as indicators of genotoxic influence.

REFERENCES

Braun, W. (1958). *German Med. Monthly. 3*, 321.
Buckton, K.E., and Evans, H.J., eds. (1973). "Methods for the analysis of human chromosome aberrations." *WHO report.*
Burgdorf, W., Kurvink, K., and Cervenka, J. (1977) *Hum. Genet. 36*, 69.
Fiertz, V. (1965). *Dermatologica. 131*, 41.
Fowler, B.A., ed. (1977). *In* "Environmental Health Perspectives". International conference on environmental arsenic. Ft. Lauderdale, Fla. Oct. 1976 (to be published).
Hueper, W.C., and Payne, W.W. (1962). *Arch. Environm. Hlth. 5*, 445.
Petres, J., Schmid-Ullrich, K., and Wolf, U. (1970). *Dtsch. med. Wschr. 95*, 79.
Rossman, T., Meyn, M.S., and Troll, W. (1975). *Mutation Res. 30*, 157.

THE USE OF HUMAN LYMPHOCYTE TESTS IN THE EVALUATION OF
POTENTIAL MUTAGENS: CLASTOGENIC ACTIVITY
OF STYRENE IN OCCUPATIONAL EXPOSURE

Tytti Meretoja †
Harri Vainio

Department of Industrial Hygiene and Toxicology
Institute for Occupational Health
Helsinki
Finland

INTRODUCTION

It has been estimated that 80-90% of human cancers are induced either directly or indirectly by environmental factors
(Boyland, 1967; Higginson, 1976; Cairns, 1975). Furthermore,
approximately 90% of the chemical carcinogens have proved also
to be mutagens (McCann and Ames, 1976). As the possibility of
new carcinogens entering the environment, *e.g.* vinyl chloride
(Heath *et al.*, 1975), has increased in recent years with the
continual rise in the number of new compounds being used in-
dustrially, short-term mutagenicity tests have become more
and more important in predicting the mutagenic and carcinogenic risks of our chemical environment.

The great interest in mutagenicity testing has resulted in
the creation of a whole battery of tests methods. As with the
time-consuming and costly animal carcinogenicity tests, the
biggest problem with these methods is extrapolating the results
to man. Though most human carcinogens have been shown to induce cancer in laboratory animals, the reverse has not always
been established. It thus seems evident that, as far as man's
well-being is concerned, the most reliable results are achieved
with tests on human cells. The human lymphocyte culture test
offers a reliable and relatively inexpensive short-term method
for examining mutagenic effects both *in vitro* and *in vivo*. The
in vivo studies are especially important in investigating the

†Deceased, December 15, 1977.

effects of various chemical exposures on industrial populations such as workers in the chemical industry, those in plastic manufacturing or in the leather and rubber industry.

The lymphocyte test system has been successfully used in the examination of several chemicals in occupational environments. For instance, the mutagenicity of benzene in occupational exposure has been demonstrated (Forni *et al.*, 1971a) and vinyl chloride has been shown to be both carcinogenic (Chreech and Johnson, 1974) and mutagenic (Ducatman *et al.*, 1975). In contrast, an intensive study on the effects of toluene on workers in a photogravure plant did not reveal chromosomal effects (Forni *et al.*, 1971b), while chromosomal examinations of workers occupationally exposed to lead have provided both positive (Schwanitz *et al.*, 1970; Deknudt *et al.*, 1973; Forni *et al.*, 1976) and negative results (Schmidt *et al.*, 1972; O'Riordan and Evans, 1974).

Styrene is one of the most widely used chemicals in the production of plastics, resins or styrene-butadiene rubber. The total number of workers exposed occupationally to styrene in Finland alone is around 2.000 persons. In mammalian metabolism styrene is converted to the epoxide styrene oxide (Leibman and Ortiz, 1969). Since both styrene and styrene oxide have been shown to possess mutagenic activity in bacteria (Milvy and Garro, 1976; Vainio *et al.*, 1976), yeast and Chinese hamster cells (Loprieno *et al.*, 1976), we undertook an epidemiologic study on the cytogenic effects of styrene among workers occupationally exposed to the chemical. The effects of styrene and its first metabolite, styrene oxide, were studied also *in vitro* in human lymphocyte culture tests.

MATERIAL AND METHODS

In Vivo Examination

The subjects for the chromosome study were 40 styrene-exposed males from three plants manufacturing polyester plastic products. The men, aged 20-55 years, had been employed in laminating work from several months to 15 years. At the end of an 8 hr workday they were given a general health examination, blood and urine samples were taken, and a complete blood count (hemoglobin level, red blood cell count, haematocrit reading, white blood cell count and platelets) was made. The occupational histories of the subjects were examined for possible exposure to other chemicals, and exposure to any agent known to induce chromosomal aberrations (recent viral diseases, recent vaccinations, cytotoxic drugs or ionizing radiation) was carefully recorded.

Urinary mandelic acid determinations were used as the biological measure of current styrene exposure (Engström *et al.*,
1976). The determinations of urinary creatinine were carried
out according to the method of Clark and Thompson (1949).

A control group of healthy men with no history of exposure
to styrene or to any agent with known clastogenic activity was
selected from outside the factory environment.

The chromosome studies were performed on peripheral blood
lymphocytes cultured according to conventional micro-methods.
A culture period of 64 hr was chosen so that a larger number
of mitoses would be available for scoring. A total of 100 metaphases for each experimental subject or control was examined
in detail for the occurrence of aneuploidy and chromosomal
aberrations. In addition, 1.000 interphase lymphocytes per
individual were examined for the presence of micronuclei and
nuclear bridges (Figs. 1 and 2). Only the micronuclei in the
near vicinity of the main nucleus were recorded.

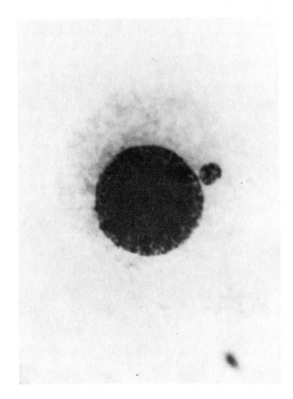

FIGURE 1. *Typical interphase cell with a micronucleus.*

In Vitro Examination

Effects of styrene and styrene oxide on human lymphocytes were examined. Whole blood microcultures were established according to conventional methods. Blood from the same healthy female donor was used throughout the experiments. An incubation time of 72 hr was chosen so that a maximal number of the stimulated, although possibly inhibited, cells would reach the mitotic stage. The chemicals were added, undiluted, to the cultures during the whole period of growth, although with styrene oxide shorter treatments were also used.

The inhibition of mitotic activity was used as the measure of cytotoxity. In the estimation of the mitotic activity or the mitotic index the number of mitoses per 10.000 cells was counted in each culture. The chromosome analysis was performed in the same manner as in the *in vivo* studies.

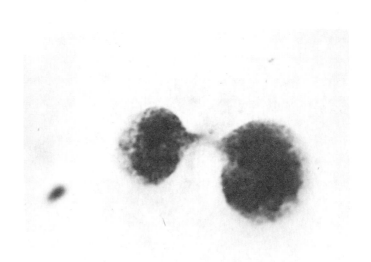

FIGURE 2. Two nuclei connected by a nuclear bridge.

RESULTS

In vivo Examination

At the time of the chromosome study the subjects felt that they were healthy. They had had no recent viral infections, nor had they received any recent vaccination or been exposed to any other agent with known clastogenic activity. The haematologic examination showed normal blood counts.

The chromosome study has thus far been completed for ten styrene exposed workers and five controls. The results are presented in Table I. As seen in the table, styrene exposure seems to correlate positively with the incidence of chromosome breakage. The frequency of micronuclei and cells connected with nuclear bridges (Fig. 2) was also increased. The chromosome aberrations were mostly breaks, but decondensation and premature condensation were also observed (Fig. 3). The results of the chromosome study of the controls agreed with the aberration frequency of control experiments made in our laboratory (an average of 2%).

In Vitro Examination

A. *Cytotoxicity of Styrene and Styrene Oxide.* The mitotic inhibition effects of the chemicals being tested were used as a measure of their cytotoxicity. Fig. 4 shows the dependence of the mitotic index on the different concentrations of styrene added to the culture for the whole period of growth. The styrene concentration of 0.008% (v/v) had no effect on the mitotic activity of the lymphocytes, but the concentration of 0.03% (v/v) lowered the mitotic index to only half of the expected value. In the cultures treated with 0.08% (v/v) of styrene, no mitoses were observed, but an increasing amount of cell debris appeared.

Styrene oxide in the same concentrations induced a complete inhibition of cell growth so that only disintegrating cells could be found on the slides. Thus styrene oxide was added in the concentration of 0.008% (v/v) to the cultures at different time periods before the cell harvest. According to the results (Fig. 5), there was no effect on mitotic activity in the cultures in which styrene oxide had been added for the last hour of growth only. However, cytotoxic effects gradually increased as the length of treatment increased. The 24 hr treatment resulted in a strong mitotic inhibition with evident death and disintegration of the cells.

TABLE I.   *The Occurrence of Aneuploidy and Chromosomal*
              from Three Plants Manufacturing Polyester
          A total of 100 metaphases were examined in each
                acid concentration (an indicator of current

Subject	Age	Years of styrene exposure	Mandelic acid concentration (mg/g of creatinine in urine)	Interphase cells with	
				Micronuclei /1000 cells	Nuclear bridges /1000 cells
Plant 1					
case 1	24	0.7	833	8	5
case 2	20	0.6	229	7	2
Plant 2					
case 3	21	1.5	3 257	14	8
case 4	37	8	23	9	3
case 5	41	2.5	219	12	0
case 6	27	2	1 452	6	3
case 7	21	1	422	6	4
Plant 3					
case 8	32	8.5	75	7	0
case 9	23	3	645	12	6
case 10	26	4	55	7	11
Controls					
case 1	32	0	0	2	0
case 2	35	0	0	0	0
case 3	33	0	0	0	2
case 4	30	0	0	2	2
case 5	30	0	0	0	0

B. *Clastogenicity of Styrene and Styrene Oxide.* A detailed chromosome analysis was performed of the cultures containing a toxic concentration of styrene (0.03% (v/v) for the whole
culture period) or styrene oxide (0.008% (v/v) for the last 8
hour of the culture). As Table II shows, both chemicals seem
to possess clastogenic activities. While the aberrations observed in the cells treated with styrene were primarily chromosome breaks, styrene oxide seemed to induce more severe chromosome destruction, which resulted in pulverized chromosomes
(Fig. 6).

Aberrations in the Lymphocytes of Styrene Exposed Workers
Plastic Products, and Unexposed Controls.
case. The duration of exposure and the urinary mandelic
styrene exposure) have been given.

Aneuploid cells	Polyploid cells	Cells with chromosomal aberrations		
		Gaps alone	Breaks	Total
2	–	–	11	11
4	–	2	9	11
2	–	–	26	26
3	–	–	25	25
5	–	–	13	13
7	–	–	15	15
6	1	–	16	16
3	1	1	16	17
5	2	–	17	17
4	–	–	15	15
1	–	–	2	2
3	–	1	3	4
3	–	–	1	1
1	2	–	1	1
2	–	–	1	1

An increase in the number of interpase cells with micronuclei or nuclear bridges was also observed. A slight effect
on the induction of aneuploidy (hypodiploidy) in the styrene
cultures was found to take place as well. In the styrene oxide
cultures the presence of aneuploidy could not be reliably
examined because the metaphases were often too damaged to
permit chromosomes to be counted.

TABLE II. The Occurrence of Chromosomal Aberrations in Cells from Cultures Containing Toxic Concentrations of Styrene or Styrene Oxide.

| | Concn. %(v/v) | Time (hr) | Interphase cells with | | Cells with chromosomal aberrations /100 metaphases | | | | |
			Micronuclei /1000 cells	Nuclear bridges /1000 cells	Aneupl.	Polypl.	Gaps alone	Breaks	Pulverized chromatin
Styrene	0.03	72	9	1	9	1	–	19	–
Styrene oxide	0.008	8	23	15	–	–	–	8	23
Control			2	–	2	–	1	1	–

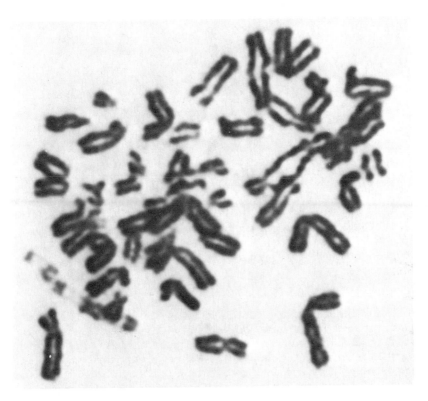

FIGURE 3. *A metaphase cell showing a dicentric chromosome and premature chromosome condensation.*

DISCUSSION

In the *in vitro* examination the cytotoxicity of styrene oxide was demonstrated to be stronger that that of styrene. The chromosome damaging effects of styrene oxide were also stronger, and this indicates that under *in vivo* conditions the agent responsible for clastogenic effects might be styrene oxide, the metabolic product of styrene.

The elimination of styrene in the form of mandelic acid makes it possible to use the urinary concentration of this acid as an indicator of styrene exposure (Engström *et al.*,

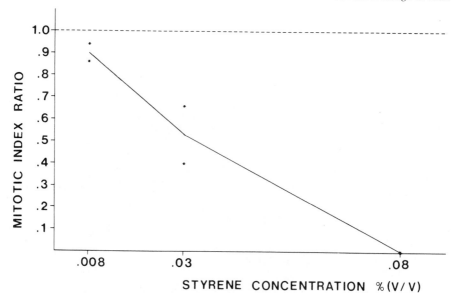

FIGURE 4. *Dependence of the mitotic index ratio (the ratio of mitotic indexes in styrene and control cultures) on the concentration of styrene added to the culture during the whole period.*

1976). We found no correlation between urine mandelic acid concentrations and chromosome damage. On the other hand, the mandelic acid concentration varies considerably with time and is an illustration of only rather recent exposure (half-life being between 9.4 and 6.4 hr). No significant correlation was observed between the incidence of chromosome aberrations and the duration of styrene exposure either. According to the results obtained so far, occupational exposure to styrene during seven months is enough to induce chromosome changes. Important information might be gained from a prospective study in which workers would be submitted to a chromosome examination before the beginning of any occupational styrene exposure and periodically checked for any changes indicating effects of the work environment. Examining the possible reversibility of clastogenic effects by following the chromosome changes of workers after occupational styrene exposure has stopped would be of great value as well. At least, it seems clear that results concerning the clastogenicity of styrene are convincing enough to demonstrate the need for large-scale mutagenicity and carcinogenicity tests so that the safe use of this economically valuable chemical can be ensured.

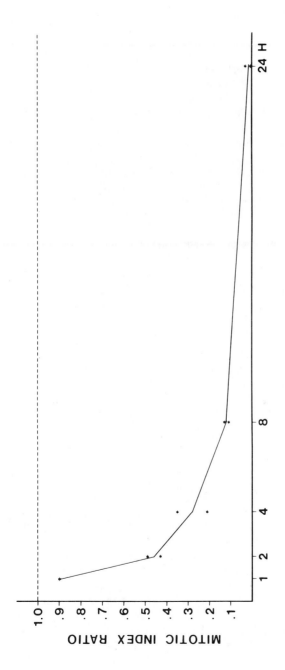

FIGURE 5. *The effects of time on the mitotic inhibition effect of styrene oxide (0.008% v/v). The mitotic index ratio shows the ratio between the mitotic indexes of the styrene oxide and control cultures.*

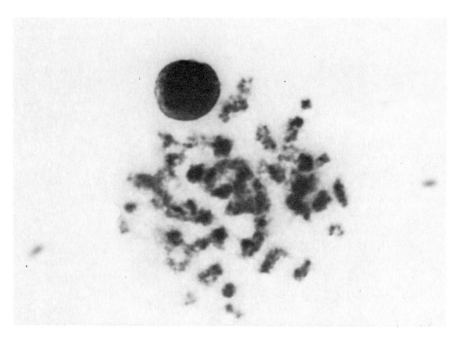

FIGURE 6. A typical damaged metaphase from a culture in which styrene oxide was added in a concentration of 0.008% (v/v) for the last 8 hours of growth.

SUMMARY

Workers from three plants manufacturing polyester plastic products were submitted to a chromosome examination for the possible clastogenic effects of styrene. The results obtained so far show a clear increase in the rate of chromosome aberrations. The incidence of aberrant cells ranged from 11 to 26% in the exposed group and was 3% or less in the unexposed control group. The frequency of interphase cells with micronuclei or nuclear bridges was also significantly increased.

In vitro experiments with human lymphocytes revealed that the cytotoxicity of styrene oxide was stronger than that of styrene. While 0.008% (v/v) of styrene in the culture for the whole period of growth had no effect on the mitotic activity of lymphocytes, the cells were able to stand the same concentration of styrene oxide only for the last 8 hr of culture or less. Examination of metaphase chromosomes from cultures containing toxic concentrations of styrene or styrene oxide showed that both chemicals possess clastogenic activity. The incidence of cells with chromosomal aberrations was 19% in styrene cultures in comparison to 31% in styrene oxide treated cultures and 2% in the control experiments.

REFERENCES

Boyland, E. (1967). *Proc. R. Soc. Med. 60*, 93.
Cairns, J. (1975). *Sci. Am. 233*, 64.
Clark, L.C., and Thompson, H.L. (1949). *Anal. Chem. 21*, 1218.
Creech, J.L., and Johnson, M.N. (1974). *Lancet 1*, 1316.
Deknudt, G., Leonard, A., and Ivanov, B. (1973). *Environ. Physiol. Biochem. 3*, 132.
Ducatman, A., Hirschhorn, K., and Selikoff, I.J. (1975). *Mutation Res. 31*, 163.
Engström, K., Härkönen, H., Kalliokoski, P., and Rantanen, J. (1976). *Scand. J. Work Environ. & Health 2*, 21.
Forni, A., Gambiaghi, G., and Secchi, C. (1976). *Arch. Environ. Health 31*, 73.
Forni, A., Cappellini, A., Pacifico, E., and Vigliani, E.C. (1971a). *Arch. Environ. Health 23*, 385.
Forni, A., Pacifico, E., and Limonta, A. (1971b). *Arch. Environ. Health 22*, 373.
Heath, C.W., Falk, H., and Creech, J.L. (1975). *Ann. N.Y. Acad. Sci. 246*, 231.
Higginson, J. (1976). *In* "The Prediction of Chronic Toxicity from Short Term Studies. Proceedings of the European Society of Toxicology. Vol. XVII." (W.A. Duncan, B.J. Leonard, and M. Brunaud, eds.), p 104. Excerpta Medica/American Elsevier.
Leibman, K.C., and Ortiz, E. (1969). *Biochem. Pharmacol. 18*, 552.
Loprieno, N., Abbondandolo, A., Baralf, R., Baroncelli, S., Bonatti, S., Bronzetti, G., Camellini, A., Corsi, C., Corti, G., Frezza, D., Leporini, C., Mazzaccaro, A., Nieri, R., Rossellini, D., and Rossi, A.M. (1976). *Mutation Res. 40*, 317.
McCann, J., and Ames, B.N. (1976). *Proc. Natl. Acad. Sci. (USA) 73*, 950.
Milvy, P., and Garro, A.J. (1976). *Mutation Res. 40*, 15.
O,Riordan, M.L., and Evans, H.J. (1974). *Nature 247*, 50.
Schmid, E., Bauchinger, M., Pietruck, S., and Hall, G. (1972). *Mutation Res. 16*, 401.
Schwanitz, G., Lehnert, G., and Gebhart, E. (1970). *Dtsch. Med. Wochenschr. 95*, 1636.
Vainio, H., Pääkkönen, R., Rönnholm, K., Raunio, V., and Pelkonen, O. (1976). *Scand. J. Work Environ. & Health 3*, 147.

SISTER CHROMATID EXCHANGE

SISTER CHROMATID EXCHANGE: THE MOST SENSITIVE
MAMMALIAN SYSTEM FOR DETERMINING THE EFFECTS OF
MUTAGENIC CARCINOGENS

Sheldon Wolff[1]

Laboratory of Radiobiology and
Department of Anatomy
University of California
San Francisco, California

In the past the most extensively studied mutagenic carcinogen has been radiation, which, at low doses, produces its
effect by damaging the genes and chromosomes in the nucleus of
cells. If the damage is in somatic cells, it can cause somatic
effects such as cancer in the individual exposed, whereas if
it is in germ cells it can cause sterility in the individual
or detrimental mutations that could possibly affect offspring
in future generations. In this modern technological age, we
now find that similar effects can be caused by many of the
plethora of chemicals to which man is exposed either because
they are useful chemicals or because they are contaminants or
byproducts of such chemicals. The magnitude of the effect is
just now becoming known.

The main tests for genetic damage have fallen into two
groups: tests at individual loci and tests in the whole genome. Included in the latter class are gross chromosomal aberrations, which because of the difference in size between a
gene and a genome, are about 10^5 times more frequent after
acute radiation than are effects at individual loci. For this
reason more than any other, the induction of ordinary chromosomal aberrations became a favored way to detect whether or
not a substance was mutagenic in higher organisms. Recently,
however, a new method has become available which is far more
sensitive for detecting the effects of the mutagens in mamma-

[1]*Work supported by the U.S. Energy Research and Development Administration*

lian cells. This is the induction of sister chromatid exchanges (SCEs) in chromosomes that have sister chromatids that
are chemically different from one another.

The breakthrough was initiated in 1972 with the publication of a paper by Zakharov and Egolina, who found that if
they allowed cells to replicate for two rounds of DNA replication in the presence of bromodeoxyuridine (BrdUrd), then the
chromosomes would contain one chromatid that. was unifilarly
substituted with BrdUrd and a sister chromatid that was bifilarly substituted. Such chromatids stained differentially with
Giemsa so that exchanges between them could be readily seen.
This observation was rapidly followed by those of Latt (1973),
who found that staining such chromosomes with the fluorescent
dye Hoechst 33258 enhanced the effect very dramatically, and
by those of Ikushima and Wolff (1974), who found that other
analogues of thymidine could cause the same effect. Other
fluorescent dyes such as acridine orange (Perry and Wolff,
1974; Dutrillaux *et al.*, 1974; Kato, 1974a) could also be used.
In many laboratories now, however, the method of choice has
been to combine staining with a fluorescent dye plus Giemsa:
the FPG, or Harlequin chromosome method of Perry and Wolff
(1974, see also Wolff and Perry, 1974), which allows the detection of sister chromatid exchanges (Fig. 1) with far greater resolution than had been possible by previously used autoradiographic techniques (Taylor, 1958). The new FPG technique
has the additional advantage that the preparations are permanent, i.e., do not fade, and do not require the use of an expensive fluorescent microscope.

The new methods for observing sister chromatid exchanges
have been very useful in studies of the basic biology of chromosome replication and structure. The resolution obtained with
the FPG technique has allowed us to resolve several long-
standing arguments about chromosomes. For instance, the presence of isolabeled chromosomes in autoradiographic preparations contributed to the idea that the chromosome consisted of
more than one DNA double helix along its length, i.e., was
binemic rather than monomemic. Isolabeling has now been found
to be an artifact of autoradiography and the fact that the
autoradiographic image cannot resolve closely spaced multiple
sister chromatid exchanges which can be clearly seen in FPG
stained preparations. Harlequin chromosomes in which the two
sister chromatids can be distinguished from one another have
also been used to disprove the hypothesis that all chromatid
deletions are the result of incomplete exchanges in small
loops along the chromosome rather than being simple chromatid
breaks. The clarity of the preparations have also confirmed
incontrovertibly that the replication pattern observed in di-

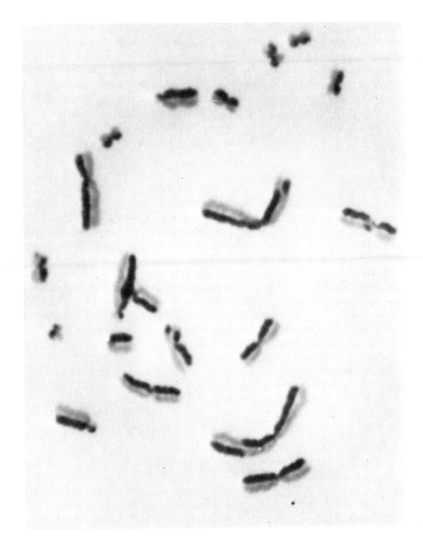

FIGURE 1. *Harlequin stained CHO cell showing sister chromatid exchanges*

plochromosomes after endomitosis shows that the newly synthesized polynucleotide strand segregates to the outside of the
chromosome.

More importantly for the topic that we are to discuss at
this conference, sister chromatid exchanges have now been
shown to be the quickest, easiest, and most sensitive mammalian system to test the genetic effects of mutagenic carcino-

gens. It had been known from the work of Kato (1974b) on auto-
radiographically determined sister chromatid exchanges that
alkylating agents could induce SCEs. The first person to use
the newer high resolution techniques for such a purpose, however, was Latt (1974), who discovered that human lymphocytes
were extremely sensitive to the crosslinking agent micomycin-
C. Solomon and Bobrow (1975) soon followed this work with
experiments in which they showed that alkylating agents induce
sister chromatid exchanges at far lower concentrations than
they did chromosome aberrations. In Figure 2 may be seen the
results of Perry and Evans (1975) who tested 14 known or suspected mutagenic chemicals for their ability to induce SCEs
in cultured Chinese hamster cells. All but two of the compounds caused a dramatic increase in the yield of SCEs. The
two that did not were maleic hydrazide, which does not affect
mammalian cells, and cyclophosphamide, or Cytoxan, a chemotherapeutic agent that requires metabolic activation before
becoming mutagenic. They further found that, with the exception of adriamycin and bleomycin, these compounds could lead
to a 10 to 20-fold increase in sister chromatid exchanges at
concentrations that did not cause an increase in chromosome
aberrations. In general, this 10 to 20-fold increase could
occur at concentrations that were 1/100th of those required
to see even small increases in the ordinary aberrations.

Because the SCE test seemed only able to pick up those
compounds that were active without metabolic activation, two
approaches have been taken to see if test systems that incorporated the ability to activate the compounds could be developed. The first of these has been to inject the compounds into
an animal so that they could pass through the liver and other
tissues containing mixed function oxidases and other enzymes
required for activation. The other has been to apply an activating system extracted from animals to the *in vitro* cultivation of the cells.

The first of these has been carried out by treating whole
animals and then culturing their blood in the presence of
bromodeoxyuridine to enable SCEs to be seen in the cultured
lymphocytes (Stetka and Wolff, 1976a) or by serially injecting
animals with bromodeoxyuridine, treating with the test chemicals, and then counting the SCEs in either bone marrow cells
(Vogel and Bauknecht 1976; Allen and Latt, 1976a) or spermatogonia (Allen and Latt, 1976a, 1976b). If the animal is
large, such as the rabbit, then the blood can be drawn serially and the time course of the induction and disappearance
of SCEs can be followed (Stetka and Wolff, 1976a). In such an
experiment, Stetka and Wollf (1976a) observed that the yield
of SCEs induced by chemicals that do not require activation
as well as those that do, first increase and then slowly decrease (Fig. 3). Adriamycin which had been reported to in-

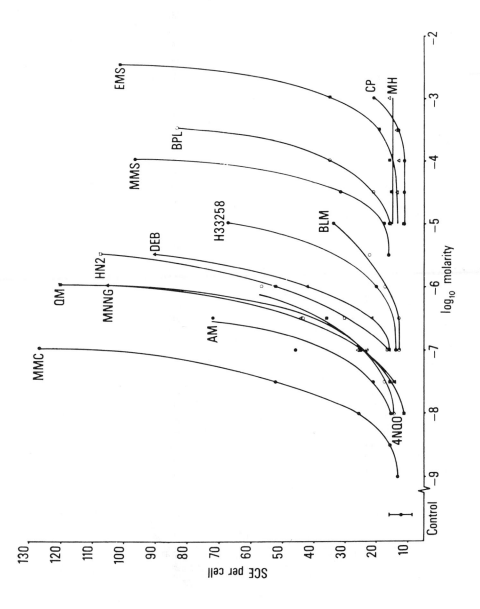

FIGURE 2. *Sister chromatid exchanges induced by 14 known or suspected carcinogens (from Perry and Evans, 1975)*

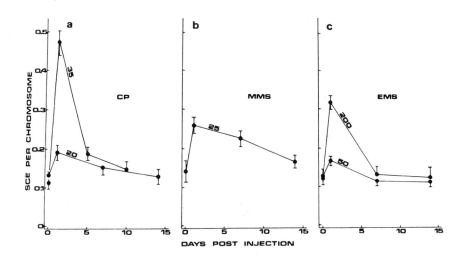

FIGURE 3. *Induction of SCEs in rabbit lymphocytes treated with mutagenic carcinogens. Note that cyclo- phosphamide causes an increase in the yield of SCEs (from Stetka and Wolff, 1976a)*

crease the yield of sister chromatid exchanges observed in blood cells cultured from patients treated with the drug also seemed to follow the same pattern, i.e., in blood cultured on successive days the yield increases (H.J. Evans, personal com- munication).

Easy as this test is, it is not nearly so easy as treat- ment of cells in tissue culture. Consequently, attempts have been made (Stetka and Wollf, 1976b; Natarajan *et al.*, 1976), to incorporate a rat liver microsome activating system with chemicals in an *in vitro* test. If Chinese hamster ovary (CHO) cells are treated with the chemicals in the presence of S-9 Mix (Ames *et al.*, 1975) as an activating system for short periods of time, then many of the compounds are activated and can induce sister chromatid exchanges. In Figure 4, we see the results of CHO cells treated for only 20 minutes with cyclophosphamide plus S-9 Mix before being cultivated in BrdUrd. A dramatic increase in SCEs is found. Figure 5 shows a cell treated in this way.

An even more sensitive test system exists, however, and this makes use of human cells in tissue culture. If cells from patients with the disease xeroderma pigmentosum are treated

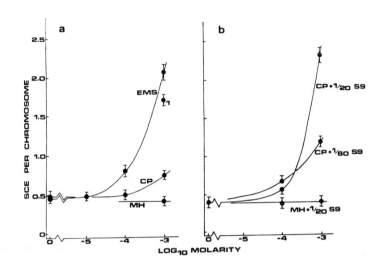

FIGURE 4. *Sister chromatid exchanges induced by ethyl-
methanesulfonate, cyclophosphamide, and maleic
hydrazide, (a) without S-9 activation, (b) with
S-9 activation (from Stetka and Wolff, 1976b)*

with low levels of mutagenic carcinogens, the yields of SCEs
are greatly increased even at concentrations of the chemicals
that do not increase the yield of SCEs in normal human cells
(Wolff *et al.*, 1977). Figure 6 presents the results of experiments that have been carried out on repair deficient XP12RO
cells from complementation group A that are deficient in excision repair. Such cells had previously been found to be more
sensitive than normal cells to UV radiation and UV-like mutagens, such as 4-nitroquinoline-1-oxide, but not more sensitive
towards X-rays and X-ray-like mutagens. This sensitivity is
manifested both as increased chromosome aberrations (Sasaki,
1973; Stich *et al.*, 1973a) and decreased survival (Stich *et
al.*, 1973b). When SCEs were used as the criterion, however,
the XP cells were sensitive to both UV-like and X-ray-like
mutagens even at concentrations that do not affect normal
human cells (Wolff *et al.*, 1977).
   This aspect of the work with SCEs not only has wide pract-
icability, it also has ramifications leading to a better
understanding of the formation of SCEs and the lesions responsible for them. Thus, by a judicious choice of the chemical
mutagens and by the use of cells that are defective in various

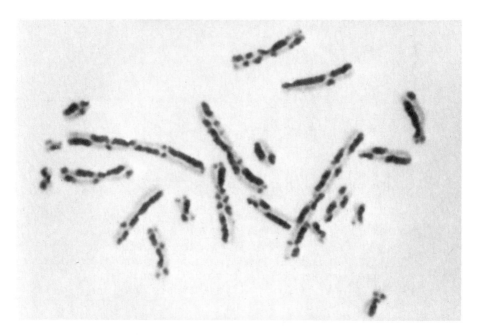

FIGURE 5. *Harlequin stained CHO cell after treatment with cyclophosphamide in the presence of activating system*

repair processes, SCE studies have been used to gather clues
about the basic mechanisms involved in SCE production. Because
the SCEs involve the exchange of partner chromatids, it had
been suggested that their formation is dependent upon DNA repair processes such as excision repair (Kato, 1974) or post-
replication repair (Kato, 1973; Bender *et al.*, 1974; Wolff *et
al.*, 1974). That the situation is not quite this simple, however, was shown in experiments in which it was found that the

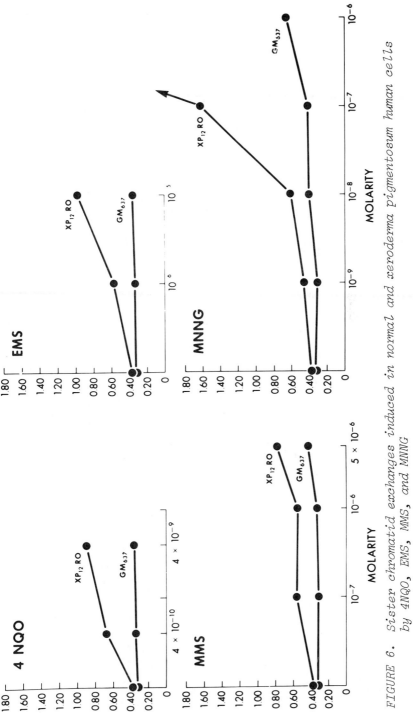

FIGURE 6. Sister chromatid exchanges induced in normal and xeroderma pigmentosum human cells by 4NQO, EMS, MMS, and MNNG

"spontaneous", or baseline, level of SCEs was normal in xeroderma pigmentosum cells that are defective in either excision
repair or postreplication repair (Wolff *et al.*, 1975).

Although all direct-acting chemicals can be detected in
either xeroderma pigmentosum or CHO cells *in vitro* , the
question still remains as to whether or not all of the chemicals that require metabolic activation can be detected *in vitro.*
This is especially true since some of the chemicals might require activation by enzymes other than those present in liver
extracts. Indeed, in a test of a series of different chemicals
by Takehisa and Wolff (1977), it was found that some are
active by themselves but increase in activity when added with
the activating system, some are active only with the activating system, and still others can be inactivated by the S-9 Mix
itself. Furthermore, some chemicals that are carcinogenic, did
not increase SCEs in CHO cells.

If CHO cells are treated with Aflatoxin B_1, which is a
potent carcinogen, the yield of SCEs increases precipitously.
In the presence of S-9 Mix the increase is even more dramatic
and occurs after a very short period of treatment with very
low concentrations (Fig.7). Another compound benzo(a)pyrene,
did not increase the yield of SCEs, even when the compound was
present for the full 24 hr culture period required for the two
rounds of DNA replication to produce harlequin chromosomes
(Fig. 8). A mere 30 min treatment of G_1 cells with the compound plus a 5-10% S-9 Mix did show that the compound could
be activated and lead to a large increase in sister chromatid
exchanges when the cell was subsequently cultured in the presence of bromodeoxyuridine (Fig. 8). Aminofluorene and acetylaminofluorene, by contrast, gave little or no increase in SCEs
whether activated or not (Fig. 9), and the highly active proximate carcinogens N-acetoxy-acetylaminofluorene and N-hydroxy-
acetylaminofluorene, which, with only a 30 min treatment, in
the absence of S-9 Mix, produce lesions that lead to SCEs, are
inactivated by its presence (Fig. 10). This observation is not
unexpected in that these compounds are very strong electrophiles that readily react with the neutrophilic proteins in
the extract of rat liver microsomes. We have subsequently
found that when acetylaminofluorene and aminofluorene are injected into the rabbit, which not only possesses the mixed
function oxidase enzymes present in liver, but also contains
other activating systems, the compounds are activated and
produce SCEs discernible when the lymphocytes are cultured.

These results indicate that suspected mutagens can be
prescreened with *in vitro* tests in the presence and absence
of S-9 Mix. Those compounds that do not require activation, as
well as some of those that do, will be quickly picked up.
Those compounds that are negative in both tests, however,

should then be screened in an *in vivo* test system such as the rabbit blood system to see if the compounds can be activated under *in vivo* conditions.

Table I presents a compilation of all chemicals that have so far been reported to be positive in the SCE test.

In conclusion, it appears that the induction of sister chromatid exchanges by mutagenic carcinogens provides a very sensitive cytogenetic test for the observations of the effects of these compounds on DNA. The induction of SCEs, particularly in xeroderma pigmentosum cells, occurs at very low concentrations, far lower than those at which effects can be noticed in mammalian systems with more ordinary mutagenic and cytological tests. The SCE tests have the advantage of being very quick. For instance, CHO cells can be seeded into flasks on one day, treated on the next, and fixed 24 hours later. The cells can be quickly stained and scored. Since such a large number of SCEs is induced per chromosome, statistically significant results can be obtained when only 20 cells are scored, as contrasted to the tedious lengthy scoring required to observe significant differences with ordinary chromosome aberrations.

Although the experiments described heretofore deal with low concentrations of the mutagenic carcinogens, they do, in fact, deal with acute treatments. Even when the chemicals are administered during the entire culture period, rather than for a mere 20-30 minutes, the actual treatment time is far shorter than the time of administration because these chemicals are generally electrophiles that have short chemical and biological half-lives. Nevertheless, the SCE system may very well have some utility for the assessment of chronic exposures. For example, Stetka and Carrano (personal communication) have carried out a series of experiments in which they exposed rabbits to successive treatments with mitomycin-C. The yield of SCEs first increased and then returned to low level, as had been found by Stetka and me for cyclophosphamide, EMS, and MMS. With successive applications of mitomycin-C, however, the yield of SCEs again increased and fell, but the fall was not great as the treatments continued. The yield of SCEs thus finally rose to a plateau level which was significantly higher than the range of SCEs found as control value in untreated rabbits. Therefore, it seems plausible that the level of SCEs found in human populations that have been chronically exposed to mutagenic carcinogens, might be higher than that found in an unexposed population.

AFLATOXIN B₁ (AFB₁)

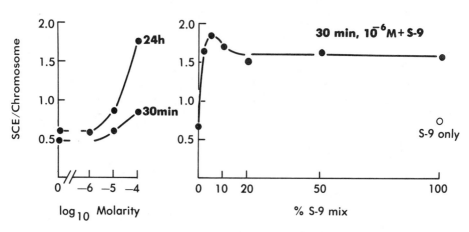

FIGURE 7.   *Sister chromatid exchanges induced by aflatoxin B₁. Left: SCEs induced after 24 hr or 30 min exposure to differing concentrations of aflatoxin B₁ without metabolic activation. Right: SCEs induced after a 30 min exposure with varying amounts of S-9 Mix (from Takehisa and Wolff, 1977).*

BENZO (a) PYRENE (BP)

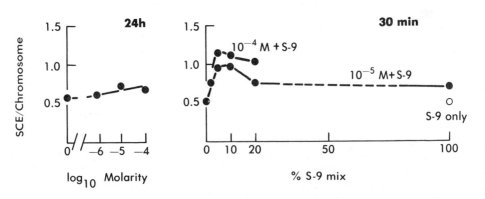

FIGURE 8.   *Sister chromatid exchanges induced by benzo(a)pyrene. Left: SCEs induced after 24 hr exposure to differing concentrations of benzo(a)pyrene without metabolic activation. Right: Effect of a 30 min exposure with varying amounts of S-9 Mix (from Takehisa and Wolff, 1977).*

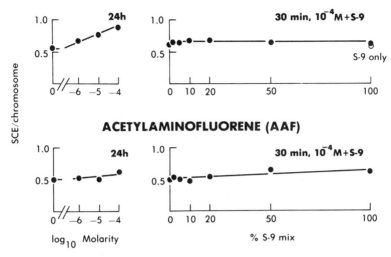

FIGURE 9. *Sister chromatid exchanges induced by aminofluorene and acetylaminofluorene. Left: SCEs found after a 24 hr exposure to the chemicals without metabolic activation. Right: SCEs found after 30 min exposure with varying amounts of S-9 Mix (from Takehisa and Wolff, 1977).*

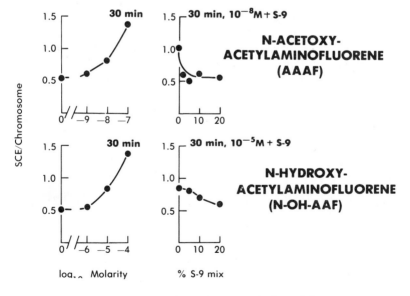

FIGURE 10. *Sister chromatid exchanges induced by N-acetoxy-acetylaminofluorene or N-hydroxy-acetylaminofluorene. Left: SCEs induced by 30 min exposure to the chemicals without metabolic activation. Right: SCEs found after 30 min exposure with varying amounts of S-9 Mix (from Takehisa and Wolff, 1977).*

TABLE I. *Positive Tests for Chemicals that Increase the Yield of SCEs (from Wolff, 1977).*

Cells	Genetic constitution	Treatment mode	Chemical	Ref.
Animal				
Human				
Lymphocytes	normal	in vitro	MMC	Latt, 1974
"	"	"	Mechlorethamine Chlorambucil Quinacrine Mustard	Solomon & Bobrow, 1975
"	"	"	Benzpyrene	Rudiger et al., 1976
"	"	"	Trenimon	Beek & Obe, 1975
"	"	"	Trenimon	Hayashi & Schmid, 1975
"	Fanconi's anemia	"	EMS, MMC	Latt et al., 1975
"	normal	in vivo	Adriamycin	Perry & Evans, 1975
Fibroblasts	"	in vitro	Trenimon	Hayashi & Schmid, 1975
"	Fanconi's anemia	"	MMC	Latt et al., 1975
"	Xeroderma pigmentosum	"	4NQO, MMS, EMS, MNNG, ENU, MMC, DMS	Wolff et al., 1977

TABLE I — Continued

Cells	Genetic constitution	Treatment mode	Chemical	Ref.
Chinese Hamster Fibroblasts	normal	in vitro	4NQO, MMC, Proflavine	Kato, 1973
"	"	"	MMC	Kato & Shimada, 1975
"	"	"	MMC, AM, 4NQO, MNNG, QM, DEB, Hoechst 33258, BLM, MMS, BPL, EMS, HN2	Perry & Evans, 1975
"	"	"	EMS, CP (with activation)	Stetka & Wolff, 1976b
"	"	"	DMN, DEN (both with activation)	Natarajan et al., 1976
Mouse Bone marrow	"	in vivo	Triaziquon, CP	Vogel & Bauknecht, 1976
"	"	"	CP	Allen & Latt, 1976b
Spermatogonia	"	"	CP	"
"	"	"	MMC	Allen & Latt, 1976a
Fibroblast	HGPRT	in vitro	MMC	Lin & Alfi, 1976
Rat Bone marrow	normal	in vitro	DMBA, TMBA	Ueda et al., 1976

TABLE I – Continued

Cells	Genetic constitution	Treatment mode	Chemical	Ref.
Rabbit Lymphocyte	normal	in vivo	CP, EMS, MMS	Stetka & Wolff, 1976a
Muntjak Fibroblasts	"	in vitro	MMC	Huttner & Ruddle, 1976
Plant Vicia faba	"	in vivo	Thiotepa	Kihlman, 1975

Abbreviations: AM, adriamycin; BLM, bleomycin; BPL, β-propriolactone; CP, cyclophosphamide; DEB, diepoxybutane; DEN, diethylnitrosamine; DMBA, dimethylbenzanthracene; DMN, dimethylnitrosamine; DMS, dimethyl sulphate; EMS, ethylmethane sulphonate; ENU, ethyl nitrosourea; HN2, nitrogen mustard; MNNG, methyl nitronitrosoguanidine; MMS, methylmethane sulphonate; 4NQO, 4-nitro-1-guanidine oxide; QM, quinacrine mustard; TMBA, trimethylbenzanthracene.

REFERENCES

Allen, J.W., and Latt, S.A. (1976a). *Nature. 260*, 449.
Allen, J.W., and Latt, S.A. (1976b). *Chromosoma. 58*, 325.
Ames, B.N., McCann, J., and Yamasaki, E. (1975). *Mutation Res. 31*, 347.
Beek, B., and Obe, G. (1975). *Humangenetik. 29*, 127.
Bender, M.A., Griggs, H.G., and Bedford, J.S. (1974). *Mutation Res. 24* 117.
Dutrillaux, B., Fosse, A.M., Prieur, M., and Lejeune, J. (1974). *Chromosoma. 48*, 327.
Hayashi, K., and Schmid, W. (1975). *Humangenetik. 29*, 201.
Huttner, K.M., and Ruddle, F.H. (1976). *Chromosoma. 56*, 1.
Ikushima, T., and Wolff, S. (1974). *Exp. Cell Res. 87*, 15.
Kato, H. (1973). *Exp. Cell Res. 82*, 383.
Kato, H. (1974a). *Nature. 251*, 70.
Kato, H. (1974b). *Exp. Cell Res. 85*, 239.
Kato, H., and Shimada, H. (1975). *Mutation Res. 28*, 459.
Kihlman, B.A. (1975). *Chromosoma. 51*, 11.
Latt, S.A. (1973). *Proc. Natl. Acad. Sci. (USA) 70*, 3395.
Latt, S.A. (1974). *Proc. Natl. Acad. Sci. (USA) 71*, 3162.
Latt, S.A., Stetten, G., Juergens, L.A., Buchanan, G.R., and Gerald, P.S. (1975). *Proc. Natl. Acad. Sci. (USA) 72*, 4066.
Lin, M.S., and Alfi, O.S. (1976). *Chromosoma. 57*, 219.
Natarajan, A.T., Tates, A.D., van Buul, P.P.W., Meijers, M., and de Vogel, N. (1976). *Mutation Res. 37*, 83.
Perry, P., and Evans, H.J. (1975). *Nature. 258*, 121.
Perry, P., and Wolff, S. (1974). *Nature. 251*, 156.
Rudiger, H.W., Kohl, F., Mangels, W., von Wichert, P., Bartram, C.R., Wohler, W., and Passarge, E. (1976). *Nature. 262*, 290.
Sasaki, M.S. (1973). *Mutation Res. 20*, 291.
Solomon, E., and Bobrow, M. (1975). *Mutation Res. 30*, 273.
Stetka, D.G., and Wolff, S. (1976a). *Mutation Res. 41*, 333.
Stetka, D.G., and Wolff, S. (1976b). *Mutation Res. 41*, 343.
Stich, H.F., Stich, W., and San, R.H.C. (1973a). *Proc. Soc. Exptl. Biol. Med. 142*, 1141.
Stich, H.F., San, R.H.C., and Kawazoe, Y. (1973b). *Mutation Res. 17*, 127.
Takehisa, S., and Wolff, S. (1977). *Mutation Res.* Submitted.
Taylor, J.H. (1958). *Genetics. 43*, 515.
Ueda, N., Uenaka, H., Akematsu, T., and Sugiyama, T. (1976). *Nature. 262*, 581.
Vogel, W., and Bauknecht, T. (1976). *Nature. 260*, 448.
Wolff, S. (1977). *In* "Annual Review of Genetics", Vol. 11. Annual Reviews, Palo Alto.
Wolff, S., Bodycote, J., and Painter, R.B. (1974). *Mutation Res. 25*, 73.

Wolff, S., Bodycote, J., Thomas, G.H., and Cleaver, J.E.
    (1975). *Genetics. 81*, 349.
Wolff, S., and Perry, P. (1974). *Chromosoma. 48*, 341.
Wolff, S., Rodin, B., and Cleaver, J.E. (1977). *Nature. 265*,
    347.
Zakharov, A.F., and Egolina, N.A. (1972). *Chromosoma. 38*, 341.

SISTER CHROMATID EXCHANGE AND CHROMOSOME ABERRATIONS
INDUCED IN HUMAN LYMPHOCYTES BY THE CYTOSTATIC
ADRIAMYCIN *IN VIVO* AND *IN VITRO*

Nils Petter Nevstad

Genetics Laboratory
Norsk Hydro's Institute for Cancer Research
Oslo
Norway

INTRODUCTION

Adriamycin (AM) is a glycoside antibiotic with an agly-
cone chromophore, adriamycinone, linked to an amino sugar,
daunosamin.

Since its isolation in Italy from a mutant of *Streptomy-
ces peucetius* (Di Marco *et al.*, 1969), AM has been used clinically as an effective chemotherapeutic agent in the treatment of solid tumors and leukemia.

Some of the most important cellular effects of AM are
probably attributable to the inhibitory action on the RNA-
and DNA syntheses. This effect has been demonstrated *in vitro*
(Silvestrini *et al.*, 1970; Meriwether *et al.*, 1972; Kim and
Kim, 1972; Wang *et al.*1972; Kitaura *et al.*, 1972; Tatsumi *et
al.*, 1974) as well as *in vivo* (Wang *et al.*, 1972; Wilmanns and
Wilms, 1972). The basic mechanism for AM effect is assumed to
be a specific complex formation with DNA. Three types of interaction are responsible for the stabilization of the complex.
The first is the electrostatic interaction between the proton-
ated amino group of the sugar residue and the ionized phosphate group. A second type of interaction is due to hydrogen
bonds. The third interaction is represented by the intercalation process, which is sustained by the weak hydrophobic stabilization taking place between the intercalating molecule and
the adjacent base pair in close Van der Waals contact.

The frequency of sister chromatid exchanges may now be determined by means of the so-called FPG technique (Perry and Wolff, 1974). It seems likely that a proportion of these SCEs are spontaneous events (Brewen and Peacock, 1969; Kato, 1974; Wolff and Perry, 1974), but the frequency increases in a dose-dependent way in the presence of clastogens.

Perry and Evans (1975) tested a series of well-known directly acting mutagens on Chinese hamster cells, and found greatly increased frequencies of SCEs. The authors concluded that SCE technique could be a very sensitive test for the pre-screening of mutagens/carcinogens.

The purpose of our experiments was to answer the following questions: 1) Is there any difference in the induction of SCEs and chromosome aberrations under *in vivo* and *in vitro* conditions? 2) Could the SCE test be a useful and sensitive tool in the examination of persons already exposed to mutagenic/carcinogenic agents? 3) Is the SCE test a more sensitive method for testing clastogens than scoring for chromosome aberrations?

MATERIALS AND METHODS

Cell Cultures

Human whole blood from males was incubated at 37°C using medium Hams F10 supplemented with 20% human serum, PHA-M (Difco) and streptomycin (100 µg/ml).

BrdU Labelling

The cultures were labelled with 1 µg/ml of 5-bromodeoxy-uridin (Sigma) (BrdU) for 48 hrs, after which the majority of the cells had been through two rounds of DNA replication, one chromatid being bifilary substituted with BrdU, and its sister chromatid unifilary substituted.

The cells were kept in total darkness to minimize effects caused by the photolysis of BrdU-containing DNA. Cultures treated with AM as described below were harvested and fixed in 3:1 methanol acetic acid. Colcemid was added for the last 2 hrs to arrest cells at metaphase. The cells were then treated with hypotonic solution (0.8% sodium citrate) for 10 min to spread the cells.

Staining

Slides with fixed cells were stained for 12 min in 0.5 μg/ml of Hoechst 33258, a bibenzimidazole derivative, in phosphate-buffered saline (pH=7.0), washed in distilled water, and mounted with a drop of water and a coverslip. They were exposed to a combination of daylight and artificial light for 24 hrs to allow the photochemical reaction to occur. Thereafter, the coverslips were removed and the slides incubated in 2 x SSC (0.6 M NaCl + 0.06 M Na-citrate) at $62^{\circ}C$ for 2 hrs. After rinsing, the slides were stained in 3% Giemsa (Gurr R60) buffered with Na_2HPO_4 pH 9.0 for 30 min, and mounted in Eukitt.

AM-treatment

In the *in vitro* experiments the cultures were treated with AM at various concentrations for the last 24 or 48 hrs before harvesting.

In the *in vivo* experiment, blood samples were taken at intervals before and after injection of a male cancer patient treated with 125 mg of AM. The AM was given intravenously as a single dose, and the patient had not received any kind of chemotherapy before. The whole blood cultures were PHA treated and incubated at $37^{\circ}C$ for 72 hrs in total darkness. BrdU labelling took place as in the *in vitro* experiment.

The plasma of all blood samples was frozen for later analyses of the concentration of AM in the plasma. Measurements of AM concentration were carried out by a fluorescence method (Bachur *et al.*, 1970).

Photography

Selected spreads were photographed and SCE were counted on the prints with occasional control in the microscope.

RESULTS

Our *in vitro* experiments indicated an increased frequency of SCEs at low doses of AM. The mean value of the controls was 4.8 per cell while treatment with 1 ng AM caused a small but significant enhancement of the SCE frequency (9.6). Very few chromosome aberrations (results of break events) could be seen at low concentrations of AM.

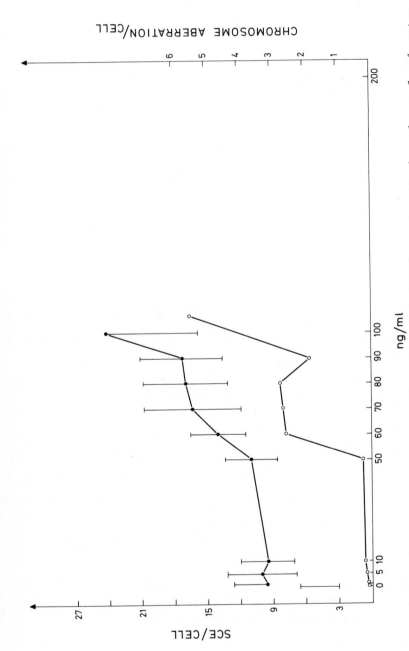

FIGURE 1. SCE (●) and chromosome aberration (o) incidence in metaphase human lymphocytes plotted against initial concentration of adriamycin. Adriamycin was added to culture medium at the time of addition of BrdU (1 μg/ml) and kept available to the cells until harvesting 48 hrs later.

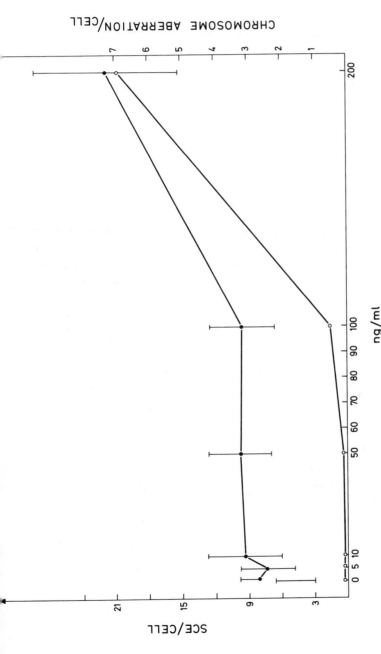

FIGURE 2. SCE (●) and chromosome aberration (○) incidence in metaphase human lymphocytes plotted against initial concentration of adriamycin. Adriamycin was added to culture medium 24 hrs after addition of BrdU (1 μg/ml) and the cells were harvested 24 hrs later. SCEs and chromosome aberrations were scored in differentially stained chromosomes of second division cells. The mean SCE frequency was scores from 25-50 cells, and the mean frequency of chromosome aberrations was scored from at least 100 cells. Vertical bars represent the standard deviation.

Maximum frequencies of SCEs and chromosome aberrations in the experiment treated with AM for 48 hrs were observed at 0.1 µg/ml AM. Higher AM concentration gave few mitotic cells, and the number of the second divisions necessary for scoring SCEs was extremely low. The results are shown in Fig. 1.

In another set of experiments the AM treatment lasted for 24 instead of 48 hrs. Also in this case the frequency of SCEs was increased by low doses of AM, but little or no increase in chromosome aberrations was observed. However, at the concentration 0.1 µg/ml the frequency of SCEs was much less increased than when treatment lasted for 48 hrs, and chromosome aberrations were still rare. Maximum frequencies of SCEs and chromosome aberrations were observed at 0.2 µg/ml AM (Fig. 2).

Measurements of the plasma concentration of AM found in the blood samples of the patient are presented in Fig. 3. Ten min after injection, the plasma concentration of AM was found to be 78 ng/ml, which is much less than expected (125 mg evenly distributed in the body would give an average of about 200 ng/ml), but after 24 hrs the plasma concentration of AM had decreased to 14 ng/ml.

In the *in vivo* experiment an increase in the frequency of SCEs was observed ten min after injection (15.5 SCE/cell with a control level of 5.0).

Six hrs after injection, however, the numbers of SCEs had decreased to 10.0, and after 1 month both SCE frequency and chromosome aberrations were down to control level (Fig. 4).

DISCUSSION

From our *in vitro* experiments it seems clear that the period of time the drug acts, is an important parameter.

Furthermore we conclude that the method for detecting SCEs in human lymphocyte cultures *in vitro* seems to be a sensitive way for testing environmental mutagens, especially at very low doses of the drug administered, when chromosome aberrations are rare or undetectable. Our observations are in agreement with the findings of other authors (Perry and Evans, 1975; Stetka and Wolff, 1976).

The frequency of SCEs on the *in vivo* experiments corresponded well with the frequencies observed in the *in vitro* experiments at similar doses (Fig. 5).

It therefore seems likely that there is no difference in the induction of SCEs under *in vivo* and *in vitro* conditions.

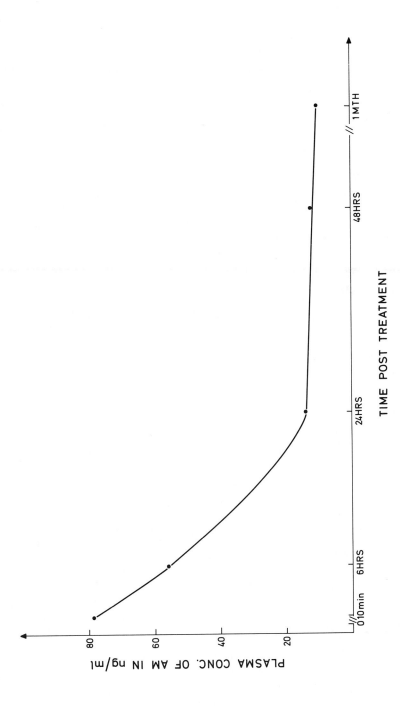

FIGURE 3. *Plasma concentration of adriamycin. Data are from a male cancer patient treated with 125 mg of adriamycin plotted against time post treatment.*

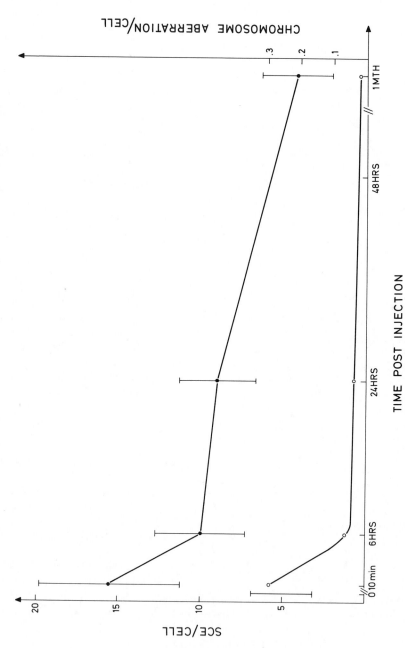

FIGURE 4. SCE (●) and chromosome aberration (○) incidence in metaphase human lymphocytes from a male cancer patient treated with 125 mg of adriamycin plotted against time post treatment.

It is generally accepted that AM penetrates the cell membrane by simple diffusion and is selectively taken up by lysosomes. Thus, we would expect the intracellular concentration of AM in the *in vivo* experiment to be much higher than the plasma concentration (78 ng/ml 10 min after injection). Even so, a low level of chromosome aberrations was observed (0.3 aberrations/cell). Comparing this to the *in vitro* results, where the maximum concentration (100 ng/ml) gave a higher number of chromosome aberrations (5.5 aberrations/cell) (Fig. 5) it seems likely that the induction of chromosome aberrations is somewhat different under *in vivo* and *in vitro* conditions.

Schinzel and Schmid (1976) examined 67 patients treated with high therapeutic doses of chemical mutagens and concluded that an increased incidence of chromatid breaks and exchanges is not a typical finding in lymphocyte cultures of persons exposed to clastogens. The authors claimed that the study of chromosome aberrations is inadequate for monitoring weak and questionable mutagens in exposed populations, and this is in agreement with our observations.

The present SCE test has several additional advantages compared to the conventional tests that involve scoring for chromosome aberrations or micronuclei. The SCE test is capable of detecting exposure to very low concentrations of chemicals, whereas the usual tests are less sensitive. Furthermore, statistically significant increases in SCEs may be detected after scoring only 10-20 cells, whereas chromosome aberration tests require the scoring of 100-200 cells. Finally the scoring of SCEs is much easier than scoring for chromosome aberrations.

The *in vivo* SCE test used to monitor humans for exposure to mutagens seems to have one disadvantage, however, because of the transient nature of the increase in the SCE frequency. This is also supported by the finding of Stetka and Wolff (1976).

Chromosome aberrations seem to decrease even faster with time after injection of AM. But if an individual is tested soon after a suspected exposure, and again at weekly intervals, the exposure of AM could be evaluated since the frequency of SCE is expected to be high at first and then to decrease to the control level within a few weeks after exposure and thereafter remain relatively constant.

On the other hand, if an individual can not be tested soon after a suspected exposure, both tests might be inconclusive, at least in this particular case.

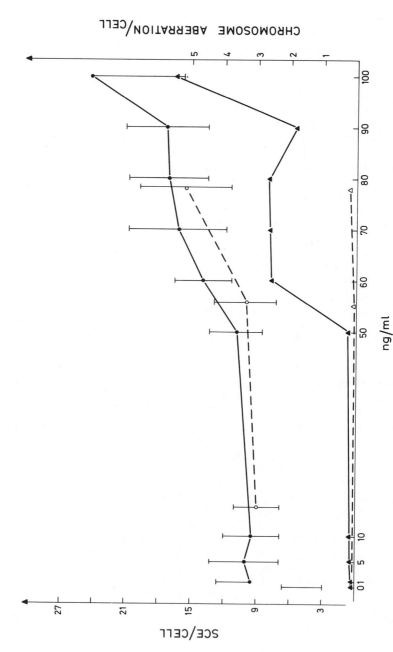

FIGURE 5. A comparison of in vitro and in vivo studies of SCE frequency (in vitro (●), in vivo (○)) and chromosome aberration frequency (in vitro (▲), in vivo (△)) in human lymphocytes after treatment with the cytostatic adriamycin.

SUMMARY

The *in vitro* results showed an increase in SCE frequency at very low doses of adriamycin (AM) (9.5 SCE/cell at 1 ng/ml with a control level of 4.8 SCE/cell). At these doses chromosome aberrations were not above control level. With a concentration of 0.1 µg/ml there was a great increase in the frequency of both SCEs (24.5) and chromosome aberrations.

The *in vivo* experiments were carried out on a cancer patient who received 125 mg of AM. The plasma concentration of AM decreased very rapidly after injection (from 0.087 µg/ml at 10 min to 0.014 µg/ml at 24 hrs).

Ten min after injection the SCE frequency increased (15.5 SCE/cell with a control level of 5.0), but six hrs later it had decreased to 10.0. After 1 month both SCE frequency and chromosome aberrations were down at control level.

The frequency of SCEs in the *in vivo* experiments corresponded well with the frequencies observed in the *in vitro* experiments at similar doses, whereas the number of chromosome aberrations did not.

Scoring for SCEs in the *in vitro* experiments seems to be a more sensitive test for the mutagenic action of AM than scoring for chromosome damage. For *in vivo* studies both methods have a time-limited value.

ACKNOWLEDGMENTS

I would like to thank Dr. Kari K. Lie, Institute of Pediatrics, University of Oslo for having carried out the measurements of the plasma concentration of AM, and Dr. Foss Abrahamsen, Dept. of Hematology, Norwegian Radiumhospital, for referring the patient. I am also grateful to Dr. philos. Anton Brøgger and Dr. med. Tobias Gedde-Dahl, Genetics Lab., Norsk Hydro's Institute for Cancer Research, Oslo for valuable advice.

REFERENCES

Bachur, N.R., Moore, A.L., Bernstein, J.G., and Lin, A. (1970). *Cancer Chemother. Rep.* 54, 89.
Brewen. J.G. and Peacock, W.A. (1969). *Mutation Res.* 7, 433.
De Duve, C., De Barsy. T., Poole. B., Trouet, A., Tulkens. P., and van Hoof, F. (1974). *Biochem. Pharmacol.* 23, 2495.
Di Marco, A. and Arcamone, F. (1975). *Arzneimittel Forsch.*

(Drug Res.) 25. 368.

Di Marco, A., Gaetani, M., and Scarpinato, B. (1969). *Cancer Chemother. Rep. 56*, 153.

Kato, H. (1974). *Nature. 251*, 70.

Kim, S.H. and Kim, J.H. (1972). *Cancer Res. 32*, 323.

Kitaura, K., Imai, R., Ishihara, Y., Yanai, H., and Takahira, H. (1972). *J. Antibiot. 25*, 509.

Meriwether, W.D. and Bachur, N.R. (1972). *Cancer Res. 32*, 1137.

Perry, P. and Evans, J. (1975). *Nature. 258*, 121.

Silvestrini, R., Gambarucci, C., and Dasdia, T. (1970). *Tumori. 56*, 137.

Schinzel, A. and Schmid, W. (1976). *Mutation Res. 40*, 139.

Skovsgaard, R. and Nissen, N.I. (1975). *Dan. med. Bull. 22*, 62.

Stetka, D.G. and Wolff, S. (1976). *Mutation Res. 41*, 333.

Stetka, D.G. and Wolff, S. (1976). *Mutation Res. 41*, 342.

Tatsumi, K., Nakamura, T., and Wakisaka, G. (1974). *Gann. 65*, 237.

Wang, J.J., Cortes, E., Sinks, L.F., and Holland, J.F. (1971). *Cancer. 28*, 837.

Wang, J.J., Chervinsky, D.S., and Rosen, J.M. (1972). *Cancer Res. 32*, 511.

Wilmans, W. and Wilms, K. (1972). *Enzyme. 13*, 90.

Wolff, S., Bodycote, J., Thomas, G.H., and Cleaver, J.E. (1975). *Genetics. 81*, 349.

Wolff, S. and Perry, P. (1974). *Chromosoma (Berl.). 48*, 341.

PSORALEN/UVA TREATMENT AND HUMAN CHROMOSOMES

Helga Waksvik
Anton Brøgger

Genetics Laboratory
Norsk Hydro's Institute for Cancer Research
Oslo, Norway

Per Thune

Department of Dermatology
University of Oslo
Norway

When irradiated with UVA light (320-380 nm) the psoralens will react with thymine residues in one DNA strand forming monoadducts (Cole, 1970) or crosslinks between thymines in two complementary DNA strands (Pathak and Krämer, 1969; Cole, 1970).

We have previously studied the clastogenic effect in human chromosomes of an alkylating and crosslinking antibiotic, mitomycin C (Brøgger and Johansen, 1972) and its interaction with caffeine in human lymphocytes (Brøgger, 1974a).

We therefore wished to compare these findings with the possible clastogenic effect of psoralen and UVA (PUVA) to see if some common denominator may be related to the crosslinking action.

Recently, PUVA has been introduced with great success in the treatment of psoriasis (Wolff *et al.*, 1976). Because of the clastogenic effect of PUVA we decided to study if chromosome damage occurs in PUVA treated psoriasis patients and chose to follow five patients very closely during the period of treatment by means of lymphocyte cultures.

MATERIALS AND METHODS

All the *in vitro* experiments were performed with lymphocytes from the same, healthy, female donor. Whole blood cultures were treated for 2 hrs with psoralen 48 hrs after PHA
stimulation, then irradiated with a Hanovia U.V.S. 500 Lamp,
and finally harvested at 72 hrs.

Chromosome damage was scored as constrictions and gaps
(folding defects), breaks and exchanges in the metaphase.

Five psoriasis patients suffering from chronic, stable
psoriasis covering at least 50% of their body surface were
investigated. They were treated with orally administered
8-methoxypsoralen (in a new formulation provided by Nyegaard
& Co A/S) and irradiation from Philips Black light lamps.
The number of treatments varied from 15 to 26 times. Four
of the patients were 100% cleared while one was about 80%
cleared. Blood samples were taken before treatment, and 2 hrs
after the 1st, 5th, 10th and 20th treatment and finally from
four of the patients when given a maintenance dose after 6
months.

Analysis of sister chromatid exchange was made by the FPG
techniqe (Perry and Wolff, 1974).

FIGURE 1.   *Chromosome Damage Induced by Psoralen/UVA:*
             Constrictions and Gaps (Bottom); Break Events
             (Top). One Exchange Is Counted as Two Break
             Events. Each Point Represents 100 Metaphases
             from One Culture, except the Last Points
             (Dotted Lines), Where Cell Death Occurred and
             a Poor Yield of Mitoses Was Obtained.

        O *25 µg/ml trimethylpsoralen + UVA*
        ● *50 µg/ml trimethylpsoralen + UVA*
        -O- *25 µg/ml methoxypsoralen + UVA + 1 mM caffeine*
        -●- *50 µg/ml methoxypsoralen + UVA + 1 mM caffeine*
        △ *UVA*
        ▲ *UVA + 1 mM caffeine*
    — + — *expected if effect of caffeine/UVA and*
             PUVA were additive

Abscissa: UVA radiation in sec; energy in 10^{-2} *J/cm*^{2}.
Ordinate: Aberrations per 100 cells.

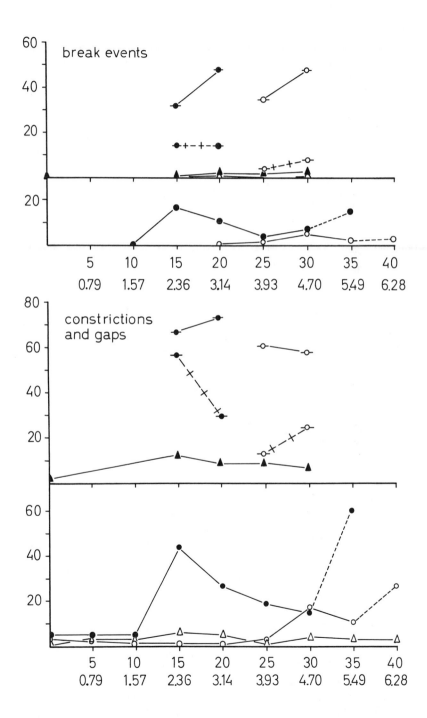

RESULTS

In vitro treatment with PUVA induced chromosome damage –
mainly constrictions and gaps, but also breaks and exchanges
(Fig. 1) – and increased the SCE frequency (Table I). The
chromosome aberrations were non-randomly distributed and
mainly localized to R bands or to R/G band junctions.

Caffeine increased in a synergistic way the chromosome
aberration yield if added after PUVA treatment (Fig. 1,
Table II), but there was no effect when caffeine was present
before and during PUVA treatment. In the psoriasis patients
there was an increase in constrictions and gaps after 10
treatments, but after 20, levels were not abnormally high
(Table III). Examinations after 6 months revealed no abnormal
findings. The SCE frequency was only examined at 6 months and
no increase was found. The average number of SCE's was 3.7
before and 3.8 after treatment.

The plasma doses of psoralen varied between 10 and 320
ng/ml at the time of irradiation.

TABLE I.   *Sister Chromatid Exchange Induced by PUVA and*
PUVA/Caffeine Treatment of Human Lymphocytes.
BrdU 1 μg/ml. Counts from 50 Cells in Each
Culture.

UVA sec	8-methoxy-psoralen μg/ml	Caffeine M	SCE/cell
0	0	0	4.58 ± 0.35
0	0	10^{-3}	5.52 ± 0.68
0	25	0	4.72 ± 0.41
0	50	0	4.68 ± 0.39
20	0	0	9.38 ± 0.69
30	0	0	16.90 ± 1.10
20	0	10^{-3}	14.54 ± 0.78
30	0	10^{-3}	21.74 ± 1.08
30	25	0	24.60 ± 1.21
30	25	10^{-3}	31.64 ± 1.09
20	50	0	14.12 ± 0.80
20	50	10^{-3}	25.14 ± 1.08

TABLE II. *Chromosome Damage from PUVA (Methoxypsoralen) and Caffeine (10^{-3} M) when Caffeine Is Added before or after PUVA Treatment. 100 Metaphases Examined from Each Culture. CG – Constrictions, Gaps; B – Breaks; Ex – Exchanges*

UVA sec	8-methoxy-psoralen μg/ml	Caffeine	Aberrations			per cell
			CG	B	Ex	
Caffeine before PUVA						
0	0	–	1	0	0	0.01
20	0	+	9	0	0	0.09
30	0	+	4	0	0	0.04
30	25	+	17	2	1	0.20
20	50	+	24	8	1	0.33
Caffeine after PUVA						
0	0	–	2	0	0	0.02
20	0	+	8	1	0	0.09
30	0	+	9	2	0	0.11
30	25	+	78	42	2	1.22
20	50	+	64	37	1	0.92
No Caffeine (for comparison, data from trimethylpsoralen, Fig. 1)						
0	0	–	3	0	0	0.03
20	0	–	5	2	0	0.07
30	0	–	4	0	0	0.04
30	25	–	17	3	1	0.21
20	50	–	27	9	1	0.37

TABLE III.   Chromosome Studies of Five Psoriasis Patients
Treated with PUVA. The Data Are Average Per
Cent Constrictions and Gaps (CG) and Break
Events (BE). 100 Cells Were Examined from
Each Patient at Each Sampling Time

	Before treatment	1	5	After 10 treatment(s)	20	6 months
CG	6.3	6.3	4.2	9.2	4.0	2.5
BE	1.7	0.6	1.2	0.8	1.0	0.0

DISCUSSION

Due to mitotic inhibition and cell death, opportunities
to observe chromosome damage are limited. The dose curves are
short, ant the last points are not very reliable, because of
the small number of metaphases. The pattern is different from
that obtained with mitomycin C (Brøgger and Johansen, 1972;
Brøgger, 1974b) since neither attentuations and lateral ex-
tensions of chromatids nor exchanges between secondary constriction regions in chromosome 1, 9 and 16 were seen. Both
agents crosslink DNA. The above mentioned effects are therefore probably not due to mitomycin C crosslinks.

The data show that there is no synergistic effect between
PUVA and caffeine when the latter is present before and during
the PUVA treatment. Caffeine has to be present after the
treatment to produce the synergistic effect. The hypothesis
that caffeine increases the number of clastogenic targets is
not supported. Caffeine added after the UVA and PUVA treatment increased also the number of SCE. This is contrary to
the findings of Kato (1973, 1974), who reports that caffeine
reduces the frequency of SCE induced by UV and mitomycin C,
clastogens which presumably act similarly to UVA and PUVA.

In our irradiation experiments two effects upon the DNA
must be considered: thymine dimers and psoralen crosslinks.
The increase in SCE frequency after irradiation alone must be
related to the dimers, whereas after PUVA both DNA changes
must be involved. Possible causes of the different effects
are summarized in Table IV. It appears that the effect from
dimers and the effect from crosslinks are about equal, and
that both effects are enhanced in the presence of caffeine.
A common denominator in all the processes which are affected

TABLE IV. *Possible Explanations for Increased SCE Frequencies Induced by UVA, PUVA, and Caffeine. The Figures are Those from Table I after Subtraction of Control and/or Caffeine Values*

Treatment		SCE/cell		Suggested explanation
Control level			4.58	Spontaneous or BrdU Effect
10^{-3}M caffeine			0.94	Caffeine effect (binding to ssDNA?)
Psoralen alone		negligible		—
UVA alone	20 s		4.80	Thymine dimers
	30 s		12.32	" "
UVA + caffeine	20 s		9.02	Caffeine effect on dimer or dimer repair
	30 s		16.22	"
PUVA alone	20 s – 50 µg/ml	9.54	4.80	Dimers
			4.74	PUVA cross links
	30 s – 25 µg/ml	20.02	12.32	Dimers
			7.70	PUVA cross links
PUVA + caffeine	20 s – 50 µg/ml	19.62	9.02	Caffeine effect/dimers
			10.60	Caffeien effect/ cross links
	30 s – 25 µg/ml	26.12	16.22	Caffeine effect/dimers
			9.90	Caffeine effect/cross links

by caffeine may be the presence of single stranded DNA, so
that all caffeine effects are in fact due to the binding of
caffeine to ssDNA (Lindahl, personal communication). This
implies that the process of SCE production involves a step
where ssDNA is present.

From our *in vivo* studies it cannot be excluded that the
lymphocytes of the psoriasis patients received a certain,
although apparently very small exposure from the PUVA treatment. Since the psoralen is circulating in the plasma, the
lymphocytes will take up the substance. The UVA irradiation
is mainly absorbed by the porphyrin groups of the hemoglobin,
which means that circulating blood will be a target for the
PUVA action. However, the circulating population of lymphocytes being exposed during the UVA treatment is unknown.

Our results are not in complete agreement with the observations of Swanbeck *et al*. (1975), who found a significant
increase in chromosome damage in some of their psoriasis
patients receiving PUVA treatment.

Schinzel and Schmid (1976) have raised certain doubts
about the sensibility of the lymphocyte culture system in
detecting low dose exposures to mutagens.

On the other hand, several investigations prove the usefulness of the method (see Brøgger, This Conference), particularly if the individuals are examined shortly after exposure (Dobos *et al*., 1974; Nevstad, This Conference). In our
psoriasis patients blood was drawn 2 hrs after PUVA treatment.
We hoped that this time would be sufficient to distribute the
exposed lymphocytes evenly among the blood cells so that they
would appear in the blood sample, but not long enough for any
effect of PUVA to disappear.

The SCE method is more sensitive than scoring for the
conventional chromosome aberrations (Wolff, This Conference).
The SCE frequency in the four patients at the maintenance
treatment after 6 months was not elevated, which is in agreement with the negative chromosome findings.

We may be wrong, but we do not feel that our results
warrant any warning against a limited use of PUVA in the
treatment of psoriasis. But, particularly since our sample
of patients was small - even if altogether they were examined
26 times - follow-up of patients is still indicated both with
respect to chromosome damage and, of course, possible carcinogenesis.

A complete report will be published elsewhere (Waksvik
et al., 1977; Brøgger *et al*., 1977).

SUMMARY

Treatment of human lymphocytes *in vitro* with trimethyl-psoralen or 8-methoxypsoralen and UVA irradiation (PUVA) induced chromosome damage - mainly constrictions and gaps, but also breaks and exchanges - and increased the frequency of sister chromatid exchange (SCE). Due to mitotic inhibition and cell death the dose curves were short: UVA irradiation energy ranging from 1.57 to 6.28 10^{-2} J/cm^2 with psoralen doses 25 and 50 µg/ml resulted in up to 0.04-0.60 constrictions and gaps/cell and up to 0.18 break events/cell. SCE frequency increased from about 5 per cell to 14 and 24.

1mM caffeine increased in a synergistic way the chromosome aberration yield and frequency of SCE.

Five psoriasis patients were studied during the course of PUVA treatment by blood samples taken on the first day of treatment prior to the administration of 8-methoxypsoralen, and 2 hrs after the UVA irradiation at the 1st, 5th, 10th and 20th treatment. There was an increase in constrictions and gaps after 10 treatments, but after 20, levels were not abnormally high.

The plasma doses of psoralen varied between 10 and 320 ng/ml at the time of irradiation.

Examinations when a maintenance dose was given after 6 months revealed no abnormal findings. At this time the frequency of sister chromatid exchanges was determined; it had not increased after therapy.

ACKNOWLEDGMENTS

We are indebted to M. Sci. Ingrid Wiik, Nyegaard & Co A/S, for determination of the plasma concentrations of 8-methoxypsoralen, and to Bitten Lunde and Reidun Björge for skilful technical assistance.

REFERENCES

Brøgger, A. (1974a). *Mutation Res. 23*, 353.
Brøgger, A. (1974b). *Hereditas. 77*, 205.
Brøgger, A., and Johansen, J. (1972). *Chromosoma. 38*, 95.
Brøgger, A., Waksvik, H., and Thune, P. (1977). *Hum. Genet.*
    (submitted).

Cole, R.S. (1970). *Biochem. biophys. Acta 217,* 30.
Dobod, M. Schuler, D., and Fekete, G. (1974). *Hum Genet. 22,*
    221.
Kato, H. (1973). *Exp. Cell Res. 82,* 383.
Kato, H. (1974). *Exp. Cell Res. 85,* 239.
Pathak, M.A. and Krämer, D.M. (1969). *Biochem. biophys. Acta*
    195, 197.
Perry, P. and Wolff, S. (1974). *Nature, 251,* 156.
Schinzel, A., and Schmid, W. (1976). *Mutation Res. 40,* 139.
Schwanbeck, G., Thyresson-Høk, M., Bredberg, A., and Lambert,
    B. (1975). *Acta derm. -venerol. 55,* 367.
Waksvik, H., Brøgger, A., and Stene, J. (1977). *Hum. Genet.*
    (in press).
Wolff, K., Fitzpatrick, T.B., Parrish, J.A., Gschnait, F.,
    Gilchrest, B., Hönigsmann, H., Pathak, M.A., and Tannen-
    baum, L. (1976). *Arch. Derm. 112,* 943.

SISTER CHROMATID EXCHANGE
IN HUMAN AND CHINESE
HAMSTER BONE MARROW CELLS
AS A METHOD FOR
STUDYING GENETIC DAMAGE

Sakari Knuutila [1]

Research Department
Rinnekoti Institution
for the Mentally
Retarded, Espoo
Finland

Hannu Norppa

Department of Genetics
University of Helsinki
Finland

Tuomas Westermark

Research Department
Rinnekoti Institution
for the Mentally
Retarded, Espoo, and
Department of Pharmacology
and Toxicology
College of Veterinary
Medicine, Helsinki
Finland

1
*Supported in part by grants from the Research Department,
Rinnekoti Institution for the Mentally Retarded, from the
Finnish Cultural Foundation and from Emil Aaltonen Foundation.*

INTRODUCTION

Recent studies of a metaphase chromosome which had re-
plicated twice in the presence of 5-bromodeoxyuridine (BUdR)
showed that bifilarly BUdR substituted chromatid after a treatment with fluorescent agent, Hoechst 33258, stains less inten-
sively with Giemsa than the unifilarly substituted chromatid
(Dutrillaux *et al.*, 1974; Kato, 1974; Perry and Wolff,
1974). This method made the detection of a sister chromatid
exchange (SCE) more rapid and reliable than the method of ^{3}H-
thymidine incorporation and autoradiography (Taylor, 1958). We
report here two methods for the determination of SCEs in human
and Chinese Hamster bone-marrow cells. Furthermore, preliminary
results concerning the frequency of SCE in hematologically
healthy subjects as well as in selenium treated hamsters are
presented.

MATERIALS AND METHODS

Human Bone Marrow Cells

Bone marrow was obtained from four slightly mentally retarded subjects (Table I, Nos. 1-4). Patient 4 had measles
whereas there was no sign of viral infection in patients 1-3.
The bone marrow was aspirated from the iliac crest under
local anesthesia (1% Lidocain^R, Orion). Heparin (Medica) was
used as an anticoagulant. For morphological studies a portion

*TABLE I. Drug Treatment and the Frequency of SCEs in
Bone Marrow Cells from 4 Hematologically Healthy Subjects.
Patient 4 Had Measles at the Time of this Study.*

Patient			Drug treatment	Mitotic index (%)	Diploid metaphases	
No.	Sex	Age yrs			Studied	SCEs/metaphase (range)
1	F	45	Orgametril^R	0.5	48	4.9 (1-9)
2	M	20	Truxal^R	0.1	19	4.5 (1-7)
3	M	36	Sennapur^R	0.1	25	4.4 (0-8)
4	M	28	Truxal^R Sordionol^R Disflatyl^R	0.5	56	5.2 (1-11)

of each sample was spread on slides and stained with May-Grünwald-Giemsa. For cytogenetic studies the rest of the aspirated material (about 2 ml) was divided into 4 to 6 portions. Each portion was incubated in the dark for 44 to 48 hours at 37°C with 8 ml MEM (Eagle), 2 ml human AB, Rh positive serum and BUdR (Sigma; 8 µg/ml). Colcemid (Ciba; 0.05 µg/ ml) was added 2 to 3 hours before the end of the incubation period. Then 0.075 M potassium chloride heated to 37°C was added, and the culture was kept for 20 min at room temperature. Ten to twenty slides were prepared per subject by the air-drying technique. The slides were stained the next day with Hoechst 33258 (0.5 µg/ml Sørensen buffer, pH 7) for 15 min, then rinsed with distilled water and subjected to UV light (Philips TUV 30 W). After UV treatment the preparations were left standing in a dish on top of a damp sponge. On the next day the preparations were incubated with 2 x SSC at 60°C for 2 hours and then rinsed with distilled water. The slides were stained for 10 min with Merck's Giemsa diluted to 5% in Sørensen buffer, pH 6.8. Nineteen to 56 well spread metaphases were scored for the number and location of SCEs (Table I). The distribution of exchanges between and within chromosomes is not reported here.

Chinese Hamster Bone Marrow Cells

The method used was a modification of that applied by Allen and Latt (1976) to mouse cells. BUdR alone, without FUdR or deoxycytidine as used by Bauknecht *et al.* (1976) and Allen and Latt (1976) was injected intraperitoneally every hour over a 9 hour period at a concentration of 240 µg (hamster 1, Table II) or 80 µg (hamsters 2-8, Table II) per gram body weight per injection. The agents under study, Selenium (Se), placebo and methyl methanesulphonate (MMS) were administered i.p. half an hour after the last BUdR injection. Fifteen to sixteen hours after the treatment injection, animals were killed by ether narcosis 3-4 hours following an i.p. injection of 10 µg colcemid per g body weight. Both femurs and humeri were removed and bone-marrow was rinsed out with a medium consisting of 3 parts MEM, 1 part human AB, Rh positive serum and colcemid (0.05 µg/ml). Hypotonic treatment, cell preparations and Harlequin staining were performed as described earlier for the human bone marrow cells.

TABLE II. The Frequencies of SCEs in Bone Marrow Cells of 8 Chinese Hamsters. The Animals Had Been Treated with Selenium, Placebo or Methyl Methanesulphonate or Were Untreated.

Hamster No.	Sex	Mutagen (μg/g body wt)	Mitotic index (%)	Metaphases studied	SCEs/one metaphase (range)
1	M	–	8.3	50	3.1 (0–12)
2	M	–	11.3	100	3.3 (0–10)
3	M	Se$^{x)}$(3)	15.8	50	6.7 (1–20)
4	M	Se (4)	3.4	50	7.0 (2–21)
5	F	Se (6)	1.0	50	11.3 (2–28)
6	M	Placebo$^{xx)}$	10.3	50	3.9 (1–7)
7	F	Placebo	10.4	50	4.8 (1–11)
8	M	MMS (100)	2.7	7	34.1 (13–47)

$^{x)}$*Composition of injected solution:*
Se 0.06 g; methyl.p.–oxybenz. 0.18 g; propyl.p.–oxybenz. 0.02 g; Aq. ster. ad. 100 ml.

$^{xx)}$*Composition of injected solution:*
Methyl.p.–oxybenz. 0.18 g; propyl.p.–oxybenz. 0.02 g; Aq. ster. ad. 100 ml.

RESULTS AND DISCUSSION

Human Bone Marrow Cells

Figures 1–2 show that the Harlequin metaphases permit reliable scoring of SCEs. The yield of the Harlequin metaphases was below 50% of all metaphases. If cells were stained soon (5–15 hours) after the starting of fixation, the non-Harlequin mitoses often gave such a good G-band staining that it was possible to analyse the karyotype.

In the present study as well as in the direct bone-marrow analysis performed previously in our laboratory (Knuutila *et al.*, 1976), the frequency of dividing cells (mitotic index) varied greatly; from less than 0.1% (patient 2) to 0.5% (patient 4). This variation may have been caused by variation in the proportion of bone marrow cells to blood cells in the aspirate. We have found that cultured aspirates with a high number of blood cells have a low mitotic index. The number of blood cells may not, however, explain why one 48 hours culture, which had been started 2 hours after the bone-marrow aspiration, had a very low mitotic index (below 0.01%, subject not reported here), whereas another culture started 4 days later

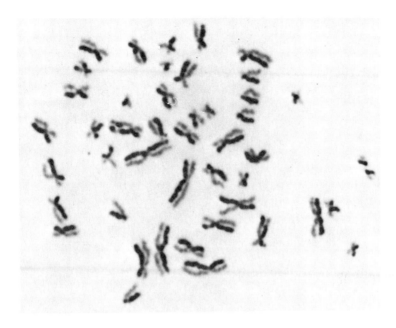

FIGURE 1.   *Harlequin Stained Human Bone Marrow Metaphase from a Measles Infected Patient (Patient 4).*

from the same aspirated material had a mitotic index of 0.2%.

Table I shows the frequencies of SCEs in all 4 subjects. The frequencies detected were more than 10 times lower than those in bone marrow cells in patients with Bloom's syndrome (Shiraishi and Sandberg, 1977). The frequencies (from 4.4 to 5.2) were also lower than those in lymphocytes cultured for 72 hours in our laboratory, from 11 healthy women, aged 19 to 44 years. In these lymphocyte studies the mean frequency of SCEs was 8.0/metaphase (range: 5.8-9.5/metaphase/subject) scored from altogether 1039 cells. The measles infected patient had a SCE frequency of 9.9 ranging from 3 to 20 per metaphase in his lymphocytes cultured for 72 hours simultaneously with his bone marrow.

Chinese Hamster Bone Marrow Cells

The mitotic index in hamster bone marrow cells was higher (1% to 15.8%) than in human bone marrow cells. The frequencies of SCEs in non-treated, in Se- and MMS-treated hamsters are given in Table II and representative findings are shown in Fig. 3a-d. According to Schwarz (1976) LD_{50} for Se is 4.1 μg per gram body weight. The results at this stage suggest that

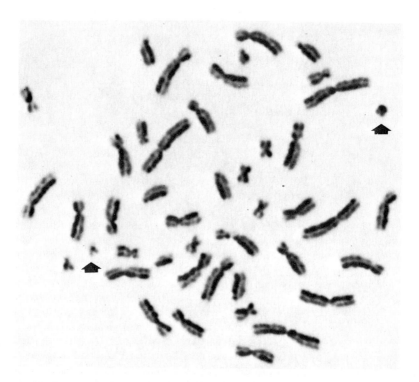

FIGURE 2.   *Harlequin Stained Bone Marrow Metaphase from a Patient with Chronic Myeloid Leukemia in Blastic Crisis (not Presented in Table 1). The Figure Shows that the Method Can also Be Used for Studying Cytogenetic Changes in Haematological Noeplasmas.*

Selenium at sublethal concentrations increases the frequency of SCEs.

We conclude that the two methods reported here are rapid, reliable and sensitive in detecting chromosomal changes caused by environmental agents. These methods can also be used for studying SCEs in hematological diseases as illustrated by the metaphase from a patient with chronic myeloid leukemia in blastic crisis (Fig. 2).

SUMMARY

Two methods for the determination of sister chromatid exchange (SCE) in human and Chinese Hamster bone marrow

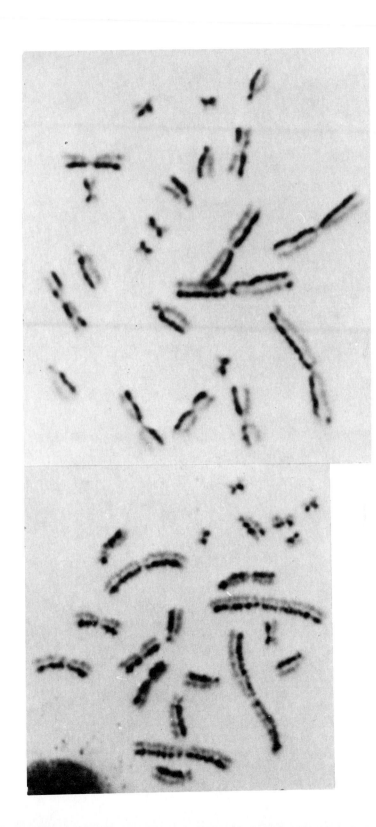

FIGURE 3a(left)-b(right). Harlequin Stained Chinese Hamster Bone Marrow Metaphases from Non-Treated Hamsters.

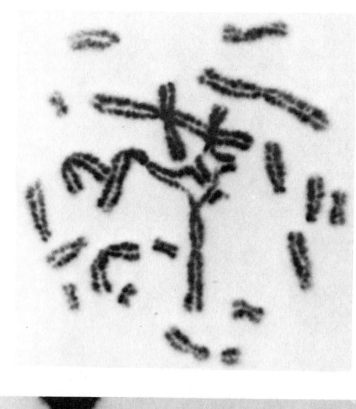

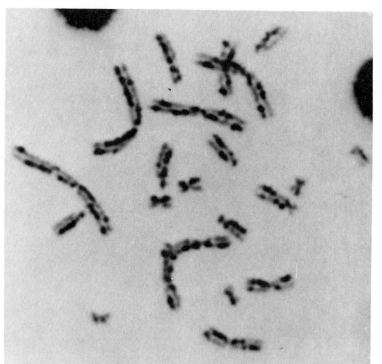

FIGURE 3c(left)-d(right). Metaphases from Methyl Methanesulphonate Treated Hamsters.

cells are reported. Bone marrow aspirates were taken from 4 hematologically healthy subjects, one of whom had measles. The aspirated material was incubated in the dark for 48 h at 37°C in the presence of BUdR. The preparations were stained with Hoechst 33258, subjected to UV light, incubated in 2xSSC and stained with Giemsa. In human bone marrow cells, the mitotic index varied greatly from person to person (0.01% to 0.5%), whereas the mitotic index in hamster cells was always higher (1.0% to 15.8%). The frequency of SCEs in human cells varied from 4.4 to 5.2/metaphase.

To study the hamster bone marrow cells BUdR was injected intraperitoneally every hour over a 9 hour period. The agents under study (Na-selenite and as a positive control methyl methanesulphonate (MMS)) were administered by the same route one half-hour after the last BUdR injection. The hamsters were killed 15 hours after the treatments. The hamsters treated with sublethal concentrations (3 to 6 μg per gram body weight) of Na-Selenite had a higher frequency of SCEs (6.7, 7.0 and 11.3) than did the 2 non-treated (3.1 and 3.3) or the 2 placebo-treated hamsters (3.9 and 4.8). The MMS treated hamster had the highest number of SCEs (34.1).

ACKNOWLEDGMENTS

We wish to thank Mrs. Marja-Liisa Koverskoi and Mr. Jorma Mäki-Paakkanen for excellent laboratory assistance. We are also grateful to Prof. Hakon Westermarck for placing at our disposal the hamsters at his Department and Cand. agronom Henning Vadgaard (Dansk Shell) for donating the Chinese Hamsters for this study.

REFERENCES

Allen, J.W., and Latt, S.A. (1976). *Chromosoma. 58*, 325.
Bauknecht, Th., Vogel, W., Bayer, U., and Wild, D. (1977). *Hum. Genet. 35*, 299.
Dutrillaux, B., Fosse, A.M., Prieur, M., and Lejeune, J. (1974). *Chromosoma. 48*, 327.
Kato, H. (1977). *Chromosoma. 59*, 179.
Knuutila, S., Simell, O., Lipponen, P., and Saarinen, I. (1976). *Hereditas. 82*, 29.
Perry, P., and Wolff, S. (1974). *Nature. 251*, 156.
Schwarz, K. (1976). *Medical Clinics of North America. 60*, 745.
Shiraishi, Y., and Sanberg, A.A. (1977). *Cytogenet. Cell Genet. 18*, 13.

A FOLLOW-UP STUDY OF PVC WORKERS TWO YEARS AFTER
EXPOSURE. PRELIMINARY RESULTS USING SISTER CHROMATID
EXCHANGE FREQUENCY AS AN ASSAY OF GENETIC DAMAGE

Inger-Lise Hansteen

Telemark Central Hospital
Laboratory of Genetics
Porsgrunn
Norway

INTRODUCTION

As a consequence of the discovery of VCM as a carcinogenic agent, all the workers at a PVC factory in Norway were
examined clinically. Cytogenetic studies were performed on 39
of these workers. The results demonstrated a slight, but significant increase in chromosome breakage frequency for the workers compared with the controls selected from the population
outside the factory. The mean chromosome breakage frequency
for the workers was 3.7% with range 0-11.9, and for the
controls 1.75% with range 0-6. An increase of chromosome aberrations caused by VCM was reported by Funes-Cravioto *et al.*,
Ducatman *et al.*, and Purchase *et al.* (1975).

At present little is known about the significance of
chromosome aberrations incurred upon workers exposed to VCM
or other chemicals. No relationship between chromosome aberrations and VCM disease could be traced for the 39 workers
under study, except for one case with a breakage frequency of
11.9%. In this one case, however, it is at present difficult
to decide if VCM is the real cause of his clinical symptoms.
Only follow-up examinations can provide the answer.

A renewed study of 39 workers with matched controls from
the office employees at the factory is underway after a period
of more than two years without exposure to VCM. The aim of this
investigation is to examine the workers with the highest breakage frequencies, and to clarify if it is VCM especially, or

the atmosphere in the chemical industry generally, which is responsible for the higher breakage frequency in the workers.

MATERIALS AND METHODS

Of the 39 workers previously examined, 38 were available for a renewed cytogenetic examination. For the present study matched controls have been chosen among employees from the office department of the factory. Of the 16 controls used in the first examination, 7 were available for reexamination.

Exact measurements of the exposure to VCM can be provided in the PVC plant from 1974. For earlier years, estimated figures indicate that the exposure was considerable (Table I). For the last two years the concentration of VCM has never exceeded 1 pp.

Lymphocyte cultures were initiated immediately after the blood samples were collected and cultured for 48 hours at 36.5 - 37°C according to conventional methods. One hundred cells per person are scored for gaps, breaks, and rearrangements. In addition the present investigation includes 72 hour cultures, with 3 µg/ml of 5-bromo-2'-deoxyuridine (BrdU) in the cultures for the last 48 hours. The slides were stained according to the method of Goto *et al.* (1975). Thirty cells per person were scored for sister chromatid exchange (SCE) figures.

TABLE I. Air Concentration of VCM in the PVC Plant. Estimates Are Based upon the Level of Production, and the Number and Types of Autoclaves used in the Respective Time Periods

Time periods	Air concentration of VCM in ppm
1950–1954	2 000
1955–1959	1 000
1960–1967	500
1968–1972	100
1973–1974	80
1974	measured 25
1975–	1

TABLE II. Sister Chromatid Exchange Frequency in PVC Workers and Their Matched Controls

PVC Case No.	Chromosome breakage % 1974	Sister chr. exchanges per cell	Range	Control case No.	Sister chr. exchanges per cell	Range
1	6	10.5	2-17	1	7.2	2-13
2	6	9.9	4-21	2	8.1	2-16
3	5	9.8	3-21	3	11.1	3-24
4	2.7	9.5	4-20	4	8.8	3-19
5	5	9.3	3-19	5	6.3	3-11
6	4	8.2	4-16	6	9.8	3-22
7	2	7.9	4-22	7	5.8	2-9
8	1	7.3	2-17	8	5.6	0-11
9	3.6	6.8	2-13	9	5.9	1-15
10	4.7	6.8	2-14	10	5.8	3-10
11	3.6	6.7	4-13	11	6.5	3-11
12	3.8	6.6	3-12	12	7.4	1-19
13	2	6.4	3-17	13	11.6	5-12
14	6	6.1	2-11	14	5.2	2-10
15	4	5.8	1-16	15	*failed*	
16	3.7	4.7	2-11	16	7.6	3-14

Mean 7.6 *Mean 7.5*

RESULTS

At present only the results from the sister chromatid exchange studies are available. Blood from 22 workers with matched controls has been cultured for SCE studies so far. The results for 16 workers with their matched controls are presented in Table II. The chromosome breakage frequency scored in 1974 is indicated for the workers. Cultures from 6 workers did not give enought cells for study, but the results for their matched controls are presented in Table III. The SCE results for 5 controls from the first investigation are presented in Table IV.

The mean number of SCEs per cell is 7.6 with range 4.7-10.5 for the workers. For the matched controls, the mean number is 7.5 with range 5.2-11.6. The mean number of SCEs for the 5 earlier controls is 8.3 with range 6.9-9.0.

SCEs were scored twice with an interval of 4 months for two control persons. SCE frequency was 9.8 and 11.1 for one person, and 5.1 and 5.8 for the other.

TABLE III.   *Sister Chromatid Exchange Frequency in PVC Workers and Their Matched Controls*

PVC case No.	Sister ch. exchanges per cell	Range	Control case No.	Sister ch. exchanges per cell	Range
1–16 mean	7.6	4.7–10.5	1–16 mean	7.5	5.2–11
17	Failed		17	11.5	5–28
18	"		18	10.3	1–22
19	"		19	7.6	1–15
20	"		20	6.5	0–15
21	"		21	5.6	2–15
22	"		22	5.1	2–10
1–16 mean	7.6		1–22 mean	7.6	

TABLE IV.   *Sister Chromatid Exchange Frequency in Controls from 1974, Reexamined Two Years Later*

Controls from 1974	Chromosome breakage % 1974	Sister ch. exchanges per cell	Range
1	0	9.0	4–18
2	2.9	8.9	1–18
3	0	8.5	2–20
4	0	8.3	2–15
5	0.9	6.9	2–17
		Mean 8.3	

DISCUSSION

A crucial point in all investigations is the selection of control material. When the matter was discussed in connection with the first investigation in 1974, it was thought important to select a control material having as little as possible exposure to industrial pollution. Another point of view is that in the chemical industry, where many different products are made, the control material should be selected from relatively unexposed persons on the factory premises; otherwise it is difficult to conclude that one special product

is the cause of the damage. Ideally there should be one
matched control per exposed worker. Deviations from the ideal
goal is a matter of practical accomplishment.

To satisfy the last point of view, the follow-up examination was planned with matched controls of office employees
as far as this was feasible. In addition, as a double control,
the previous control material available was examined again.

The only results we have got so far have been obtained
with the SCE technique. The results indicate that there is
probably no difference in the frequency of SCEs between
PVC workers (7.6, Table II) and their matched controls (7.5)
two years after exposure. There is no difference in SCEs
between the matched controls (7.5) and the earlier control
material (8.3) either, indicating that the previous control
material can be used.

The SCE frequency is fairly stable in one person, judging
from the results of two controls cultured twice with an interval of 4 months.

The mean SCE frequency of 7.5 per metaphase found in
this investigation is in agreement with the SCE frequency of
7.9 found by Morgan and Crossen (1976), who examined 50 normal
individuals and scored 50 cells per person.

The SCE technique was not used for the first examination
of the PVC workers. Only chromosome breakage frequency and
number of abnormal cells were recorded then. It is therefore
difficult to compare the results we have got so far using the
SCE technique with chromosome breakage frequency. Evidence
from the literature (Kato and Shimade, 1975; Solomon and
Bobrow, 1975; Perry and Evans, 1975) suggests, however, that
SCE is a more sensitive assay for chromosome damage than the
earlier available techniques, and that agents causing an increase in SCEs also produce visible chromosome damage at a
higher concentration. As we found no increase in SCE frequency for the workers compared to their controls two years
after exposure, it is reasonable to expect no difference with
regard to chromosome damage either. A decrease in chromosome
damage is further in accordance with the results of Yoder et
al., 1973, who demonstrated for pesticides a high chromosome
breakage frequency during exposure, with a fall to normal
values when the exposure ceased. In hamsters, Stetka and
Wolff (1976) demonstrated an increase in SCE one day after
acute exposure to different chemical agents, with a return to
normal values two weeks after exposure. Results presented by
Dr. D. Anderson at this meeting suggest, however, that
chronic exposure to VCM can give SCE frequencies within the
normal ranges. SCE frequencies scored for acute exposure may
therefore not be comparable with results from chronic exposure.

Forni *et al.* (1971) found, in the case of benzene exposed workers, that the frequency of stable chromosome aberrations increased with time after exposure, while the unstable chromosome aberrations decreased with time.

No definite conclusions can therefore be drawn before all the results from the present follow-up study are available.

SUMMARY

About two years after the first cytogenetic study of 39 PVC workers, a control study is now beind carried out. The concentration of PVC in the plant during these two years has not exceeded 1 ppm. For the first examination, controls were selected from the population outside the factory. For the control study, matched controls have been chosen among office employees of the factory, and scoring of sister chromatid exchanges is included. Preliminary results for 16 workers give a mean of 7.6 sister chromatid exchanges per cell with a range of 4.7-10.5. The comparable figures for the matched controls are 7.5 with a range of 5.1-11.1. This indicates that there is probably no difference in the frequency of sister chromatid exchanges between PVC workers and their matched controls two years after exposure.

ACKNOWLEDGMENTS

I wish to thank Drs. L. Hillestad and S. Storetvedt Heldaas for medical advice, and K.O. Clausen, V. Haugan and J. Bakken for skilful technical assistance.

REFERENCES

Ducatman, A., Hirschorn, K., and Selikoff, I.J. (1975). *J. Mutation Res. 31*, 163.
Forni, A., Cappellini, A., Pacifico, E., and Vigliani, E.C. (1971). *Arch. Environ. Health. 23*, 385.
Funes-Cravioto, F., Lambert, B., Lindsten, J., Ehrenberg, L., Natarajan, A.T., and Osterman-Golkar, S. (1975). *Lancet. i*, 459.
Goto, K., Akematsu, T., Shimazu, H., and Sugiuama, T. (1975). *Chromosoma (Berl.). 53*, 223.
Kato, H., and Shimada, H. (1975). *Mutation Res. 28*, 459

Morgan, W.F., and Crossen, P.E. (1977). *Mutation Res. 42*, 305.
Perry, P., and Evans, H.J. (1975). *Nature. 258*, 121.
Purchase, I.F.H., Richardson, C.R., and Anderson, D. (1975). *Lancet. ii.* 410.
Solomon, E., and Bobrow, M. (1975). *Mutation Res. 30*, 273.
Stetka, D.G., and Wolff, S. (1976). *Mutation Res. 41*, 333.
Yoder, J., Watson, M., and Benson, W.W. (1973). *Mutation Res. 21*, 335.

EPIDEMIOLOGICAL APPROACHES

EPIDEMIOLOGIC APPROACHES FOR SURVEILLANCE OF
GENETIC HAZARDS WITH PARTICULAR REFERENCE TO
ANESTHETIC GASES

Peter F. Infante[1]

Industry-wide Studies Branch
Division of Surveillance, Hazard
Evaluation and Field Studies
National Institute for Occupational
Safety and Health
Center for Disease Control
Cincinnati, Ohio

Since the early 1940's there has been a tremendous proliferation of man-made chemicals into the environment. Of increasing concern is the fact that a majority of these chemicals has not been evaluated for potential danger to this or future generations either as mutagens, carcinogens or teratogens. At this "Conference on Genetic Damage in Man Caused by Environmental Agents," it seems only appropriate to review some background and some methodological approaches for assessing the biological activity and reproductive hazards of some chemicals to man.

Because increasing numbers of women in the United States are now seeking employment outside the home, Government, Industry and Labor recently have directed attention toward the effects of occupational exposure to toxic agents on women, and specifically on the embryo and fetus. In lieu of adequate industrial engineering control, women of childbearing ages in the United States are being transferred from jobs located in areas of sustained exposure to toxic agents because of concern for adverse effects on the unborn. The alternative to transfer of women out of high exposure areas for some companies has been the requirement that women present medical evidence

[1]

The views expressed in this paper do not necessarily represent those of the National Institute for Occupational Safety and Health.

that they cannot bear children. Such approaches, however,
ignore possible toxic effects on reproduction through male
occupational exposure, i.e. possible mutagenic effects.

Although population surveillance of potential genetic
hazards might possibly be helpful for the identification of
toxic work environments, as was the case for operating rooms,
it should be kept in mind that the currently employed epidemiological method may be too insensitive in many situations,
because of the dilutions of findings resulting from under-
reporting, either from memory lapse or from complete unaware-
ness of the event being studied. Even when spontaneous abortion is recognized during early gestational development, at
times there is no reporting to the physician and hence no
record for confirmation.

In addition, the lack of a national policy in the United
States for retention of personnel and medical records in the
occupational setting often makes it impossible to have available a large enough sample size for adequate evaluation of a
potential genetic risk, the result being an unacceptably high
probability for beta error, i.e. the reporting of an absence
of increased risk when indeed one is present. (This is not to
say that the positive identification of an adverse reproductive event in association with an agent or work environment
is not real). Further, epidemiologic study, being *post hoc*
in nature, does not allow for the practice of preventive
public health. For these reasons, by necessity, we must
depend and rely upon the use of laboratory testing as our
primary means of identifying agents which may pose a genetic
risk to humans.

In some cases, however, a wide variety of agents may pol-
lute the work environment, making the duplication of this
environmental insult difficult or impossible for some laboratory testing procedures. In these situations, epidemiologic
assessment by necessity must play a major role in the evaluation of potential germ cell damage.

When questionnaire and/or interview-survey is to be used
for the assessment of reproductive effects of toxic agents or
work environments, it is important to standardize the approach
and to be aware of some of the variables which may contribute
significantly to the accuracy or inaccuracy of the data
ascertained.

Assuming that the respondent has access to the information wanted, there are still two main obstacles to accurate
reporting: 1) memory, the ability to recall accurately, and
2) motivation, the willingness of the respondent to report
accurately. As would be expected, studied have shown that
the rate of reporting is high for episodes which occurred
close to the date of interview, but dropped consistently with

increasing rapidity the longer the elapsed time between the
interview and the episode being reported (Cannell, 1965).
(One would assume that the severity of the event also would
effect recall). Now, to give an example of where memory could
possibly influence study results.

Reports showing effects on reproduction among female
operating room personnel are shown in Table I. Pharoah *et al.*
(1977), concluded that their results were at variance with all

TABLE I. *Reports of Pregnancy Outcome Among Female
Anesthetists*

PARAMETER MEASURED	SIGNIFICANT DIFFERENCE (+ OR −)	PERIOD BETWEEN EVENT AND ASCERTAINMENT (YEARS)	INVESTIGATORS
Spontaneous Abortion	+	?	Askrog and Harvald, 1970
Spontaneous Abortion	+	?	Cohen et al., 1971
Spontaneous Abortion	+	?	Knill-Jones et al., 1972
Birth Defects	−	all pregnancies	
Spontaneous Abortion	+	0−10	Cohen et al., 1974
Birth Defects	+		
Spontaneous Abortion	+	0−40	Knill-Jones et al., 1975
Birth Defects	+		
Spontaneous Abortion	−	0−25	Pharoah et al., 1977
Birth Defects	+		
Stillbirths[1]	−		
Birth Weight	+		

[1]*Stillbirth rate:* 1.73% (10/578) *in exposed women vs.* 0.83%
(46/5546) *in controls was not statistically significant.*

other studies (Cohen *et al.*, 1971; Knill-Jones *et al.*, 1972;
Askrog and Harvald, 1970; Cohen *et al.*, 1974; Knill-Jones
et al., 1975) and particularly the results of the study conducted by the American Society of Anesthesiologists (Cohen *et
al.*, 1974) (ASA), because the latter study concerned only
those pregnancies occurring within ten years of the survey.
They stated that this procedure introduced an age bias because
it excluded the early pregnancies and that this procedure could
have accounted for the higher spontaneous abortion rate in
the ASA (Cohen *et al.*, 1974) study.

However, their own data do not support this hypothesis as
evident by a u-shaped distribution for spontaneous abortion
rates by five-year maternal age intervals. These data are
shown in Table II. The abortion rate is 24% for mothers 24

TABLE II. Spontaneous Abortion Rates

MATERNAL AGE	APPOINTMENT			ALL
	Anaesthetist	*Other*	*None*	
≤ 24	..	24.6%	17.4%	24.2%
25–29	11.7%	11.6%	10.5%	11.5%
30–34	11.3%	12.8%	10.4%	12.2%
35–39	18.3%	17.5%	14.7%	17.1%
40+	45.8%	31.5%	34.0%	33.6%
All ages	13.8%	13.8%	12.0%	13.5%

(From Pharoah et al. (1977) Lancet 1,1).

years of age or younger, and only 11%, 12% and 17% for
mothers between 25 and 39 years. Thus, it would appear
unlikely that the differences in results between the two
studies are due to age bias; more likely they are due to
memory lapse. This might especially be true when spontaneous
abortions occurring in the early period of gestational development are being investigated up to 25 years later. Because
of difficulties in recall, one would assume that you would
have less dilution in results if the event being studied is
closer to the period of ascertainment.

It is noteworthy that the study by Pharoah *et al.* (1977)
did demonstrate an excess of stillbirths, an event occurring
later in gestational development. They also observed an
excess of congenital anomalies as well as a significant
reduction in birth weight.

With regard to whom should be interviewed, a past survey
has shown that when an individual was asked to report about
his own hospitalization, the under-reporting was 7%. Reporting for close relatives was somewhat worse, but rose to 22%
for distant relatives (Cannell, 1965). Thus, when logistically
feasible, it would seem preferable to interview women for
pregnancy outcome rather than the husband. Because of memory
lapse, under-reporting rather than over-reporting is usually
the anticipated bias.

Although birth defects may be confirmed from medical
records, the reporting of spontaneous abortion, particularly
those occuring early in gestational development, may not be
so complete. Estimates for the percentages of recognized
abortions being reported to physicians are not readily available. Thus, in terms of corroborating information obtained
from respondents, medical records cannot be expected to
provide data on undiagnosed or unreported conditions.

The second major concern, respondent motivation, is also
an important factor in determining the amount and accuracy
of data available to the interviewer. Conditions which are
not embarrassing or detrimental to the individuals' self-image
are expected to be reported more accurately. For example, a
subject may report an appendectomy but be more reluctant to
report her husband's impotence or sterility (Cannell, 1965).
In some situations, the respondent may be motivated to over-
report. Such could be the case if workers and their families
had prior knowledge of a specific agent or work environment
being associated with the event being measured. Conversely,
fear of job loss resulting from concern about possible dis-
continuation of a process related to an excess of spontaneous
abortion might result in a bias toward under-reporting.

These are just a few of the factors to consider in epidemiologic studies of this type. Too often, investigators sit
down, think up a few questions and send untrained auxiliary
personnel into the field without adequate consideration of
possible bias on the part of the respondent or the interviewer. At times, adequate attention might not be given to
the sample size needed to the method of data analysis.

I will confine the remainder of my presentation to spontaneous abortion and birth defects as a reflection of possible germ cell damage to man. Three illustrations serve as
examples of mutagenic or carcinogenic agents which also have
been associated with adverse effects on reproduction resulting from male occupational exposure (Cohen *et al.*, 1974;
Knill-Jones *et al.*, 1975; Cohen *et al.*, 1975; Infante *et al.*,
1976; Sanotskii, 1976). These reports are summarized in Table
III.

In 1974, Cohen *et al.* (1974) reported a significant excess of congenital abnormalities among the unexposed wives of
male operating room personnel. These data were standardized
for maternal age and smoking habits at the time of pregnancy.
No increased risk of spontaneous abortion was observed among
the wives of exposed male respondents compared with the wives
of unexposed male respondents. No increase in cancer was
found among the exposed males, but an increased incidence of
hepatic disease, including serum hepatitis was reported. One
year later, Knill-Jones *et al.* (1975) also reported a significant association between male operating room personnel and
total congenital abnormalitites among their offspring. Again,
male exposure was not associated with an increase in the frequency of spontaneous abortion.

In contrast to these observations, Cohen *et al.* (1975)
reported a significant excess of spontaneous abortion among
the wives of dentists exposed to anesthetic gases, whereas
birth defects were not in significant excess, *i.e.*, the p
value was 0.26. (Liver disease excluding serum hepatitis
was also reported to be in significant excess for exposed
dentists as compared to unexposed dentists). Perhaps the
observation of a significant excess of abortion rather than
birth defects among the wives of dentists is a reflection of
competitive risks whereby conceptuses in the former group do
not complete gestational development to be born with congenital anomalies. If this be the case, the differences might be
the result of exposure to different anesthetic gases or different proportions or levels of the same waste anesthetic
gases. Average concentrations of nitrous oxide in dental
operating rooms are reported as high as 6700 ppm (Miller and
Corbett, 1974), which is higher than in most hospital operating rooms (DHEW Publication No. (NIOSH) 77-140, 1977).

TABLE III. Mutagens or carcinogens associated with adverse effects on human reproduction

AGENT	RESPONSE IN SUB-HUMAN TEST SYSTEM	REPRODUCTIVE EFFECTS IN HUMANS FROM MALE EXPOSURE ONLY	
		Observed Effects	Investigators
Waste Anesthetic Gases	mutagenic and carcinogenic	1. Excess of congenital abnormalities	Cohen et al., 1974. Knill-Jones et al., 1975.
		2. Excess of spontaneous abortions	Cohen et al., 1975.
Vinyl Chloride	mutagenic and carcinogenic	Excess of spontaneous abortions	Infante et al., 1976.
Chloroprene	mutagenic	1. Decrease in motility of sperm and shape of sperm	Sanotskii, 1976.
		2. Excess of spontaneous abortions	

In a study of chronic exposure of rats to nitrous oxide at
subanesthetic levels, injury to the seminiferous tubules was
observed, the toxic effect being confined to spermatogenic
cells with consequent reduction in mature spermatoza and the
appearance of multinucleated forms. Recovery of spermatogenesis occured after return to room air for more than three
days (Kripke *et al.*, 1976). Synergism between nitrous oxide
and halothane in the production of nuclear abnormalities in
fibroblasts has also been reported (Sturrock and Nunn, 1976).

In the study of pregnancy outcome among the wives of
dentists, some have suggested that exposure to dental x-radiation may be the contributing factor to the observation of
adverse effects on pregnancy. This explanation seems unlikely
in view of the fact that the control group was made up of
general dentists who also use x-radiation. With regard to
observations of excessive congenital anomalies among the
wives of male operating room personnel, some have suggested
that the women over-reported because of prior knowledge of
reproductive hazards associated with operating room exposures.
However, prior study had indicated that the reproductive
hazards were associated with female employment in the
operating rooms. Therefore, if the wives of males employed
in operating rooms did have prior knowledge of such hazards,
they may have thought of themselves as a control group for
the females exposed and it is conceivable they could have
been influenced to under-report. It also seems unlikely that
the wives would have had sophisticated enough information to
know that they were to report an excess of birth defects, but
not an excess of miscarriages, as both studies demonstrate,
since prior studies have demonstrated both of these parameters
to be in significant excess among females employed in
operating rooms (Askrog and Harvald, 1970; Cohen *et al.*, 1971;
Knill-Jones *et al.*, 1972).

In terms of correlation between chemical carcinogenesis
and mutagenesis, it is noteworthy that the three inhalation
anesthetic agents tested for carcinogenecity – namely,
trichloroethylene (DHEW Publication No. (NIOSH) 77-140, 1977),
chloroform (DHEW Publication No. (NIOSH) 77-140, 1977) and
isoflurane (Corbett, 1976) – have all induced hepatic tumors
in experimental animals. Known human carcinogens are also
structurally similar to certain inhalation anesthetic agents
(Corbett, 1976). Bis(chloromethyl) ether and chloromethyl
methyl ether are structurally similar to the anesthetic agents
isoflurane, methoxyflurane and enflurane, while vinyl chloride
is structurally similar to trichloroethylene.

In terms of mutagenecity testing of inhalation anesthetic
agents, trochloroethylene has yielded a positive response in
S. Typhimurium (Baden *et al.*, 1976) and *E. Coli* (Greim *et al.*

1975) while nitrous gas has been positive in *Tradescantia* (A.H. Sparrow, personal communication). Further mutagenecity testing of anesthetic agents would be helpful in understanding the mechanisms underlying the epidemiologic observations.

A significant excess of spontaneous abortion among the wives of workers exposed to vinyl chloride (VC) has also been reported (Infante *et al.*, 1976). Prior to the husband's exposure to VC, the abortion rate in their wives was similar to that experienced by the control group consisting of the wives of rubber and polyvinyl chloride (PVC) fabrication workers, prior to the latter group's respective exposures. However, among pregnancies occurring subsequent to the husband's exposures, the wives of VC workers experienced a significant excess of spontaneous abortion as contrasted with similar data for the wives of rubber and PVC workers.

In terms of methodology, the choice of rubber and PVC workers as a control group would tend towards a bias of underestimating the true risk of spontaneous abortion in the wives of VC workers, because the "controls" were undoubtedly exposed to other industrial mutagens.

The third example of a chemical mutagen which has been associated with adverse effects on human reproduction is 2-chlorobutadiene, more commonly referred to as chloroprene. This compound is used extensively in the chemical industry in the U.S. as the starting material for the synthetic rubber, polychloroprene, which is marketed under the trade name Neoprene. Table IV indicates the evidence for cytogenetic, mutagenic or adverse reproductive effects of chloroprene as contrasted with similar information for VC. The data for chloroprene have previously been reported in more detail (Infante, In press). As early as 1936, Von Oettinger *et al.*, (1936) demonstrated damage in the reproductive organs of male mice and rats following exposure to chloroprene. At chloroprene inhalation concentrations of 12 ppm for male mice and 120 ppm for male rats, degenerative changes in the reproductive organs and sterility in up to 60 to 70 per cent of the exposed animals were observed. Chloroprene has caused mutation in several strains of *S. Typhimurium*, has induced sex-linked recessive lethal mutations in *Drosophila*, and has demonstrated a dominant lethal effect in the rat. Chromosomal aberrations were observed in bone marrow cells of these same animals. Several reports have indicated excesses of chromosomal aberrations in the circulating lymphocytes of workers exposed to chloroprene (Infante, In press). Excesses of miscarriages among the wives of male workers exposed to chloroprene and a decrease in the motility and number of sperm in chloroprene workers themselves have also been reported (Sanotskii, 1976).

TABLE IV. *Evidence for Cytogenetic, Mutagenic or Reproductive Effects of Vinyl Chloride and Chloroprene*

TEST SYSTEM	AGENT	
	Vinyl Chloride	Chloroprene
Laboratory Observations		
Microbial (E. Coli,[1] S. Pombe,[1] S. Typhimurium)	+	+
Insect (Recessive Lethal in Drosophila)	+	+
Plants (Tradescantia)	+	?
Mammals		
Metabolites in Hamster Cells	+	?
Chromosomal Aberrations in Male Rats	?	+
Reproduction Interference Following Exposure to Male Mice and Rats	−	+
Decrease in Motility and Number of Sperm in Rats	?	+
Human Observations		
Chromosomal Aberrations in Male Workers	+	+
Excess Miscarriage in Wives of Male Workers	+	+
Decrease in Motility and Number of Sperm in Workers	?	+

[1]*Mutagenic Activity of Chloroprene has not been tested in these Organisms.*

Thus, the spectrum of evidence for the mutagenicity of chloroprene is even greater than the evidence for vinyl chloride being a mutagen.

The data presented for the mutagenic effects of waste anesthetic gases, vinyl chloride and chloroprene raise the question regarding the amount and type of data needed to allow the scientific community to classify a compound as an agent which poses a potential mutagenic hazard to humans. In terms of public health and practice of preventive medicine, it is hoped that geneticists can soon arrive at an acceptable battery of tests in sub-human systems which could be used for an adequate estimate of potential genetic damage in man caused by environmental agents. The alternative may be to implement regulatory control of environmental mutagens on the basis of *post hoc* enumeration of abortuses and congenital anomalies.

This latter approach is particularly disquieting in light of the insensitivity of the epidemiologic method, and in light of the fact that adequate data bases on which to conduct such studies are often lacking or intentionally destroyed as a consequence of lack of a national policy for personnel record retention in the industrial setting.

REFERENCES

Askrog, V. and Harvald, B (1970). *Nord. Med. 83*, 498.

Baden, J., Wharton, R., Hitt, B., Brinkenhoff, M., Simmon V., and Mazze, R. (1976). *Fed. Proc. Amer. Soc. Exp. Biol. 35*, 410.

Cannell, C.F. (1965). *In* "Genetics and the Epidemiology of Chronic Diseases" (J.V. Neel, M.W. Shaw and J.W. Schull, eds.), p. 395. DHEW, Pub. Health Service Pub. No. 1163, Washington.

Cohen, E.N., Bellville, J.M. and Brown, B.W. (1971). *Anesthesiology 35*, 343.

Cohen, E.N., Brown, B.W., Bruce, D.L., Cascorbi, H.F., Corbett, T.H., Jones, T.W. and Whitcher, C.E. (1974). *Anesthesiology. 41*, 321.

Cohen, E.N., Brown, B.W., Bruce, D.L., Cascorbi, H.F., Corbett, T.H., Jones, T.W. and Whitcher, C.E. (1975). *J. Am. Dent. Assoc. 90*, 1291.

Corbett, T.H. (1976). *Ann. NY Acad. Sci. 271*, 58.

DHEW Publication No. (NIOSH) 77-140 (1977)

    U.S. Department of Health, Education and Welfare. (1977).

Greim, H., Bonse, G., Radwan, Z., Reichert, D. and Henschler, D. (1975). *Biochem. Pharm. 24*, 2013.

Infante, P.F., McMichael, A.J., Wagoner, J.K., Waxweiler, R.J. and Falk, H. (1976). *Lancet 1*, 734.

Infante, P.F. (1977). *Env. Health Perspectives*, In press.

Knill-Jones, R.P., Moir, D.D., Rodrigues, L.V. and Spence, A.A. (1972) *Lancet 1*, 1326.

Knill-Jones, R.P., Newman, B.J. and Spence, A.A. (1975) *Lancet 2*, 807.

Kripke, B.J., Kelman, A.D., Shah, N.K., Balogh, K. and Handler, A.H. (1976). *Anesthesiology. 44*, 104.

Millard, R.I. and Corbett, T.H. (1974). *J. Oral Surgery. 32*, 593.

Pharoah, P.O.D., Alberman, E., Doyle, P. and Chamberlain, G. (1977). *Lancet 1*, 1.

Sanotskii, I.V. (1976). *Env. Health Perspectives. 17*, 85.

Sturrock, J.E. and Nunn, J.F. (1976). *Anesthesiology. 44*, 461.

Von Oettinger, W.F., Hueper, W.D., Duchmann-Gruebler, W. and Wiley, F.H. (1976). *J. Ind. Hyg. Tox. 18*, 240.

APPROACHES TO THE EVALUATION OF GENETIC DAMAGE AFTER A MAJOR HAZARD IN CHEMICAL INDUSTRY: PRELIMINARY CYTOGENETIC FINDINGS IN TCDD-EXPOSED SUBJECTS AFTER THE SEVESO ACCIDENT

M.L. Tenchini
C. Crimaudo
G. Simoni
L. De Carli[1]

Istituto di Biologia Generale
Facoltà di Medicina
Università di Milano
Italy

R. Giorgi

Istituto di Genetica
Università di Pavia
Italy

F. Nuzzo

Laboratorio di Genetica Biochimica ed Evoluzionistica del C.N.R.
Pavia
Italy

[1]*The present work is part of a Special Project to investigate TCDD-exposed pregnancies, directed by Prof. G.B. Candiani of the Clinica Ostetrica 1° and Prof. L. De Carli of the Istituto di Biologia Generale, Facoltà di Medicina, Università di Milano (Italy). The project is supported by the Assessorato alla Sanità, Regione Lombardia (Italy).*

INTRODUCTION

In addition to raising an immediate health problem, the
Seveso accident also posed vital questions about the long term
effect of dioxin, one of the most serious being potential
genetic damage. Few data are available on the mutagenic properties of the compound, which is known mostly as a teratogenic
agent (Hussain *et al.*, 1972; Neubert *et al.*, 1973; Green and
Moreland, 1975; Smith *et al.*, 1976). Moreover the choice of
adequate test procedures and approaches for detecting the
mutagenic effects of the chemical in exposed subjects creates
difficulties. The main issues in the evaluation of potential
genetic damage caused by dioxin were thus both theoretical and
methodological.
  From a theoretical standpoint, there were two major questions:
  a) what kind of genetic damage should be looked for and
what criteria should be used to evaluate it?
  b) would it be possible to identify specific categories
of exposed subjects with differing levels of genetic risk?
  On the methodological side the following questions were
the most relevant:
  a) how could laboratory tests and epidemiological investigations be combined and coordinated in evaluating the potential genetic damage?
  b) which were the most reliable and feasible methods for
determining point mutations and chromosome aberrations both
in laboratory tests and in the population studies?
  We restricted our attention to the cytogenetic aspects,
trying to find a proper answer, in this area, to each of the
above questions. In principle, a genetic damage by dioxin
might well be detectable on chromosomes. It therefore seemed
appropriate to look for this kind of alteration on both somatic and germ cells. At somatic level the chromosomal effect
should ideally be assessed on cells from different tissues.
However, as a first approach, the peripheral blood lymphocytes were the most readily accessible material. Cytogenetic
alterations in germ cells can be revealed on meiotic preparations performed on biopsies of gonadal tissues. The examination of meioses does not appear justified until definite evidence has been gathered that dioxin actually increases the frequency of chromosome aberrations. Another more practicable way
of detecting this type of damage is to identify individuals
with anomalous karyotypes in the progeny of subjects exposed
to the chemical. Hence, one should consider the possibility
of prenatal diagnosis performed on amniotic fluid cells. However, in our opinion, before undertaking any systematic program in this direction, more precise information should be

gathered on the cytogenetic risk. Such information can be gained both from results of tests on laboratory animals and mammalian cultured cells and from epidemiological studies performed on human populations with the lymphocyte test.

In order to make any assumption on the degree of genetic risk to which the people of the contaminated area may have been exposed, and select samples of individuals to be examined, it is helpful to know the contamination levels of the soil and buildings and the degree of individual exposure. As to contamination levels, the area affected by the toxic cloud was divided into three zones, namely Zone A, Zone B and the security zone. This classification was made on the basis of arbitrary threshold TCDD concentrations (250 $\mu g/m^2$ for zone A, 5 $\mu g/m^2$ for zone B and <5 $\mu g/m^2$ for the security zone). It is difficult, if not impossible, to establish any correlation between the contamination level of the soil and the actual dose absorbed by the residents in each zone. Estimates of this correlation depend both on physical factors, which influence the distribution of the chemical in the environment, and on the chances of different individuals of coming in contact with contaminated material.

The clinical manifestations and, particularly, dermatological symptoms do not necessarily indicate the degree of contamination of the organism by the chemical. These symptoms are likely to reflect either the individual's susceptibility to the toxic effect of dioxin or his ability to eliminate the chemical. With regard to the mode of exposure, a distinction can be made between:

a) subjects with acute exposure, *i.e.* individuals present in the area when the toxic cloud was released and/or during the period immediately following, and

b) subjects with chronic exposure, *i.e.* workers at the ICMESA plant, not residing in the contaminated area.

As to the methodological approach to be adapted in evaluating cytogenetic damage, the population studies to be performed in the contaminated area cannot be considered sufficient to provide conclusive data. There is no doubt that these studies should be supplemented with experiments carried out on animals and on cultured mammalian cells. The decisive value of laboratory experiments lies in the fact that they provide evidence of dose-response relationships, which should be considered a prerequisite for demonstrating a mutagenic effect at gene or chromosome levels.

Among the laboratory tests those performed on treated animals, such as the bone marrow test, the micronucleus test and the dominant lethal test are certainly the most reliable, inasmuch as they reflect most of the conditions in which chromosome damage is produced in a mammalian organism and are

also more sensitive than tests on cultured cells. The latter
tests, based on treatment of cells *in vitro*, are only partial-
ly informative since they make use of systems that are highly
selective in cellular functions; moreover their reduced sensitivity is due to a high background of spontaneous chromosome
damage.

In cytogenetic studies on human populations, the peripheral blood lymphocyte test, though subject to a number of limitations, is presently the only test applicable to large samples. While a negative result of a cytogenetic analysis cannot
rule out the possibility of chromosome damage by a mutagenic
agent, as recently shown by Schinzel and Schmid (1976), a
positive result is valuable as it calls for more detailed investigation by extending the analyses to other tissues, including bone marrow, skin and gonadal tissues. Such analyses
may not be justified *a priori* and in any case they could
hardly be performed on all the individuals in a large sample.

A class of subjects exposed to TCDD to whom special attention was paid because requests for chromosome analysis were
more pressing and a decision could not be delayed, were women
living in the Seveso area and in the first trimester of their
pregnancy when the accident occurred at the ICMESA plant. However, chromosome analysis on peripheral blood and foetal cells
was not instrumental for deciding whether or not to interrupt
the pregnancy. Therefore we did not accept requests to carry
out amniocentesis for a cytogenetic prenatal diagnosis in
cases of pregnancies initiated before the TCDD exposure.
Thirty-one women who were in this condition applied for a
therapeutic abortion in view of the teratogenic risk due to
TCDD exposure.

The teratogenic properties of dioxin had been well documented in animals prior to the Seveso accident (Neubert *et
al.*, 1973; Smith *et al.*, 1976). However, the availability of
aborted material gave us an opportunity to undertake a comparative study on the frequency of chromosome lesions in maternal and foetal tissues after TCDD exposure.

The usefulness of performing amniocentesis and a cytogenetic prenatal diagnosis on women who became pregnant after the
Seveso accident is questionable. Uncertainty on this point
will eventually be resolved, when more data are available on
the incidence of chromosome aberrations in TCDD-exposed subjects. Four women who were in this condition elected to
interrupt their pregnancy. The cytogenetic analysis has been
performed on the aborted tissues.

The whole project for cytogenetic investigation of the
Seveso population living in the contaminated area includes the
analysis of a sample of 400 individuals from the general population and of a sample of 25 pregnant women submitted to
therapeutic abortion.

The first sample has the following composition:

200 subjects with presumed acute exposure to TCDD, taken from the residents of zone A.

100 subjects with presumed chronic exposure, taken from non-resident workers at the ICMESA plant.

100 control subjects taken from inhabitants of the area surrounding the contaminated area.

A total of 245 peripheral blood cultures from TCDD-exposed subjects and 95 from control subjects have so far been set up. Chromosome preparations have been performed both after 48 and after 72 hours incubation in order to check the effect of the culture conditions on the incidence of chromosome aberrations. Examination of the chromosome preparations of the whole sample of the TCDD-exposed and control subjects is under way. The results of the analyses are not yet available.

Data are, however, available on chromosome analyses in cases of TCDD-exposed pregnancies which were terminated by therapeutic abortion. Whenever possible cultures have been set up both from maternal peripheral blood and from foetal tissues including the placenta, embryo and umbilical cord.

MATERIALS AND METHODS

Tissue Specimens

The abortions were induced in a period ranging between 8 and 16 weeks of gestation. Tissue fragments were placed in complete culture medium supplemented by serum and kept at room temperature for a period not exceeding 4 hours until the cultures were set up. Samples of tissues were obtained from the placenta, embryo and umbilical cord. At the same time a sample of peripheral blood was withdrawn from the mother.

Culture Procedure

Peripheral blood cultures were performed according to the standard micromethod. Cultures of specimens under coverslips in plastic Petri dishes were set up from solid tissues. The medium used in all cultures was reinforced Eagle (aminoacids and vitamins 4x) supplemented with 20% foetal serum.

Cytological Preparations

Chromosome preparations from peripheral blood were
made after 72 hours, following Colcemid treatment with 0.04
µg/ml for 3 hours, hypotonic shock with KCl 0.075 M and fixation with methanol-acetic acid 3:1. The chromosome preparations of fibroblast cultures were made *in situ* after culture
times varying from 9 to 40 days, following the treatment with
Colcemid 0.04 µg/ml for 7 hours, hypotonic shock with 1%
sodium citrate and fixation as for the blood. The preparations
were regularly performed on outgrowing cells in the primary
explants. Only in two cases were trypsinized cells used from
early passages. Staining was made according to the conventional Giemsa technique.

RESULTS

    Table I reports the results of the chromosome counts on
peripheral blood of the mothers. As can be seen, there were no
significant deviations from the normal diploid mode.

*TABLE I.   Chromosome Counts of Blood Culture from Women
            at the Time of Interruption of Pregnancy*

Subject	No. of chromosomes/cell					
	<44	44	45	46	47	>47[b]
S.V. (1)	2	1	–	102	–	–
C.I. (2)	–	2	9	139	–	4
F.V. (3)	–	2	3	100	1	2
P.M. (5)	2	2	–	83	1	2
D.P.N. (7)	3	2	14	80	4	–
M.G. (8)	4	–	4	96	–	–
C.L. (9)	3	5	12	86	2	1
B.A. (15)	5	3	5	88	–	–
M.C. (16)	2	1	5	92	1	5
T.F. (17)	–	2	5	92	–	2
P.S. (19)	9	5	5	81	–	3
C.V. (21)	9	2	–	93	–	3
D.A. (25)[a]	6	4	5	86	1	–

[a] *Pregnancy initiated after the Seveso accident*
[b] *Polyploids included*

The frequencies of various types of structural chromosome aberrations in the maternal peripheral blood are reported in Table II.

The values are within the normal range.

Chromosome preparations on foetal tissues were made after varying culture times depending on growth efficiency. The outcome of the cultures is summarized in Table III. A marked individual variability was noted in the growth potential of the various tissues. No correlation was observed between different tissues belonging to the same aborted material.

The results of chromosome counts on fibroblasts derived from placental, embryonic and umbilical cord tissues are reported in Tables IV and V. No significant deviations are observed from the normal diploid mode. Aneuploid cells are mostly metaphases with missing chromosomes; this type of variation may be due to artifacts of the preparation.

Various types of structural chromosome aberrations found in foetal tissues are shown in Fig. 1. The frequencies of these aberrations are reported in Tables VI and VII.

A relatively high frequency of chromatid and chromosome breaks, either including or excluding gaps, can be observed. The frequencies are consistently higher in the embryonic tissues than in placental and umbilical cord tissues. Two values are prominent, namely those of cases 16 and 21. A marked variability in the frequency of aberrations can also be seen.

The results of chromosome counts and scoring for structural chromosome aberrations on cultures of aborted tissues, which showed poor growth, are reported in Tables VIII and IX. These data are given separately as they were not considered uniform with the others, due to possible effects of the prolonged culture. This might have resulted in a higher incidence of chromosome aberrations.

DISCUSSION

The significance of cytogenetic findings on peripheral blood lymphocytes as indicators of chromosomal damage has recently been questioned. There are instances of *in vivo* treatment with, or exposure to high doses of classical mutagenic agents or agents known to cause chromosome lesions which did not result in any appreciable increase in chromosome aberrations, as detected in peripheral blood (Schinzel and Schmid, 1976). The relevance of a positive finding depends on the types of aberration observed. Gaps, chromatid and isochromatid breaks and also chromatid exchanges are poor indices of herit-

TABLE II. Frequencies of Chromosome Structural Aberrations in Blood Cultures from Women at the Time of Interruption of Pregnancy

| Subject | Cells analysed (No.) | Total No. of aberrations | Types of aberrations No. | | | | Aberrant cells % | | Aberrations/ damaged cell |
			Gaps	Chromatid breaks	Chromosome breaks and fragments	Rearrangements[a]	Including gaps	Excluding gaps	
S.V. (1)	105	9	6	3	–	–	7.61	2.85	1.12
C.I. (2)	158	7	2	4	–	1	4.43	3.16	1.00
F.V. (3)	110	3	1	2	–	–	2.72	1.81	1.00
P.M. (5)	93	5	3	–	2	–	5.37	2.15	1.00
D.P.N. (7)	102	1	–	1	–	–	0.98	0.98	1.00
M.G. (8)	102	–	–	–	–	–	<1.00	<1.00	–
C.L. (9)	109	3	1	1	1	–	2.75	1.83	1.00
B.A. (15)	100	5	3	2	–	–	5.00	2.00	1.00
M.C. (16)	111	2	1	–	–	1	1.80	0.90	1.00
T.F. (17)	103	3	1	2	–	–	2.91	1.94	1.00
P.S. (19)	104	6	3	3	–	–	3.84	2.88	1.50
C.V. (21)	106	3	2	1	–	–	2.83	0.94	1.00
D.A. (25)[b]	100	3	2	1	–	–	3.00	1.00	1.00

[a]Dicentrics, chromatid exchanges, rings, abnormal chromosomes
[b]Pregnancy initiated after the Seveso accident

TABLE III.   Outcome of Cultures of Foetal Tissue

Tissue	No. of specimens	Outcome of culture				No. of successful chromosome analyses
		Growth rate high	Growth rate low	Growth failure	Bacterial contamination	
Embryo	19	11	4	1	3	14
Placenta	18	9	4	2	3	13
Umbilical cord	6	3	2	–	1	5
Total	43	23	10	3	7	32

TABLE IV.   Chromosome Counts in Fibroblasts from Placental and Umbilical Cord Tissues (Cultures with High Growth Rate)

Case	Tissue	No. chromosomes/cell					
		<44	44	45	46	47	>47[b]
S.V. (1)	Placenta	4	3	8	109	–	14
C.I. (2)	Placenta	–	–	1	61	1	–
F.V. (3)	Placenta	–	–	4	49	2	–
P.M. (5)	Placenta	1	–	3	95	2	–
M.G. (8)	Placenta	4	–	–	39	–	1
M.A. (10)	Placenta	–	2	–	48	–	–
T.M.C. (11)	Placenta	–	–	1	54	–	–
6741[a](22)	Placenta	6	7	9	292	3	7
	Umbilical cord	1	3	6	137	1	10
6742[a](23)	Umbilical cord	5	3	7	217	–	7
D.A.[a](25)	Placenta	3	4	1	37	–	5
	Umbilical cord	4	9	23	398	4	3

[a]Pregnancies initiated after the Seveso accident
[b]Polyploids included

TABLE V.   *Chromosome Counts in Fibroblasts from Foetal*
Tissues (Cultures with High Growth Rate)

Case	No. of chromosomes/cell					
	<44	44	45	46	47	>47[b]
M.G. (8)	1	2	–	207	1	1
M.A. (10)	–	2	1	77	–	–
T.M.C. (11)	4	–	–	100	–	1
C.C. (12)	1	–	1	111	1	–
M.C. (16)	2	2	5	80	2	–
5428 (18)	7	–	3	127	–	3
P.S. (19)	11	4	6	172	1	3
C.V. (21)	2	4	–	96	–	2
6741[a](22)	3	4	6	227	1	7
6742[a](23)	8	1	4	162	2	10
D.A.[a](25)	4	10	17	356	4	12

[a]*Pregnancies initiated after the Seveso accident*
[b]*Polyploids included*

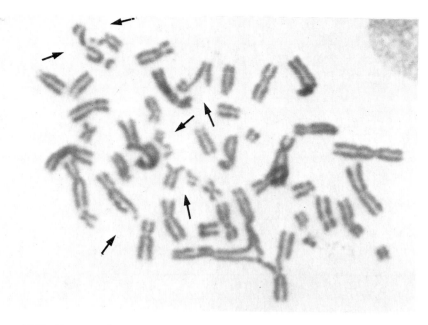

FIGURE 1.   *Different types of structural chromosome*
aberrations in a metaphase from a culture
of embryonic tissue

TABLE VI. Frequencies of Chromosome Structural Aberrations from Placental and Umbilical Cord Tissues (Cultures with High Growth Rate)

Case	Tissue	Karyotype	Cells analyzed	Total No. of aberrations	Types of aberrations No.				Aberrant cells %		Aberrations/damaged cell
					Gaps	Chromatid breaks	Chromosome breaks and fragments	Rearrangements[a]	Including gaps	Excluding gaps	
S.V. (1)	Placenta	46,XX	150	16	9	2	3	2	10.66	6.66	1.00
C.I. (2)	Placenta	46,XX	63	3	1	2	–	–	4.76	3.17	1.00
F.V. (3)	Placenta	46,XX	55	2	1	–	1	–	3.63	1.81	1.00
P.M. (5)	Placenta	46,XX	101	11	4	3	3	1	7.92	4.95	1.37
M.G. (8)	Placenta	46,XY	42	12	9	3	–	–	19.04	7.14	1.50
M.A. (10)	Placenta	46,XX	50	4	3	–	1	–	4.00	2.00	2.00
T.M.C. (11)	Placenta	46,XY	55	4	2	2	–	–	7.27	3.63	1.00
6741[b](22) {	Placenta	46,XY	328	43	26	13	2	2	6.68	3.17	1.76
	Umbilical cord	46,XY	165	38	12	19	5	2	12.55	6.66	1.99
6742[b](23) {	Umbilical cord	46,XY	244	31	16	8	5	2	7.37	4.50	1.72
	Placenta	46,XX	54	13	7	4	–	2	14.81	5.55	1.62
D.A.[b](25) {	Umbilical cord	46,XX	443	67	39	16	10	2	8.57	4.06	1.76

[a] Dicentrics, chromatic exchanges, rings, abnormal chromosomes
[b] Pregnancies initiated after the Seveso accident

TABLE VII. Frequencies of Chromosome Structural Aberrations in Fibroblasts from Foetal Tissues (Cultures with High Growth Rate)

Case	Karyotype	Cells analyzed No.	Total No. of aberrations	Gaps	Chromatid breaks	Chromosome breaks and fragments	Rearrangements[a]	Aberrant cells % Including gaps	Aberrant cells % Excluding gaps	Aberrations/ damaged cell
M.G. (8)	46,XY	212	18	10	5	2	1	7.07	2.83	1.20
M.A. (10)	46,XX	80	26	9	15	2	–	15.00	13.75	2.16
T.M.C. (11)	46,XY	106	22	8	7	6	1	16.03	12.26	1.29
C.C. (12)	46,XY	114	31	11	14	5	1	18.42	11.40	1.47
M.C. (16)	46,XX	91	49	10	28	7	4	31.86	25.27	1.68
5428 (18)	46,XY	140	18	4	6	6	2	7.85	6.42	1.63
P.S. (19)	46,XY	197	64	46	16	1	1	19.41	7.58	1.78
C.V. (21)	46,XX	104	69	26	38	4	1	36.51	24.57	1.84
6741b (22)	46,XY	248	55	27	21	4	3	16.86	8.06	1.22
6742b (23)	46,XY	191	21	11	5	5	–	8.37	3.14	1.31
D.A.b (25)	46,XX	415	77	32	31	11	3	13.01	8.67	1.42

[a] Dicentrics, chromatid exchanges, rings, abnormal chromosomes
b Pregnancies initiated after the Seveso accident

TABLE VIII.   *Chromosome Counts in Fibroblasts from Abortive Tissues (Cultures with Low Growth Rate)*

Case	Tissue	No. of chromosomes/cell					
		<44	44	45	46	47	>47[b]
D.P.N. (7) { Placenta		–	–	–	5	–	–
Foetus		1	1	–	14	–	–
C.C. (12) Placenta		–	–	–	22	–	–
M.C. (16) { Umbilical cord }		1	–	–	5	–	–
T.F. (17) { Umbilical cord }		1	–	–	5	–	–
Foetus		1	–	–	8	–	–
6742[a] (23) Placenta		–	–	1	5	–	–
7006[a] (24) { Placenta		1	–	2	27	–	1
Foetus		3	–	4	8	–	3

[a] *Pregnancies initiated after the Seveso accident*
[b] *Polyploids included*

able chromosome damage, except when there is a very high increase in the frequency of these aberrations. More indicative are chromosome rearrangements such as dicentrics, rings, or translocations detectable by the presence of atypical chromosomes.

Another general remark worth making concerns the frequency of chromosome aberrations in the control cultures performed in different laboratories. There are appreciable differences from one laboratory to another in the background values of chromosome lesions found in cultured cells, presumably due to (A) factors inherent in the culture media; (B) the varying frequency and degree of contamination by PPLO, and (C) differences in the procedures used for chromosome preparations (Schneider *et al.*, 1974). These sources of variability must be taken into consideration in evaluating cytogenetic findings.

Owing to the preliminary nature of our results, any discussion on the significance of data on chromosome aberrations in relation to TCDD exposure is necessarily limited.

An essential piece of information, that is still missing, is control data on aborted tissues. Abortions without any clinical indication or pharmacotherapeutic history, possibly induced in women resident in the area surrounding the contaminated one, *i.e.* in the same physical environment, would pro-

TABLE IX. *Frequencies of Chromosome Structural Aberrations in Fibroblasts from Abortive Tissues (Cultures with Low Growth Rate)*

Case	Tissue	Karyo-type	Cells ana-lyzed	Total No. of aberra-tions	Types of aberrations No.				Aberrant cells No.		Aberra-tions/ damaged cell
					Gaps	Chromatid breaks	Chromo-some breaks and frag-ments	Rearran-gements[a]	Inclu-ding gaps	Exclu-ding gaps	
D.P.N. (7) {	Placenta	46,XX	5	2	2	–	–	–	1/5	–	2.00
	Foetus	46,XX	15	6	3	2	–	1	5/15	3/15	1.20
C.C. (12)	Placenta	46,XY	22	1	1	–	–	–	1/22	–	1.00
M.C. (16) {	Umbili-cal cord	46,XX	6	7	3	4	–	–	3/6	2/6	2.33
T.F. (17) {	Umbili-cal cord	46,XY	5	9	1	7	–	1	3/5	3/5	3.00
	Foetus	46,XY	8	–	–	–	–	–	–	–	–
6742[b] (23)	Placenta	46,XY	6	1	–	1	–	–	1/6	1/6	1.00
7006[b] (24) {	Placenta	46,XX	32	7	3	3	1	–	5/32	5/32	1.40
	Foetus	46,XX	20	–	–	–	–	–	–	–	–

[a]Dicentrics, chromatid exchanges, rings, abnormal chromosomes
[b]Pregnancies initiated after the Seveso accident

vide the most reliable controls. Unfortunately, such controls are difficult to obtain owing to the legal restriction still existing in Italy.

The absence of any evident effect either on the distribution of chromosome numbers or on the frequency of structural chromosome aberrations in the peripheral blood of the women who applied for abortion, does not rule out the possibility of an effect on the frequency of chromosome lesions in the foetus. Such an effect might be due to increased sensitivity of foetal tissues to the toxic effect of dioxin. As a matter of fact the frequency of structural chromosome aberrations is significantly higher in the abortive tissues than in the peripheral blood of the mothers. The highest values are observed in embryonic tissue.

The alterations are predominantly of the chromatid type and rearrangements are rare. A marked degree of variability is, however, noticeable among individuals. This is the salient feature of these data. We are inclined to believe that a large portion of this variability is due to factors inherent in the culture. In this connection it is worth noting that marked differences occurred in the efficiency of growth among the various tissue samples. However, it was not possible to establish any simple relationship between growth capacity and frequency of chromosome aberrations. In the absence of adequate controls at the present stage of our investigations it is not possible to give any definite answer to the question as to whether the higher frequency of chromosome aberrations found in the foetal tissue reflects chromosome damage induced by dioxin. A comparison can be made between the present data and those obtained on foetal cells suspended in amniotic fluid withdrawn by amniocentesis at the end of the first trimester of gestation in pregnancies not exposed to potential mutagenic or teratogenic agents (Simoni *et al.*, 1977). An average frequency of chromosome aberrations (excluding gaps) of 13% was found by analyzing 25 samples of amniotic fluid cells with a normal diploid karyotype. This would seem to indicate that in foetal cells the frequency of spontaneous chromosome damage is generally higher than that observed in peripheral blood and fibroblasts from adult individuals. We would therefore conclude that the chromosome aberrations found in abortuses from TCDD-exposed women do not appear to be more frequent than those spontaneously occurring in cultures of a comparable type of cell.

SUMMARY

In the course of cytogenetic investigations on TCDD-exposed subjects from the population of Seveso, special attention was paid to women who had started their pregnancy immediately prior to or after the explosion at the ICMESA plant and who had subsequently elected to abort in view of the possible teratogenic effects of the chemical.

Chromosome analyses were aimed at revealing differences between the frequency of chromosome lesions in the maternal peripheral blood and in foetal tissues. The chromosome preparations were scored for the presence of numerical and structural variations.

No significant change was found in the chromosome number in any sample analysed. However, marked fluctuations in the frequencies of chromosome lesions were observed in different subjects and different tissues. The overall frequency of chromosome lesions found in the foetal cells appear to be higher than that usually found in fibroblasts from normal adult tissues, but is closely comparable to that observed in amniotic fluid cells during the same gestation period, in non-exposed pregnancies.

REFERENCES

Green, S. and Moreland, F.S. (1975). *Toxicol. Appl. Pharmacol.* *33*, 161.

Hussain, S., Ehrenberg, L., Loefroth, G., and Gejvall, T. (1972). *Ambio.* *1*, 32.

Neubert, D., Zens, P., Rothenwellner, A., and Merker, H.J. (1973). *Environ. Health Perspect.* *5*, 67.

Schinzel, A. and Schmid, W. (1976). *Mutation Res.* *40*, 139.

Schneider, E.L., Standbridge, E.J., Epstein, C.J., Golbus, M., Abbo, G., and Rogers, G. (1974). *Science.* *184*, 477.

Simoni, G., Della Valle, G., Larizza, L., Sacchi, N., and Rosella, F. (1977). Oslo Symposium on Clinical Genetics, Oslo, May 14-15, 1977.

Smith, F.A., Schwetz, B.A., and Nitschke, K.D. (1976). *Toxicol. Appl. Pharmacol.* *38*, 517.

STUDIES OF SPONTANEOUS ABORTION,
MALFORMATIONS AND BIRTH WEIGHT IN
POPULATIONS EXPOSED TO ENVIRONMENTAL
POLLUTANTS; THEORETICAL CONSIDERATIONS
AND EMPIRICAL DATA[1]

Gunhild Beckman
Lars Beckman
Ingrid Nordenson
Stefan Nordström

Department of Medical Genetics
University of Umeå
Umeå
Sweden

INTRODUCTION

The exposure to environmental pollutants is a problem of
increasing concern in the modern industrialized society. The
consequences of such exposure in terms of genetic damage and
morbidity are often difficult to evaluate. In the evaluations
of genetic risks of environmental chemicals one must take into
considerations the fate of the chemical in the human metabolism, synergistic effects (*e.g.* interaction with smoking), the
type of genetic damage and finally various types of bias in
the epidemiological-statistical analysis. Extrapolations from
experimental data on organisms other than man are difficult
and must always be made with great caution. Furthermore since
experiments with man cannot be made our insight into these
problems has to be based on studies of special groups of
people exposed to environmental pollutants. Of particular
interest are populations in highly polluted areas and special

[1]*Supported by the Swedish Work Environmental Fund and the
Swedish Medical Research Council*

occupational groups. Most of the empirical data discussed in this paper are derived from a study of the population around the Rönnskär smelter in northern Sweden. Fig. 1 shows the location of the Rönnskär smelter.

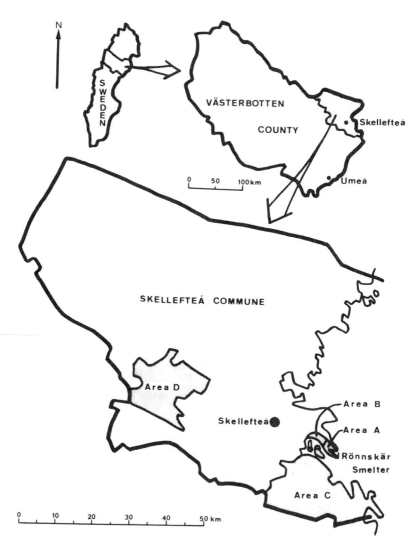

Figure 1. *Location of the Rönnskär smelter and investi-gation areas.*

TYPES OF DATA AND EXPECTED RESULTS

If a certain environmental pollutant causes genetic damage (point mutations and chromosome breaks), what kinds of effects can we expect, and what types of data are available to demonstrate such effects? The rate of spontaneous abortions is expected to increase due to an increased frequency of dominant, lethal point mutations and chromosome aberrations. In newborns we expect increased rates of dominant and sex-linked recessive diseases, chromosome aberrations and congenital malformations. Studies of sex ratio and birth weight may also provide auxiliary information (see below). In adults we expect increased rates of chromosome aberrations and diseases which may arise as a result of somatic mutation, *e.g.* cancer.

Studies of rare monogenic diseases are difficult from any points of view; autosomal recessive mutations are not informative at all, and as regards the dominant and sex-linked disorders the diagnostic problems are serious and the numbers in general too small for a meaningful statistical analysis.

In this paper we have chosen to evaluate the usefulness of studies of congenital malformations, spontaneous abortion and birth weight in populations exposed to environmental pollutants. These parameters are rather "soft" and complex, but they have the advantage of being feasible to study in samples of moderate size; this is a prerequisite for applicability in most situations.

Congenital is not synonymous with heritable and an increased rate of congenital malformations may therefore reflect genetic damage as well as non-genetic effect on the developing fetus. The same substance may be involved in both types of mechanisms. Spontaneous abortions may be caused by point mutations, chromosome abnormalities and environmental factors. The same agent may cause germinal mutations in the parents and direct somatic damage to the fetus.

Birth weight is a complex parameter, which may be influenced by a series of genetic and environmental factors. Thus birth weight is higher in males and increases with parity, specially between the first and second births (Karn and Penrose, 1951; Fraccaro, 1958; Camilleri and Cremona, 1970). The effect of smoking is well documented, but the causality is debated (see Hickey *et al.*, 1973). Low birth weight in a group may indicate impaired intrauterine growth caused by either genetic or environmental factors. Low birth weight may also be due to an increased rate of prematurity which in turn may be dependent on other factors *e.g.* the rate of previous abortion (see Papaevangelou *et al.*, 1973). Low birth weight in exposed populations is perhaps more likely to indicate non-genetic influence on the fetus rather than genetic damage.

Nevertheless, together with studies of malformations and
abortions, birth weight may provide useful auxiliary information.

An increased mutation rate should theoretically lead to a
decreased sex ratio. Large numbers or very drastic effects
are, however, needed in order to demonstrate a statistically
significant decrease in male births.

The types of human materials available and their usefulness have been discussed in detail in the symposium "Evaluation of genetic risks of environmental chemicals" (Ramel,
1973).

SPONTANEOUS ABORTIONS

A number of methodological difficulties are encountered
in studies of spontaneous abortions. One obvious difficulty
is that early, non-hospitalized cases of abortions cannot be
studied. The reliability of diagnosis and registration is
often questionable. Furthermore it is sometimes difficult to
discriminate between spontaneous and induced (criminal) abortions. Thus in Sweden (Uppsala) the rate of criminal abortions
was estimated to be between 2 and 4% of all pregnancies in
1966 (Petterson, 1968). In studies of spontaneous abortions
one should try to randomize the effects of the above-mentioned
confounding factors.

In Sweden the frequency of spontaneous abortions has increased from about 6% to 14% over the past 50 years (Petterson, 1968). A variety of genetic and environmental factors
could have influenced this trend. Furthermore, the reliability of the observations might have improved with time. Most
studies are based on retrospective data which may be biased
in several ways. Estimates of abortion rates in recent years
are usually between 10 and 15%. In a prospective study from
Uppsala, Sweden, Petterson (1968) estimated the spontaneous
abortion rate from the 7th week of pregnancy to 14%. The
above mentioned time trend makes it desirable that series
to be compared, *e.g.* exposed groups and controls, are matched
with respect to time.

It is of importance that the type of ascertainment of
abortion data is known and corrected for. Control groups often
consist of women coming to maternity clinics for delivery. In
such materials the last-born child, which by definition cannot be an abortus, should of course not be included in the
material on which the estimate of abortion is based. If this
type of simple proband correction is neglected, the abortion
frequency in the control series may be underestimated. If
the abortion frequency in this type of control series is

compared with the frequency of abortion in all pregnancies of a series of occupationally exposed women, one may erroneously conclude that the exposed group has an increased abortion rate.

The risk for spontaneous abortion increases after previous legal or spontaneous abortions (Warburton and Fraser, 1964; Petterson, 1968). There is also a tendency towards an increased abortion rate in higher pregnancy orders (Petterson, 1967; Freire-Maia, 1970). This trend may reflect an increased effect of deleterious environmental factors with exposure time and chronological time. Freire-Maia (1970) compared the abortion rates in an irradiated group (medical personnel) and a control group. The abortion frequencies were 10.72% and 7.87% respectively. This difference was statistically significant (t= 2.63; P<0.01). When the data were presented by pregnancy order it was found that the pregnancy effect on abortion rate was stronger in the irradiated group. Assuming linear regression, the regression coefficients (per cent abortion on pregnancy order) were 2.39 in the irradiated group and 0.84 in the control group. The difference between them was highly significant (t= 3.78; P<0.001). The results of Freire-Maia suggest that in studies of exposed groups the amount of information may be increased by taking birth order into consideration.

Table I shows frequencies of spontaneous abortions in four different areas (A-D) located at different distances from the Rönnskär smelter (Fig. 1). All pregnancies in women from the four areas born 1930 and later and attending the Skellefteå hospital were included. The data for the four areas are comparable, since they have been collected under identical conditions from the files of the same hospital (Skellefteå).

TABLE I. *Frequency of Spontaneous Abortion in Four Different Areas A-D (See Fig. 1) at Different Distances from the Rönnskär Smelter*

Area	Abortions		Number of pregnancies
	n	*%*	
A Skelleftehamn	149	11.0	1,358
B Klemensnäs-Ursviken	73	9.2	791
C Bureå	79	8.2	969
D Boliden	85	7.6	1,118

The frequency of spontaneous abortion showed a significant heterogeneity between areas (χ^2= 9.85, 3 d.f., P<0.05). The highest frequency (11 per cent) was found in area A, which is close to the smelter and the lowest frequency (7.6 per cent) was found in area D, which is the most distantly located area. The difference between areas A and D shows a high degree of statistical significance (χ^2= 10.38, 1 d.f., p<0.005).

In the study of Freire-Maia (1970) the abortion frequencies in the exposed group and the control were 10.7 per cent and 7.9 per cent respectively, and the regression coefficients (per cent abortion on pregnancy order) were 2.4 and 0.8, respectively. In our study the difference in abortion frequency between areas A and D was of the same order of magnitude (11 per cent vs. 7.6 per cent). However, when we calculated the regression coefficients, we found no significant difference between A and D; the coefficients were 3.2 and 3.0 respectively. It is also notable that both estimates were rather high compared to those by Freire-Maia.

The observed differences in abortion frequency between areas located at different distances from the Rönnskär smelter may reflect an effect of environmental pollution from the smelter. Other explanations are, however, also possible since the areas differ in other aspects besides distance from the smelter, *e.g.* degree of urbanization, and most probably social patterns and genetic composition. Definite conclusions should await the results of further investigations including studies of women working inside and close to the industrial area.

CONGENITAL MALFORMATIONS

If environmental pollutants cause congenital malformations we should expect increased malformation rates not only around polluting industries, but also to some extent in urbanized areas generally. Beckman and Nordström (1976) examined the frequencies of five common malformations in about 60,000 live born births in Västerbotten county in northern Sweden. The malformations were: cleft palate, cleft lip with or without cleft palate, polydactyly, syndactyly and club foot. The frequency of all malformations together was in the total material 4.59 per 1,000 births. The county was divided into 14 regions composed of parishes (see Fig. 2). Significantly increased malformation rates were found in two regions (1 and 6). Region 8, where the Rönnskär smelter is located had a lower than average malformation frequency (4.4 per 1,000). Three areas (Lycksele, Skellefteå and Umeå on the map) have a higher degree of urbanization than the other regions. Contrary to expectation the malformation rate (4.23 per 1,000) in these

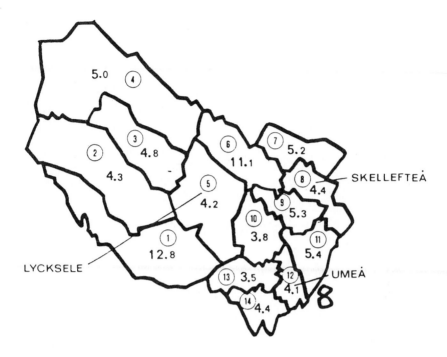

Figure 2. *Frequency of five common malformations (per 1,000 live births) in 14 different regions in Västerbotten county. Numbers of regions within circles. The locations of the three cities (and hospitals) are indicated (from Beckman and Nordström, 1976).*

urbanized areas was significantly lower (χ^2= 5.12, 1 d.f., P<0.05) than in the rural areas (5.65 per 1,000).

The frequency of cleft palate is known to be higher in northern than in southern Sweden (Beckman and Myrberg, 1972). Beckman and Nordström (1976) noted that the frequency of oral clefts (cleft lip and cleft palate) was significantly higher (P<0.01) in the Skellefteå area than in the Umeå and Lycksele areas. The frequency of cleft palate in this area was as high as 1.27 per 1,000 (see Table II). Cleft defects are well-known indicators of teratogenic agents and we therefore asked the obvious question whether the high frequency of cleft palate in the Skellefteå area could be traced to the Rönnskär smelter. Would *e.g.* the frequency be still higher close to the smelter? We therefore studied the frequencies of cleft defects among the births in the St. Olof and St. Örjan parishes during 1954-1976 (see Table II). This study is a part of an ongoing detailed study of all malformations in the

TABLE II. *Frequencies of CL(P) (Cleft Lip with or without Cleft Palate) and CP (Cleft Palate) in Different Swedish Populations*

Population	CL(P)		CP		Number of births
	n	per 1,000	n	per 1,000	
[a]Malmöhus county	59	1.34	16	0.36	44,109
[b]Västerbotten county Total	69	1.13	59	0.97	61,061
[b]Skellefteå hospital area	34	1.44	30	1.27	23,614
[c]St. Olof and St. Örjan parishes	10	1.18	7	0.82	8,503

[a]*Böök, 1951*
[b]*Beckman and Nordström, 1976*
[c]*This investigation; these parishes correspond roughly to areas A and B close to the smelter*

Skellefteå area during the same period. The St. Olof and St. Örjan parishes correspond rather closely to areas A and B on Fig. 1.

Table II shows that the frequencies of both cleft defects were in fact lower in the two parishes located close to the smelter. This result does not seem to justify the conclusion that the increased rates of cleft defects in the Skellefteå area are due to environmental pollution from the Rönnskär smelter. Neither does this result justify any general conclusions concerning the rates of congenital malformations in the area surrounding the smelter. These problems are currently being investigated.

BIRTH WEIGHT

Birth weights were studied in the populations in the four different areas in the Skellefteå commune (Fig. 1). These data were compared with the birth weights in the Umeå population (Table III). Umeå, which is more urbanized than the four areas, is located in the same county (Fig. 1), but at a fairly long distance from the Rönnskär smelter. In the presentation in Table III the data are divided according to pregnancy order. In the two areas close to the smelter (A and B) the birth weight was significantly lower than in

TABLE III. *Birth Weights in the Populations in Four Different Areas (A-D) in the Skellefteå Commune and in Umeå*

Area	Pregnancy	Mean	SD	n
A	1	3332.97	488.95	525
	2	3443.62	545.91	400
	3 and >3	3454.35	567.95	232
	Total	3394.70	528.45	1157
B	1	3379.15	492.46	320
	2	3453.02	555.54	228
	3 and >3	3424.89	596.50	141
	Total	3411.50	536.40	689
C	1	3399.51	567.08	374
	2	3567.47	500.16	261
	3 and >3	3575.75	527.96	212
	Total	3495.49	543.53	847
D	1	3410.40	560.93	446
	2	3523.46	559.97	336
	3 and >3	3515.58	636.26	213
	Total	3470.31	580.21	995
Umeå	1	3371.74	532.78	1182
	2	3495.35	530.68	846
	3 and >3	3568.30	596.10	672
	Total	3460.14	554.31	2700

areas C and D and in Umeå. It is noteworthy that the decreased birth weights in areas A and B concern almost exclusively higher pregnancy orders (3 and >3). The meaning of this result may be that environmental pollutants have affected fetal growth in areas A and B and that this effect becomes more pronounced with increased number of pregnancies. There are, of course, other possible explanations. In highly industrialized areas the way of life is probably more stressing than in rural areas. As far as we know detailed studies of variations in birth weights between small geographical areas have not been done before. Furthermore we do not know whether the low birth weight is due to retarded fetal growth, in which case the gestation time may be normal, or whether women in areas A and B have a decreased average gestation time. Low birth weight in the presence of normal gestation time would perhaps point more towards a toxic environmental effect, while low birth weight *and* gestation time may indicate other factors like maternal stress.

CONCLUDING REMARKS

In this presentation we have tried to apply studies of abortion rates, malformations and birth weights in studies of populations located at different distances from a known source of environmental pollution, the Rönnskär smelter. We found an increased abortion rate and a decreased birth weight in the population which was closest to the smelter. The smelter emits a variety of toxic or potentially toxic substances to air and water, of which the ones that have caused most concern are arsenic, lead and sulphur dioxide. The general problem in these kinds of investigations is that a variety of explanations apart from the "pollution hypothesis" are possible. If, however, the results point in the same direction *e.g.* a population close to a polluting industry shows increased abortion rate and lowered birth weight, the suspicion of a toxic environmental effect increases.

Finally we should like to emphasize that definite conclusions will have to await the study of employees at Rönnskär as well as the detailed study of malformations.

REFERENCES

Beckman, L., and Myrberg, N. (1972). *Human Hered.* *22*, 417.
Beckman, L., and Nordström, M. (1976). *Hereditas.* *84*, 35.
Böök, J.A. (1951). *Acta Genet.* *2*, 289.
Camilleri, A.P., and Cremona, V. (1970). *J. Obstet. Gynaec. Brit. Cwlth.* *77*, 145.
Fraccaro, M. (1958). *Human Biol.* *30*, 142.
Freire-Maia, N. (1970). *Social Biol.* *17*, 102.
Hickey, R.J., Clelland, R.C., and Harner, E.B. (1973). *Lancet. i,* 270.
Karn, M.N., and Penrose, L.S. (1951). *Ann. Eugenics.* *16*, 147.
Petterson, F. (1968). "Epidemiology of Early Pregnancy Wastage". Norstedts, Stockholm.
Papaevangelou, G., Vrettos, A.S., Papadatos, C., and Alexiou, D. (1973). *J. Obstet. Gynaec. 80,* 418.
Ramel, C. (ed.) (1973). "Ambio Special Report", No. 3. Göteborgs Offsettryckeri, Stockholm.
Warburton, D., and Fraser, F.C. (1964). *Human Genet.* *16*, 1.

ALCOHOL AS A MUTAGEN *IN VIVO* AND *IN VITRO*

Olli Halkka
Kalervo Eriksson

Department of Genetics, University of Helsinki, and
Research Laboratories of the State Alcohol Monopoly (Alko),
Helsinki, Finland

INTRODUCTION

The possibility that ethanol is a teratogenic or mutagenic agent in small mammals or other vertebrates was investigated in several laboratories from about 1917 to about 1933. During this period, the mood of most of the groups working on this problem gradually turned from optimism to confusion and uncertainty. This seems to be evident in the series of studies by MacDowell and his collaborators (cited in Bluhm, 1930, Hanson and Cooper, 1930, and Badr and Badr, 1975). Another long series of studies was performed by F.B. Hanson and his co-authors. In the last of a series of papers, Hanson and Cooper (1930) remarked: "this is the last of a long series of failures to modify the germ plasm by external agents".

According to Bluhm (1930), the studies by Pearl (1917), Stockard and Papanicolaou (1918), Stockard (1922), Gyllensvärd (1923), Rost and Wolf (1925) all suffered either from too small a starting material or from other methodological weaknesses.

In a series of contributions Bluhm (1922, 1926, 1930a, b, 1933) herself finally reached the conclusion that ethanol is mutagenic in the mouse. The total material investigated by Bluhm comprised 32.000 albino mice. But the results achieved by Bluhm were soon subjected to severe criticism by a number of geneticists.

In the period between 1933 and 1959, the number of studies published on the mutagenic or clastogenic effects of ethanol was practically nil. These 26 silent years were followed by a period during which attention was increasingly directed towards alcohol as a potential mutagen. The years 1959-1977 saw the publication of at least 37 papers on this subject. The present general acceptance of the role of ethanol as a clastogenic agent or a mutation-producer rests mainly on studies performed in the 1960s and 1970s.

EVALUATION OF THE RESULTS OF RECENT STUDIES ON ETHANOL MUTAGENESIS

In vivo investigations

Being a ubiquitous component of the chemical environment of plants and animals, ethanol is frequently able to invade somatic or even germ line cells. Considerable amounts of ethanol may be present in plant roots growing for some time under anaerobic conditions. Michaelis *et al.*, (1959) treated growing roots of *Vicia faba* with varying concentrations of ethanol up to the molar concentration 0.5, and noted a break frequency of about 2 breaks per cell (see also Rieger and Michaelis, 1960a, b, c, d, 1961a, b, 1962, Michaelis *et al.*, 1963). This result contrasts with the negative outcome of earlier studies by Loveless and Revell (1949) on *Vicia*.

The chromatid aberrations induced by ethyl alcohol show preferential distribution in certain chromosome regions containing constitutive heterochromatin or at least regions close to such heterochromatin (Michaelis and Rieger, 1968; Rieger and Michaelis, 1970; Rieger *et al.*, 1975).

The observation by Rieger, Michaelis and their collaborators that alcohol is able to induce radiomimetic effects in plant roots has been confirmed by Sax and Sax (1966a, b) in experiments with *Allium cepa*. Ethanol treatment also affected the mitotic cycle in *Allium* (Arcara and Ronchi, 1967).

The *in vivo* experiments with plants also include studies on meiotic divisions of *Triticum* (Kabarity, 1968) and *Zea* (Maguire, 1975). Both authors noted abnormal meiotic chromosome behaviour after ethyl alcohol treatment.

Recently, Harsanyi *et al.*, (1977) treated conidia of *Aspergillus nidulans* with ethyl alcohol in concentrations ranging from 0.25% to 20% (v/v). Ethanol was demonstrated to have a distinct mutagenic effect, although no such effect was noted in earlier experiments with *Neurospora* by H. Döring (see Stubbe, 1940) (cf. Fabian and McCullough, 1934).

Experiments performed *in vivo* on animals, with the purpose of revealing possible mutagenic properties of ethanol, are rather few in number. The early work by Mann (1923) on *Drosophila* yielded a negative result. Manna and Mazumder (1964) were successful in producing chromosome breakage in the spermatocytes of the grasshopper *Phloeba antennata*. The breaks were concentrated in a certain region of the X-chromosome of the species.

Schöneic (1966), working with mouse ascites tumours *in vivo*, was unable to produce any increase in chromosome breaks even with alcohol concentrations very close to those having a lethal effect. In studies with normal male mice receiving alcohol through a gastric tube, Badr and Badr (1975) noted increased occurrence of lethal mutations. In later studies with mice, Badr *et al.*, (1977) scored an increased number of chromosome aberrations after ethanol treatment.

Experimenting with rats, Kohila *et al.*, (1976) administered ethanol *per os* and scored aberrations in spermatogonial mitoses. No increase in aberration frequency was observed. In further experiments, it was found that in alcoholic rats receiving a diet with abnormally low levels of thiamine, the frequency of aberrations was as low as in controls (Halkka and Eriksson, 1977). Nor do rats seem to be sensitized to the clastogenic effects of ethanol by an abnormally large supply of retinol (preliminary results of works in progress; cf. van Thiel *et al.*, 1974).

In vivo studies with human beings must necessarily rely on material obtained under less stringent conditions than those possible in experiments with mice or rats. Some indication that ethanol had a teratogenic effect was given by Jones and Smith (1973) and Jones *et al.*, (1973). After analysing the blood cells of alcoholic subjects, de Torok (1972a, b) claimed that the frequency of aneuploid mitoses was far higher than in non-alcoholic controls. In this connection, it may be noted that multipolar mitoses have been observed in various organisms after alcohol treatment (Barthelmess, 1957; Barthelmess and Einlechner, 1959; Brenner, 1949; see also Barthelmess, 1970).

Perhaps the most interesting *in vivo* results with human beings have been obtained by G. Obe and his collaborators (Obe and Herha, 1975; Obe *et al.*, 1977). Exchange aberrations of the chromatid and chromosome type were found to be significantly more frequent in chronic alcohol users than in the non-alcoholic control subjects.

In vitro investigations

Meisner *et al.*, (1970) added alcohol at the toxic concentration of 1.2% to continuous cultures of human skin. This treatment resulted in a five-fold increase in breaks at two days after addition of the drug. These results agree with Bregman's report (1971) that lymphocytes showed an increase in break and gap frequency after treatment with 1.2% ethanol. He found, however, that a lower, and non-lethal concentration (0.3%) had no effect. Using concentrations up to about 0.5 on lymphocytes, Cadotte *et al.*, (1973) observed the same frequency of aberrations as that scored from the control cultures.

Obe *et al.*, (1977) treated human lymphocyte cultures with 0.25% and 0.5% ethanol (v/v). This treatment was found to have no effect on the chromosomes. Similarly negative results were obtained in investigations with Chinese hamster ovary cells (Obe and Ristow, unpublished).

With this same test system, CHO cells, acetaldehyde was found to be able to cause sister chromatin exchanges when administered at very low concentrations (0.0005-0.001 (v/v)) It thus seems possible that the chromosomal aberrations observed after alcohol administration are caused by this primary product of ethanol oxidation, and not by ethanol itself.

DISCUSSION

Experiments and observations aimed at detection of mutagenic or clastogenic effects of ethanol have yielded conflicting results. The fact that, in mammalian cells, lethal concentrations *in vivo* are necessary to produce aberrations *in vitro* may perhaps be explained by assuming that at high concentrations of ethanol, oxidation of acetaldehyde does not proceed normally.

Badr and Badr (1975) suggest that, in mice, dominant lethal mutations are produced primarily in epididymal spermatozoa and late spermatids. This would indicate that alcohol (and its metabolic products) invades germ line cells at concentrations high enough to produce point mutations. But although plant and invertebrate test system have shown ethanol to be an effective clastogen at low concentrations, no clastogenic effects can be found in the spermatogonia of heavily alcoholized rats (Kohila *et al.*, 1976). Obviously, data on alcohol metabolism in various parts of the testis tissue are necessary for an understanding of the basis of germ line mutagenesis. The germ line effects of alcohol in general need further clarification.

A difficult problem is posed by the increased incidence of chromosome aberrations observed in the blood cells of alcoholic human beings (Obe and Herha, 1975). It appears possible that the mechanisms underlying these clastogenic effects are partly connected with the indirect effects of alcoholism.

SUMMARY

Alcohol (ethanol) has been claimed to be mutagenic or clastogenic *in vivo* in somatic cells of man and mouse, in meiosis of the mouse and grasshopper, in meiosis of wheat and maize, in root tips of onion *(Allium)* and bean *(Vicia)*, and in germinating conidia of the ascomycete fungus *Aspergillus*. No mutagenic effect could be demonstrated with the spermatogonia of the rat and with ascites tumour of the mouse.

Experiments on the mutagenicity of alcohol *in vitro* in human cell cultures have yielded results conflicting with those obtained *in vivo*. It seems that the concentrations necessary to produce clastogenic effects in cell cultures are far higher than those lethal *in vivo*.

The mechanism of mutation induction by ethanol remains to be clarified. In some organisms, heterochromatin appears to be specifically affected. The results of recent experiments with Chinese hamster ovary cells suggest that mutagenesis observed after ethanol administration may be due to the first product formed during ethanol oxidation, acetaldehyde.

ACKNOWLEDGMENTS

We are grateful to Mrs. Tarja Kohila, B.Sc., for technical assistance. The study has been financed by grants from the Foundation for Alcohol Research and from the National Research Council for Sciences (Academy of Finland).

REFERENCES

Arcara, P.G., and Ronchi, V.N. (1967). *Caryologia. 20,* 229.
Badr, F.M., and Badr, R.S. (1975). *Nature. 253,* 134.
Badr, F.M., Badr, R.S., Asker, R.L., and Hussain, T.H. (1977).
In "Alcohol Intoxication and Withdrawal: Experimental Studies,
    III. Advances in Experimental Medicine and Biology"
    (M.M. Gross, ed.), Plenum Press, New York and London, in
    Press.
Barthelmess, A. (1957). *Protoplasma. 48,* 546.
Barthelmess, A., and Einlechner, J. (1959). *Protoplasma. 51,*
    325.
Barthelmess, A. (1970). *In* "Chemical mutagenesis in mammals
    and man" (F. Vogel and G. Röhrborn, eds.), p. 69. Berlin,
    Heidelberg and New York, Springer.
Bluhm, A. (1922). *Ztschr. f. Abst.-u. Vererbungslehre. 28,* 75.
Bluhm, A. (1926). *Biol. Zentralbl. 46,* 651.
Bluhm, A. (1930a). *Biol. Zentralbl. 50,* 102.
Bluhm, A. (1930b). *In* "Zum Problem "Alkohol und Nachkommen-
    schaft". Eine experimentelle Studie". (Festschrift für
    A. Ploetz; *Arch. Rassenbiol. 24*) p. 87. Lehmanns Verlag,
    München.
Bluhm, A. (1933). *Arch. Rassen- und Gesellschaftsbiologie.*
    27, 353.
Bregman, A.A. (1971). *Environmental Mutagen Society Newsletter.*
    4, 35.
Brenner, S. (1949). *Nature. 164,* 495.
Cadotte, M., Allard, S., and Verdy, M. (1973). *Ann. Génét. 16,*
    55.
Fabian, E.W., and McCullough, N.B. (1934). *Journ. of Bact. 27,*
    583.
Gyllensvärd, C. (1923). "Beitrag zur Frage der Erblichkeit
    der Alkoholwirkungen". Schwed. Inaugdiss. Stockholm.
Halkka, O., and Eriksson, K. (1977). *In:* "Alcohol Intoxication
    and Withdrawal: Experimental Studies, III. Advances in
    Experimental Medicine and Biology" (M.M. Gross, ed.),
    Plenum Press, New York and London, in Press.
Hanson, F.B., and Cooper, Z.K. (1930). *Journ. Exp. Zool. 56,*
    369.
Harsanyi, Z., Granek, I.A., and Mackenzie, D.W.R. (1977).
    Mut. Res. 48, 51.
Jones, K.L., and Smith, D.W. (1975). *Teratology. 12,* 1.
Jones, K.L., Smith, D.W., Ulleland, C.N., and Streissguth,
    A.P. (1973). *Lancet. 1,* 1267.

Kabarity, A. (1968). *Acta Biol. Acad. Sci. Hung. 19*, 49.
Kohila, T., Eriksson, K., and Halkka, O. (1976). *Med. Biol. 54*, 150.
Loveness, A., and Revell, S. (1949). *Nature. 164*, 938.
Maguire, M.P. (1975). *Genetics. 80*, Suppl., 54.
Mann, M.C. (1923). *Journ. Exp. Zool. 38*, 213.
Manna, O.K., and Mazunder, S.C. (1964). *Naturwissenschaften. 51*, 646.
Meisner, L.F., Inhorn, S.L., and Nielsen, P.M. (1970). *Mammalian Chromosomes Newsletter. 11*, 69.
Michaelis, A., Ramshorn, K., and Rieger, R. (1959). *Naturwissenschaften. 46*, 381.
Michaelis, A., Nicoloff, H., and Rieger, R. (1962). *Biochem. Biophys. Res. Commun. 9*, 280.
Michaelis, A., and Rieger, R. (1963). *Nature. 199*, 1014.
Michaelis, A., and Rieger, R. (1968). *Mut. Res. 6*, 81.
Obe, G., Ristow, H., and Herha, J. (1977). *In* "Alcoholic Intoxication and Withdrawal: Experimental Studies, III. Advances in Experimental Medicine and Biology" (M.M. Gross, ed.), Plenum Press, New York and London, in Press.
Obe, G., and Herha, J. (1975). *Humangenetik. 29*, 191.
Pearl, R. (1917). *Journ. Exp. Zool. 22*, 125.
Rieger, R. (1963). *Proc. 11th Intern. Congr. Genet.* (Hague 1963), *1*, 88.
Rieger, R., and Michaelis, A. (1960). *Abh. Dtsch. Akad. Wiss.* (Berl.), *Kl. F. Med. 1*, 54.
Rieger, R., and Michaelis, A. (1960). *Biol. Zbl. 79*, 1.
Rieger, R., and Michaelis, A. (1960). *M.-Ber. Dtsch. Akad. Wiss.* (Berl.), *2*, 290.
Rieger, R., and Michaelis, A. (1961). *Naturwissenschaften. 48*, 139.
Rieger, R., and Michaelis, A. (1961). *Chromosoma* (Berl.), *11*, 573.
Rieger, R., and Michaelis, A. (1962). *Kulturpflanze. 10*, 212.
Rieger, R., and Michaelis, A. (1970). *Mutation Res. 10*, 162.
Rieger, R., Michaelis, A., Schubert, I., Döbel, P., and Jank, H.-W. (1975). *Mutation Res. 27*, 69.
Rost, X., and Wolf, X. (1925). *Arch. Hygiene. 95*,
Sax, K., and Sax, H.J. (1966). *Science. 152*, 676.
Sax, K., and Sax, H.J. (1966). *Proc. Natl. Acad. Sci. 55*, 1431
Schöneich, J. (1966). *Humangenetik. 3*, 84.
Stockard, C., and Papanicolaou, X. (1918). *Journ. Exp. Zool. 26*,
Stockard, C. (1922). *Brit. Med. Journ. 2*,
Stubbe, H. (1940). *Biol. Zentralbl. 60*, 113.
van Thiel, D.H., Gavaler, J. and Lester, R. (1974). *Science. 186*, 941.

de Torok, D. (1972). *Ann. N.Y. Acad. Sci. 197*, 32.
de Torok, D. (1972). *International Symposium Biological Aspects of Alcohol Consumption*, (27-29 September, Helsinki) *The Finnish Foundation for Alcohol Studies. 20*, 135.

ON THE PHILOSOPHY BEHIND THE NEW ICRP
RECOMMENDATIONS, IN RELATION TO
ENVIRONMENTAL MUTAGENS

Per Oftedal

Institute of General Genetics
University of Oslo
Oslo
Norway

The International Commission on Radiological Protection
(ICRP) is a relevant subject at a conference on chemical genetic hazards, since the end points of radiation and chemical
hazards - cancer and mutation - are the same. These effects
are stochastic, *i.e.* frequencies increase but not the degree
of affliction, in contrast to the normal industrial hazards
where effects are largely non-stochastic, and reversible and
curable. The ICRP has been active since the mid-twenties. The
period from 1925 to 1975 may be characterized as dominated by
a "philosophy of safe doses". True enough, the recommended
dose limits are decreased at each turn of the road, but basically they were viewed as acceptable limits indicating a level
of negligible effect (ICRP 1966a,b). This philosophy has
changed during the last few years. The present communication
gives a brief summary of the new recommendations (ICRP 1977).
Generally speaking, it is now accepted that radiation
carries a risk, and that the aim of the recommendations is to
reduce this to an acceptable level. This level is not negligible and certainly not zero, but should compare with risk in
a safe industry, which is taken to be 10^{-4} deaths per working
year. In reality a radiation risk at this level is lower than
a corresponding industrial risk, because the former refers to
only death and not to other damage leading to loss of working
time, and also there is a long latency period between insult
and manifest damage. It is reckoned that a dose limit of 5 rem
per year for whole body exposure for occupationally exposed
personnel leads to this order of risk of one in 10,000. This is
the same dose contained in the present recommendations. Expo-

sure of single organs - *e.g.* only the thyroid - allow corre-
spondingly higher doses, to give the same total risk.

This dose limit of 5 rem is to be regarded as a red light,
a stop. If below 5 rem, light is green, but two further conditions should be fulfilled: a) The dose should be minimal,
i.e. there is no justification for laxity even if the exposure
appears to be far below the limit, and (or but) b) the procedure should be optimal, *i.e.* the work procedure should not
lead to higher risks from other causes in order to reduce the
risk due to radiation.

The recommendations are extensive and fairly detailed on
many points, and may be used to derive concentration limits
for radioactive substances, and other working rules. However,
it is characteristic that radiation risk is seen in line with
and in comparison to other normal occupational risks (Pochin,
1974). In oil industry in the North Sea, such risk is maybe
10^{-2} deaths per year among deep sea divers, while among cler-
gymen or university professors it is maybe 10^{-5} - 10^{-6} deaths
per working year. Genetic risks due to radiation are deemed to
be of the same order as stochastic somatic risks. The total
somatic and genetic risk level is taken to be about $100 \cdot 10^{-6}$
per rad for occupational exposure.

In the early years ICRP's aim was limited to protection
of those exposed in their work. Only after 1955 did the general
population receive attention, due to fallout from bomb tests
and releases from the atomic industry. For the general population, the formerly advocated 5 rem/30 years is not found in
the new recommendations. There is in fact given *no* limit. The
argument is that every dose increment - and dose commitment
increment - should be justified by the benefit derived from
the activity. (Individual members - small numbers - of the
public may as before be allowed 1/10 of the occupational dose,
i.e. 0.5 rem, per year). For the population in general, doses
of the order of 1-10 mrem/year are foreseen. Another way of
putting it is 1 manrad per MW energy used per year (WHO 1977).
If a person uses 1 kW/year, the acceptable dose should then be
1 mrad/year, leading to a genetic risk of about 10^{-7} per year.
It is estimated that at equilibrium, the level will be 3 x
higher. The somatic risk should be added to this. Since only
about one half of the population is genetically significant,
and since damages expressed early or late in life are difficult to compare or sum, numerical calculations obviously can
not provide a complete picture.

This philosophy of justification by benefit is of course
difficult to carry to its full logical conclusion, due to
geographical separation of source and exposure, and time lag
between exposure and effect. Releases of radioactivity in one
country may come to be limited by damage to the next generation in a different part of the world.

The estimates of genetic risk used by the ICRP are almost the same, though with a narrower range, as those in the NAS-NCR BEIR (1972) Committee and the UNSCEAR (1972) report, and the UNSCEAR (1977) report which will appear in the fall of 1977. This is to be expected since the data base is the same, though interpretations will vary.

Recently, a genetics Task Group of ICRP (Oftedal *et al.*, 1977) has presented a report. In the Task Group were represented expertise on "classical" mouse radiation genetics (Ehling) as well as on clinical genetics (Carter), *ad hoc* survey insight (Vogel) and new registry information (Trimble). This constitutes a shift in emphasis from the at least formerly more experimental radiation data oriented BEIR and UNSCEAR committees. The ICRP Task Group also had the support of the population geneticists Edwards (Birmingham) and Neel (Michigan), as well as the experimental mouse geneticist Searle (Harwell), as corresponding members. Remarkably good consensus was reached in the committee, ending up with the risk estimates mentioned above, for the explicit purpose of radiation protection. But it should be stated that several members, in particular Neel and Edwards, recognized a problem in the induction of recessive mutations under the very long perspectives. However, quantification was not deemed possible.

So much for philosophy and recommendations.

What has emerged with great clarity both during the Task Group work and in recent discussions in the Committee I of the ICRP is that the risk is low and acceptable dose limits are feasible, but that there is a great and deeply felt need for more information and an extended data base with regard to prevalence and heritability of genetically determined ill health in human populations (Newcombe 1973, Stevenson 1959, Vogel and Rathenberg 1973, Edwards 1974, Trimble and Doughty 1974).

Obviously, Scandinavia has much to offer in this area: $20 \cdot 10^6$ homogeneous population, good demography, fairly high medical standard (but few epidemiologists) and uniform and stable social and economic conditions. In my opinion, there is much useful information for us to fetch here, and this is information which no one else can gather. I think the Scandinavian human geneticists have a specific responsibility here. Another equally valuable initiative would be to start a continued and systematic analysis of chromosome aberrations in abortuses, as a basis for a surveillance, similar and in parallel to the already organized European malformation survey, sponsored by the medical birth registries. Those two surveys could be tied together.

We have today excellent possibilities to identify suspect substances as mutagens/carcinogens in our environment by microbial and chromosomal methods, or using *Drosophila*

(Bridges, 1974, Sobels, 1976). This can provide us with warnings to which we can react by elimination, by improving working or application conditions, or by regulation. Our reaction
follows generalized patterns, however refined our investigation may be. An exact cost-benefit analysis is a distant ideal
even for radiation, and will hardly ever be needed for chemicals.

I would like to mention two cases, both of current interest in Norway. The first example is *vinyl chloride* (Hansteen,
this conference). Working limits were formerly determined
largely by dangers of explosion and of non-stochastic effects.
Then, in Norway we had *one case* of the specific liver cancer
associated with vinyl chloride - in about 25 years and about
600 workers. This is less than one death per 10^4 working
years. In other words an industrial situation which is normally considered safe, and used by ICRP as a norm. Nevertheless
the situation is unacceptable due to the presence of a risk
which may be identified as such. The solution, as we all know,
is the introduction of changes in technology leading to 2-3
orders of magnitude reduction in exposure, but without any
exact cost-benefit analysis. Just how extensive investments
were made to save a small number of lives is not known.

Another example is *amitrole:* A small but suspiciously
high number - hardly significant but difficult to explain, and
still increasing - of cancer of various types has occurred in
a group of railroad workers in Sweden (Sundell, 1972). The
workers were exposed to amitrole during herbicide spraying.
The working conditions were bad and the exposure probably very
high. The group is so small and the work situation so difficult to analyze that no definitive answer will probably ever
emerge. Nevertheless, the use of amitrole in agriculture was
stopped in Norway on the basis of a hypothesis of dose-effect
linearity, and the product "man x working days x ppm". This
is an obvious carry-over from the ICRP philosophy to chemical
hazards. Whether this is justified remains to be sorted out.
Amitrole has not been found to be a mutagen in any of a
battery of tests (Laamanen *et al.*, 1976). But this negative -
or non-positive - evidence is invalidated by the suspicion
based on epidemiology. Again no attempt at a true cost-benefit analysis has been made.

Chemical mutagens and carcinogens are now in the process
of being drawn into a regulatory system more or less of the
ICRP pattern. A new international commission is being set up
by the International Association of Environmental Mutagen
Societies, and will in the coming years apply itself to the
extremely difficult task of creating a set of recommendations
for regulatory action by the various national health authorities. It has received the long and somewhat cumbersome name
of The International Commission for the Protection against

Environmental Mutagens and Carcinogens (ICPEMC) (Sobels, 1976). The work is just beginning, but assuming that 4 out of 5 cancers may be environmentally caused, and a completely unknown fraction of the genetic load is to be ascribed to environmentally caused mutation, the need for a regulatory framework of general applicability is obvious. Recommendations and regulations must be conservative to the extent that they provide protection even for the more sensitive members of the public.

The recommendations as regards genetic radiation damage are based on dose to gonia. Referring to chemicals, it should be remembered that the increased sensitivity in post meiotic stages may be so great ($\gg$ x 10) that short exposures of sensitive stages may become more important than chronic exposures of less sensitive stages. The need for both a norm for occupational exposure, and for population exposure is obvious, as exemplified in Ramel's (this conference) discussion of vinyl chloride, which is probably mainly an occupational problem, and the waste from the vinyl chloride industry, which is a population exposure problem in the form of tar.

In conclusion, there can be no question about the need for testing and mapping the situation and identifying the mutagenic agents in our environment. But important gains may be won already today out of improvement in production technology, refinement in application and use of unavoidable mutagens, and above all through a restrictive policy on the part of the responsible authorities.

REFERENCES

BEIR (1972). "The Effects on Populations of Exposure to Low Levels of Ionizing Radiation". Report of the Advisory Committee on the Biological Effects of Ionizing Radiations. National Academy of Sciences, National Research Council, Washington, D.C.

Bridges, B. (1974). *Mutation Res.* 26, 335.

Edwards, J.H. (1974). *Progr. Medical Genet.* 10, 1.

ICRP (1966a). "The Evaluation of Risks from Radiation". ICRP Publication 8, Pergamon Press, New York.

ICRP (1966b). "Recommendations of the International Commission on Radiological Protection" (adopted September 17, 1965). ICRP Publication 9, Pergamon Press, New York.

ICRP (1977). "Recommendations of the International Commission on Radiological Protection" (adopted January 17, 1977). Pergamon Press, New York (in the press).

Laamanen, I., Sorsa, M., Bamford, D., Gripenberg, U., and Meretoja, T. (1976). *Mutation Res.* 40, 185.

Newcombe, H.B. (1973). *In* "Chemical Mutagens", Vol. 3 (A. Hollaender, ed.), p. 57, Plenum, New York.

Oftedal, P., Carter, C.O., Ehling, U.H., Trimble, B.K., and Vogel, F. (1976). "Task Group on Genetically Determined Ill Health". Report to Committee I of the ICRP (working paper, to be published).

Pochin, E.E. (1974). *Community Health.* *6*, 2.

Sobels, F.H. (1976). *Mutation Res.* *38*, 361.

Stevenson, A.C. (1959). *Radiation Res. Suppl.* *1*, 306.

Sundell, L. (1972). Private communication.

Trimble, B.K. and Doughty, J.H. (1974). *Ann. Hum. Genet.* *38*, 199.

UNSCEAR (1972). "Ionizing Radiation: Levels and Effects". United Nations, New York.

UNSCEAR (1977). "Ionizing Radiations: Levels and Effects". United Nations, New York (in the press).

Vogel, F., and Rathenburg, R. (1973). *Adv. Hum. Genet.* *5*, 223.

WHO (1977). "Health Implications of Nuclear Power Production". Report of a Working Group, Brussels 1-5 December 1975. Regional Office for Europe, WHO, Copenhagen.

BIOCHEMICAL METHODS FOR MONITORING GENETIC DAMAGE

CHEMICAL MUTAGENESIS AND
INDUSTRIAL MONITORING

Marvin S. Legator
Stephen J. Rinkus

Department of Preventive
Medicine and Community
Health
Division of
Environmental Toxicology
and Epidemiology
University of Texas
Medical Branch
Galveston, Texas

I. INTRODUCTION

In less than ten years, a new area of toxicology, often
referred to as genetic toxicology, has emerged. Long before
the emergence of this area of toxicology, there was the realization that as the incidence and severity of parasitic
diseases declined, impairments in man resulting from gene and
chromosomal mutations would exert a more serious impact on
the health and economy of the human community. The finding
that some man-made chemicals in widespread use in the population are mutagenic in sub-human experimental systems implies
the possibility that such chemicals constitute a potential
genetic hazard to man. This possibility assumes greater importance when one considers the unprecedented reliance on
synthetic chemicals which man has developed in this century.
The overwhelming majority of these chemicals are uncharacterized with respects to their mutagenicity. Prominent categories of these synthetics include prescription and non-
prescription drugs, pesticides, food additives, and industrial chemicals. It was only with the development of procedures in animals in the latter part of the 1960's that a

feasible approach was made toward identifying and characterizing mutagenic agents. Since then, much progress has been
made in developing and testing new methodologies; strategies
for the mass screening of chemicals have been proposed; and
the field has even received recognition in the recently passed
Toxic Substances Control Act in the U.S. One of the aspects
that remains to be studied more intensely and is the very
heart of genetic toxiology is the actual quantitation of the
genetic risk posed to man by the conditional use of a vital
mutagenic substance. Such studies necessarily involve the
monitoring of human populations. This paper will discuss the
problem of genetic disease, including cancer, and the present
state of art of detecting mutagenic activity in non-human
systems and in human populations.

II. CHEMICAL MUTAGENESIS IN MAN

     With the design to upgrade the standard of living,
science and industry develop and introduce on the market new
chemicals at a rate of as many as 1000 new compounds a year in
commercial quantities in the U.S. (Train 1976). That these
chemicals, to combat sickness, to increase the food supply,
etc., have proved effective is hardly questionable. The disappearance of bacterial and parasitic diseases as the great
killers that they once were at the turn of the century, the
decline in infant mortalities, and the rise in the life expectancy can be attributed in part to modern medicine which readily employs synthetic drugs. The great strides that have
been made in the rate of production of stable foods owe some
of their success to the wide-scale employment of chemical
pesticides. Similarly, cases can be argued for the use of
industrial chemicals.
     However, that these chemicals have exerted other effects
on the population is likewise without doubt. In retrospect,
some of these substances have caused chronic, adverse effects
such as cancer and genetic disease. The mutagenic risk poses
itself to both present and future generations: to the former,
as the production of genetic lesions in somatic cells,
possibly leading to cancer; while to the latter, as the production of transmissible gene mutations and alterations in
chromosome structure and number. In its report, the U.S.
Council on Environmental Quality (1975a) advised the President
that a high percentage of all cancers are related to various
environmental factors, including exposure to some chemicals.
In socioeconomic terms in the U.S., the estimated 1.3 million
cancer patients require $1.8 billion per year in hospital care
alone, or tens of billions of dollars if all direct expendi-

tures incurred are considerd (DHEW 1974 a); as much as 1.7
million work years are lost to the national economy and to
family income as a result of cancer (Murray 1974).

Unfortunately, comprehensive data on the incidence,
variety, and etiology of actual gene and chromosomal mutations
have not been compiled. Estimates on the occurrence of true
genetic disease in the U.S. population would indicate that
12 million people carry defective genes or chromosomes that manifested themselves with birth defects (DHEW 1974b). Based
on chromosome examinations of 31,801 newborn children in seven
metropolises in Europe and North America, indications are that
from one in 117 to one in 285 newborns exhibit major cytogenetic abnormalities (Friedrich and Nielsen 1973). Concerning
the prenatal period, the occurrence of chromosomal aberrations
in spontaneous abortions has been reported at one in 33 aborted fetuses for the last five months of gestation, one in four
for the fourth month, and one in three for the first trimester (Carr 1972). The overall incidence of conceptuses that
are spontaneously lost could be as high as 45% (Carr 1971).
It is hard to believe that the human race has endured successfully such rates over all the generations since its emergence
on earth som 40,000 years ago. It must be kept in mind that
discussion only of those genetic lesions whose phenotypic
expression allows detection during gestation and birth would
be a gross understatement of the magnitude of mutations and
genetic disease. It would not account for the other dominant,
recessive, and sex-linked disorders that manifest themselves
later in the lifetime of the individual; likewise, it would
overlook the multifactorial, genetic interplay in some forms of
heart disease, arthritis, diabetes, cancer, hypertension, and
schizophrenia. Furthermore, *Drosophila* studies would suggest
that mere heterozygosity for a lethal, or mildly deleterious,
recessive mutation involves an incurred disadvantage on viability, despite the phenotypic dominance of the other, unmutated
gene (Temin *et al.* 1969).

Again, the socio-economic impact of genetic diseases is
massive. Estimates have placed the monetary burden of genetic
diseases at 25% of the national expenditure for health services in the U.S. (Lederberg 1971). The loss in life-years as
the result of birth defects, 80% of which are thought to be
genetic in origin (DHEW 1974c), dwarfs the other major diseases (Fig. 1).

This is not to insinuate that all nor necessarily the
majority of chemicals cause cancer or induce mutations. The
fact is that not enough information on the entire health consequences of the chemicals is known. Some are definitely
detrimental, and the overwhelming majority have been released
on the market with inadequate testing for mutagenicity. For
example, a survey of the top 200 prescription drugs for the

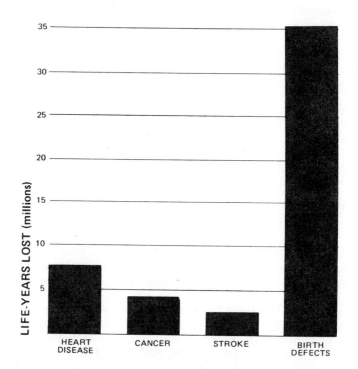

FIGURE 1.   *Life-years lost due to heart disease,*
cancer, stroke, and birth defects
(reproduced from DHEW 1974c)

U.S. in 1974 (69% of the market) revealed that of the 190
separated entities comprising these top 200 drug products,
only 53% (101 chemicals) had at least one mutagenicity-related
study reported in the literature; the vast majority of the
testing that was performed was in systems of little or no
toxicological relevance (Rinkus unpublished).

The scope of the problem goes beyond intentional, mass
exposure to uncharacterized chemicals. The U.S. Environmental
Protection Agency (1975) has identified collectively 253 separate organic chemicals in drinking water samples across the
country, including halogenated hydrocarbons: vinyl chloride,
carbon tetrachloride, and chloroform; and pesticides: DDT,
aldrin, dieldrin, heptachlor, and chlordane -- all confirmed
or suspected carcinogens (IARC 1972; IARC 1974a; IARC 1974b;
Epstein 1976). Industrial and municipal waste, urban and
rural runoff, natural sources, and water and sewage treatments
practice are among the confirmed sources of these pollutants.
Again, the potential carcinogenicity of many of the other
chemicals in the water supply is simply not known. However

the fact that some of these chemicals do appear in blood
(Dowty 1975) and tissues (U.S. Council on Environmental Quality 1975b), and are implicated in increased cancer rates has
been reported (Environmental Defense Fund 1974).

The circumstances that led to this predicament in general are not as elusive as their resolution. In a 20th century
continuation of the Industrial Revolution, society embarked on
an era of unprecedented use of synthetic chemicals. It is an
age much in step with the Nuclear and Space Ages and part of
the same technicological, social phenomena. Mass exposure to
these substances began in a time that lacked an appreciation
of carcinogenesis and mutagenesis. Since then chemicals have
become as enwebbed in and essential to our life style as com-
puters; and their production has grown into a multi-million
dollar affair -- all this in less than the span of one human
lifetime. On the basis of the great number of chemicals,
their ubiquitousness in society, what is already known about
the adverse effects of some of the chemicals, and undetermined
about the others, it would appear that the miracles of the
chemical age have an unforeseen price. The latency period
following the induction of a cancer has been estimated at 15
to 40 years (Selikoff and Hammond 1972), and without national
surveys on the incidence of pregnancies spontaneously aborted
due to chromosome anomalies and the incidence of genetic diseases in general, one can only speculate on the detrimental
effects to the genetic pool from injurious chemical exposure.
In terms of ourselves as a population of living organisms,
we are suffering chemical shock -- the extent of which is only
beginning to be gauged.

III. NON-HUMAN MUTAGENICITY TESTING

The growing awareness that it is no longer possible to
introduce new chemicals into the environment without toxicological testing in the area of mutagenesis makes it imperative
that the relevance of existing procedures be re-examined. Mutagenicity testing should not be considered simply as a means to
identify a potential carcinogen, but also as a means of preventing genetic abnormalities whose importance to man may well
eclipse all other areas of toxicological testing. Several
strategies, so-called tier approaches, have been developed for
the systematic screening of chemicals for mutagens. Proposals
(Flamm 1974; Bora 1976) to utilize exclusively *in vitro* systems to develop priorities for further testing in long-term
animal studies are toxicologically naive and inadequate for
evaluating potential mutagens. The use of *in vitro* microsomal
preparations combined with microorganisms or other indicators

as an exclusive primary screen is an approach whose deficiencies must be recognized.   An *in vitro* microsomal activating
system cannot reflect the complex dynamic processes that are
carried out in the intact animal.   Indeed, it is not possible
to devise a standard *in vitro* activation system that can be
used generally to screen potential mutagens and carcinogens.
Even if such an *in vitro* system could activate all compounds
that are metabolized by the host's enzyme systems, the fact
that many materials are either potentiated or detoxified by
other routes, *e.g.* intestinal flora, would argue against the
use of the system as an initial screen test.   Thirdly, an
important class of chemicals that induce nondisjunction by
affecting the spindle mechanisme, one of the most important
categories of cytogenetic abnormalities, would be missed by
microbial studies.

In the literature, one can find studies and editorials
suggesting that the actual correlation between carcinogenicity
and mutagenicity in microsomal activation systems is as high
as 93% (Perchase *et al.* 1976).   Such reportings have been
offered in support of adoption of a tier approach that would
utilize only *in vitro* systems as primary screens.   These correlations must be seen in their proper perspective.   The determination of a meaningful correlation, one that if high enough
would justify the exclusive use of *in vitro* testing in a tier
approach, presumes an understanding of the set of all carcinogens: of their chemical classes, of the frequencies of these
classes, and of future-yet-contrived arrangements of carbon,
nitrogen, oxygen, etc. atoms.   Such understanding simply does
not exist, and may not ever exist.   These hitherto reported
correlations are products of the sets of carcinogens tested
and, more importantly, of the manner in which they were chosen
In most cases, selected rather than randomly chosen compounds
have been evaluated.   Given such circumstances, one must
seriously question why the higher correlations so far reported
are any more credible than the lowest reported one, 44%
(Odashima 1976).

All responsible scientists share the goal of characterizing human mutagens and replacing them with non-mutagenic
substitutes or otherwise supervising their use in society.   It
is essential, however, that indefensible toxicological procedures should not be adopted to achieve this goal.   The crux of
the matter is not the identification of active compounds by
in vitro procedures, with or without activation, but the fact
that *a priori* potent carcinogens and mutations will be missed.
Industrial toxicologists and other interested individuals
should not be lulled into a false sense of security and assume
that the chemicals are not mutagenic when the compound is
found not to be active by existing microbial procedures
(Kolata 1976).   In principal, with a test chemical, one would

start with the best available animal systems, including those
tests that evaluate metabolic products in the intact host.
Since there is no single test for detecting chemical mutagens,
the testing protocol would utilize a battery of tests carried
out in the intact animal. The procedures for detecting and
characterizing various types of genetic lesions include the
following (Legator and Zimmering 1975):

A. *Indirect Detection of Gene and Chromosomal Mutations:*
host-mediated assay, and body-fluid analysis in experimental
animals (discussed later) using various indicator systems as
well as *in vitro* studies, with and without microsomal activation.

B. *Detection of Chromosomal Mutations in vivo.*
1. Dominant lethal test.
2. Translocation studies.
3. Micronuclei test.
4. Cytogenetic analysis of both meiotic and
mitotic cells.

C. *Detection of "Premutational Lesions": Repair
Studies in vivo.*

D. *Alkylation of Macromolecules* (discussed later).

While collaborative studies have rarely been conducted
in the field of toxicology, it is noteworthy that already in
the field of genetic toxicology the dominant lethal test and
in vivo cytogenetic analysis have been subjected to collaborative studies (Kilian *et al.* 1975). The integration of the
results from all of these systems should offer the optimum
opportunity for identifying mutagens that are potential hazards to man. The subsequent studies of an active compound
would rely on refining procedures to isolate and identify the
active compound and, subsequently, to characterize the genetic
lesions induced by the chemicals under study. In the field of
chronic toxicology, available methods are used to evaluate new
compounds before and after they are introduced on the market as
well as to evaluate currently used materials. Although, in
many instances, these procedures are time consuming and expensive, no one would suggest that they be abandoned until more
suitable methods are developed. In like manner, to postpone
the evaluation of chemicals for mutagenic activity, or to
settle for procedures that may yield misleading information, is
to diminish the importance of this area of toxicology. Existing procedures are as good as, if not better than, existing
methods in the field of chronic toxicology. Indeed, if one
employs a battery of tests that would detect compounds which
cause gene and chromosomal mutations, including nondisjunction,

the total cost would be only one-third of the some $100,000
that is presently allotted for a single carcinogenicity study.

IV.   INDUSTRIAL MONITORING

      The industrial environment provides a unique set of cir-
cumstances for detecting and characterizing chemical mutagens.
In many instances, employees can be identified who are exposed
to a variety of chemicals, some of which are unique to various
industries, while others represent exaggerated cases of what
may be occurring in the general population. It is encouraging
to note that some of our more progressive corporations have
embarked on a comprehensive program to detect chemical mutagens.
      In the context of an industrial program to characterize
mutagenic agents, there are three aspects that one can consider. First, there are the experimental compounds that have
yet to be introduced on the market. These chemicals should
be thoroughly investigated by a variety of available mammalian
procedures (as previously discussed) prior to their introduction on the market. A second aspect would be the monitoring of
workers exposed to specific chemicals; this will be discussed
later. Finally, classical epidemiological studies can be conducted on the workers. However, it should be kept in mind
that while epidemiological studies have been successful in
demonstrating sound associations between cause and effect when
the effect has been as obvious, as lung cancer from smoking and
angiosarcoma from vinyl chloride exposures, they have been less
successful in detecting effects which are not as easily dis-
tinguised from the background rate. The limitation of epidemiological studies as discussed by Comfod and Van Ryin
(unpublished manuscript) regarding saccharin and bladder cancer is an illustrative case:

            Armstrong and Doll (1975), motivated by reports
      of animal experiments in which bladder cancer resulted
      from exposure to high doses of saccharin, undertook a
      case-control study of 18,733 patients dying from blad-
      der cancer and 19,709 patients dying from other forms
      of cancer. They also studied the saccharin consump-
      tion of 200 diabetics and 200 matched controls. In
      view of the more than tenfold excess consumption of
      saccharin by male diabetics one would have expected,
      if saccharin is carcinogenic at the levels ingested,
      1.26 mg/kg/day, a higher risk of bladder cancer for
      diabetics. In fact, none was found. The risk of bladder
      cancer for male diabetics relative to non-diabetics

was estimated as 1.00 -- but with 95% confidence limits of 0.62-1.60. A downward extrapolation from the
Canadian animal tests on saccharin by the Food and Drug
Administration ... indicates a bladder cancer risk of
$0.4x10^{-3}$ for humans consuming the equivalent of one
bottle of diet soda daily or about 1.5 mg/kg/day. Of
the one million male deaths in the United States in
1969 almost six thousand were from bladder cancer ...
indicating a life-time from bladder cancer of $6x10^{-3}$.
The Food and Drug Administration result thus indicates
that a diabetic consuming 1.5 mg/kg/day, or slightly
more than reported by the diabetics in the Armstrong-
Doll study, would have a life-time risk of $6.4x10^{-3}$,
or a relative risk of 1.07. This is well within the
Armstrong-Doll confidence limits, so that their results
are consistent with both the hypothesis suggested by
the animal results and with the hypothesis of no effect.
One might add that even with human samples large enough
to have yielded confidence limits that excluded either
1.00 or 1.07 but not both, the effects of confounding
variables, such as cigarette smoking, on effects this
small would have made the results difficult to interpret.

Given the severe limitations of epidemiological studies
and the need for monitoring populations such as those found in
industry, major emphasis should be placed on those techniques
that assess the potential for a chemical to induce mutations
in human subjects. Cytogenetic surveillance of workers in
industry is well established and is discussed elsewhere in
this monograph. Two additional procedures that, together with
cytogenetic studies, can form the basis of a meaningful surveillance program in industrial and other potentially high
risk populations are subsequently discussed.

A. *Body-Fluid Analysis*

The analysis of body fluids in animals and man by the use
of various indicator systems is an important procedure that
provides a greater assessment of the mutagenic potential of a
compound. Its flexibility in terms of the choice of indicator
and the ability to carry out the analysis in a variety of
species, including man, makes it a unique test for chemical
mutagens. It allows for the indirect measurement of gene and
chromosomal mutations in the host and permits this determination for the parent compound and its metabolities that appear
in the body fluids (urine, blood, and semen) regardless of
where the metabolites are produced in the host. Chemicals
that cannot be detected *in vitro* by standard activiation procedures can be potentially identified by this method.

In the initial publication of the host-mediated assay
(Gabridge *et al.* 1969), the analysis of blood and urine of
treated animals for mutagenic activity was described as a modification of the host-mediated assay procedure. Since that
initial report, several papers have appeared on urine analysis
of treated animals for mutagenic activity using a variety of
indicators. Of particular significance are the reports by
Siebert and Simon (1973a; 1973b) and Siebert (1973) that urine
from patients treated with the anti-cancer drug cyclophosphamide
exhibits mutagenic activity. However, the true strength of this
procedure became evident when chemicals other than anti-cancer
agents were shown mutagenic by this method. These findings
have established the utility of this procedure for population
monitoring.

Legator *et al.* (1975) and Connor *et al.* (1977) found that
the urine of patients who received two widely used drugs, the
schistosomacide niridazole and the trichomonacide metronidazole
(Flagyl), contained mutagenic substances. Urine collected from
a single patient who had been treated in the United States with
a single oral dose of 25 mg/kg of niridazole demonstrated mutagenic activity with *Salmonella* tester strain TA1538 four days
after the treatment. The availability of patients receiving
metronidazole permitted a study of eight human subjects. Control urine obtained from individuals prior to therapy showed
no mutagenic activity whereas mutagenic activity was found in
every individual who received the drug. Fig. 2 illustrates the
results for six patients whose urine was examined on days 0, 1,
6,8, or 9 of drug treatment. A daily analysis was performed on
two additional patients, and the results of 10 days of treatment are recorded in Fig. 3. The total amount of metronidazole
in the urine was determined spectrophotometrically, and chro-
matographic analysis was performed to determine the specific
metabolites according to a procedure described by Stambaum *et*
al. (1968). The activity found in the urine of treated subjects
was far higher than that anticipated from *in vitro* dose-response
results with metronidazole as illustrated in Fig. 3. Two of
the principal metabolites present in the urine, 1-acetic acid-
2-methyl-5-nitroimidazole and 1-(2-hydroxyethyl)-2-hydroxyme-
thyl-5-nitroimidazole, were obtained in pure form and were tested against tester strain TA1535 (Fig. 4). The latter was 10
times more mutagenic than metronidazole, and the former was totally inactive. On the basis of these studies, it is likely
that the hydroxymethyl metabolite is responsible for the high
degree of mutagenic activity found in human urine. Our inability to find enhanced activity after exposure to microsomal activation would indicate that the hydroxymethyl metabolite is
not formed in the *in vitro* activation procedure. Since this
compound is found in the urine of animals and man after administration of the drug, these results suggest that the *in vitro*

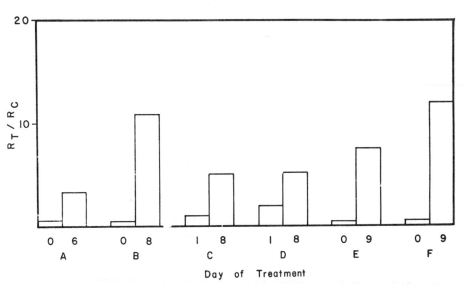

FIGURE 2. *Six patients (A, B, C, D, E, and F) whose urine*
were tested against Salmonella tester strain
TA 1535 before (days 0 or 1) and during (days 6,
8 or 9) treatment with metronidazole, 750 mg/day.
R_t/R_c is the number of histidine revertants on
treated plates per number of histidine revert-
ants on control plates.

activation system lacks the enzymes, components, or conditions
necessary to carry out the metabolism seen in the host and
what positive response that is observed may be coincidental.

Of special relevance to industrial monitoring has been the
recently reported measurement of mutagenic activity in the
urine of two workers highly exposed to epichlorohydrin during
an industrial spill (Kilian *et al.* 1977). Epichlorohydrin
(1,2-epoxy-3-chloropropane) is a known alkylating agent that
is widely used in industry as a solvent for resins, gums,
cellulose, joints, and especially as a raw material for the
manufacture of epoxy resins. Sram *et al.* (1976) have reviewed
and extended mutagenicity studies with this chemical: it is
active *in vitro* against *Salmonella* tester strains, in the host-
mediated assay, in inducing chromosomal abnormalities in mice
and in *in vitro* chromosome studies with human lymphcytes; the
chemical is not active in the dominant lethal test.

Studies in our laboratory were initiated to determine the
presence of mutagenic substances in urine of workers exposed
to epichlorohydrin using *Salmonella* tester strain TA1535 as
the indicator organism. Six individuals exposed to 0.8-4.0 ppm

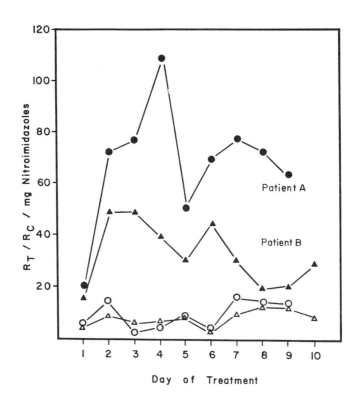

FIGURE 3. *Mutagenic activity detected with S. typhimurium*
TA 1535 in urine samples of 2 patients treated
with metronidazole. Patients received a dosage
of 750 mg/day, and the urine was collected 1 hr
after the morning dose on the day indicated.
●, 0, Patient 1; ▲, Δ, Patient 2. Actual detect-
ed mutagenic activity (solid symbols); expect-
ed mutagenic activity was based on equivalent
amount of metronidazole (open symbols), R_t/R_c,
the number of histidine revertants on treated
plates per number of histidine revertants on
control plates (Connor et al. 1977).

epichlorohydrin per day were monitored for 5 days, as well as
5 non-exposed workers. In the exposed and non-exposed group,
there were no indication of mutagenic substances in the urine.
During the course of this study, two workers were inadvertently
exposed to levels greater than 25 ppm of the chemical. The

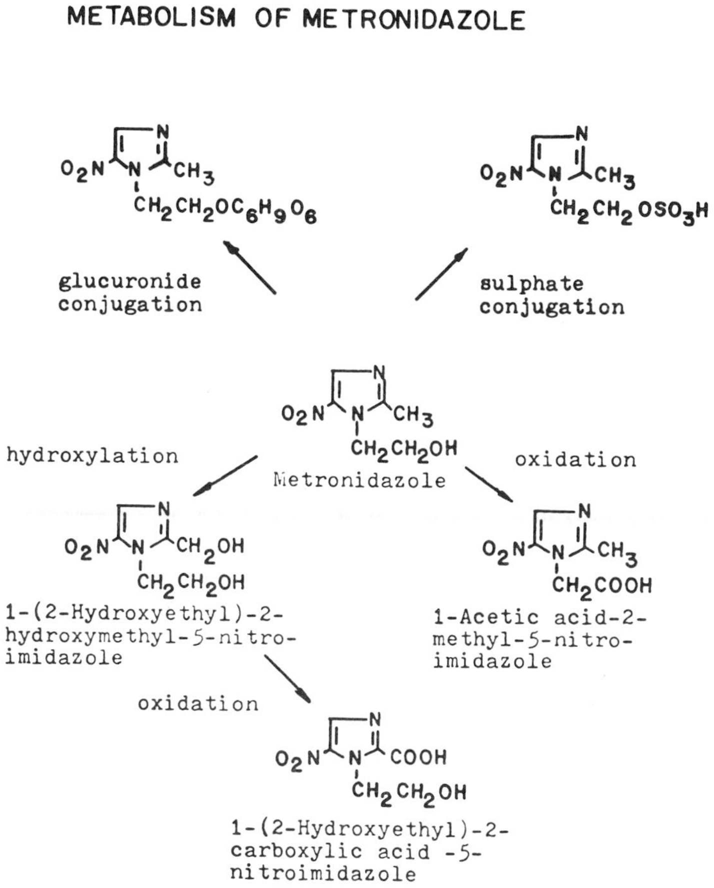

METABOLISM OF METRONIDAZOLE

FIGURE 4. Metabolism of metronidazole (modified from Ings et al. 1975).

urine of the two workers was obtained nine hours after the accident; detectable levels of mutagenic activity were present. Fig. 5 illustrates the results of this study. This is the first example of detecting activity in the urine of workers following exposure to an industrial chemical. Further studies with this procedure should add a new dimension to industrial monitoring. Although the *Salmonella* tester strains have been traditionally used in this procedure, use of other indicators will extend the utility of this method. Another major contribution to expanding the capabilities of this procedure was the development of a general concentration procedure by Connor and Legator (1977) which should enhance our ability to detect potential mutagens in the urine or blood.

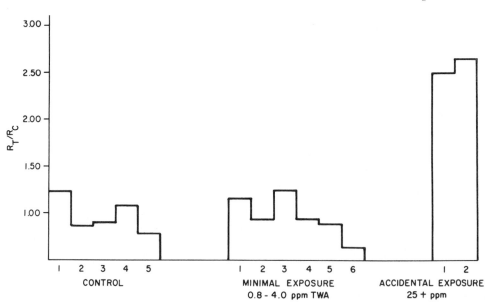

FIGURE 5. *The mutagenic activity in Salmonella tester strain TA 1535 of urine of chemical workers exposed to epichlorohydrin, and of controls. See text of the paper for further description.*

B. *Alkylation of Macromolecules – A Technique of Great Potential for Monitoring Human Subjects*

A characteristic of a large group of carcinogenic and mutagenic chemicals is their ability to alkylate or otherwise chemically react with nucleophilic atoms in macromolecules. The more commonly recognized alkylating agents include alkyl sulfates, alkyl sulfonates, alkyl halides, dialkyl nitrosamines, diazo compounds, lactones, and epoxides. In fact, such agents constitute in numbers the most significant group of mutagens. Chemicals that react in such a way with the nucleophilic atoms of DNA can also similarly react with RNA and amino acids. It is therefore possible to use the alkylation of specific amino acids in proteins as an indication of mutagenic activity. Recognition of this fact prompted Osterman-Golkar *et al.* (1976) to suggest the use of radioactive reagents. Truong and Legator (1977) have recently developed this approach into a more practical procedure that can be utilized in experiments with animals as well as human subjects. The use of radioactive material is eliminated by monitoring the disappearance of normal amino acid residues in hemoglobin with an amino acid analyzer. The knowledge of the number and sequence of amino acids in this

thoroughly investigated protein allows for the selection of
specific amino acids for analysis and standardization of the
assay. In contrast to proline whose nitrogen atom is not available for reaction due to its position in the ring and linkage
with the carbonyl group of the adjacent amino acid, histidine,
methionine, cysteine, serine, and other amino acids have nucleophilic atoms capable of reacting with electrophilic agents.
The purity and standardization of the assay is accomplished by
calculating the ratio of the reactable amino acids to unreact-
able proline. Of the reactable amino acids, histidine is the
preferred amino acid for the quantitative determination of
alkylation due to its more frequent occurrence in the human
hemoglobin tetramer (38 histidine molecules per tetramer),
its stability, and the ease of detection with the amino acid
analyzer.

The method of preparing hemoglobin for the analysis consists of the following steps:

1. Blood obtained from treated subjects is centrifuged at
3000 rpm to remove plasma.

2. 1 to 2 ml of red blood cells are washed 3 times with
3-4 volumes of saline.

3. Lyse with cold distilled water.

4. Remove lipoproteins by shaking with one half volume of
CCl_4 and centrifuge at 7000 rpm for 10 minutes.

5. Precipatate hemoglobin with 15 volumes 1.5% HCl in
acetone; wash 3 times with acetone and dry.

6. To 0.3 - 0.4 mg of hemoglobin, add 1 ml 6 N HCl.

7. Seal under vacuum.

8. Hydrolyze for 24 hours at 110°C.

9. Dry over NaOH.

10. Apply 150 µl to Beckman model 191 Amino Acid Analyzer.

In rodent testing, the known alkylating agent EMS and the
schistosomacide hycanthone were investigated. Fig. 6 illustrates the results of this study. Of special interest is the
finding with hycanthone which has previously been shown to
cause frameshift mutations in *Salmonella* (Hartman *et al.* 1971).

Although studies to date have been carried out in rodents,
this analysis can be carried out with equal ease in human subjects. Since the life span of the erythrocyte is 4 months, one
would anticipate finding a record of chemical insults that may
have occurred during this 4-month period with this procedure.
It may therefore be possible to obtain precise dosimetric measurements in man after chemical exposure to alkylating agents.

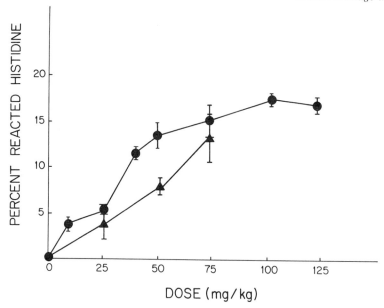

FIGURE 6. *In vivo alkylation of histidine in rat hemoglo-*
bin by MMS (●) and hycanthone methanesulfonate
(▲). Twenty-four hours after a single, intra-
peritoneal injection of the chemical, albine
rats were killed with a blow to the head; 2 cc
blood were obtained by cardiopuncture with a
heparinized syringe. Preparation of the blood
for analysis is described in the text of the
paper. Percent reacted histidine is calculated
from the disappearance of N'(2) - and N'(4)-
unsubstituted histidine (e.g. 10% reacted histi-
dine is inferred when the quantity of unsubsti-
tuted histidine is found to be 90% of the quan-
tity of unsubstistuted histidine obtained with
controls). Each point represents the mean and
standard deviation of three treated animals.

V. CONCLUSION

It was predictable, if indeed not inevitable, that a new
area of toxicology such as genetic toxicology would encounter
difficulties and setbacks during its early developmental phase.
While there is lettle debate about the importance of this area
in light of the consequences of mutagenic and carcinogenic
agents to the population, it is quite another matter to interpret mutagenicity findings and to agree on benefit-risk assess-

ments based on such findings. At one end of the testing spectrum, unjustified importance is placed on negative findings in simplistic *in vitro* systems; while at the other end, completely inadequate epidemiological data are offered to attest to the safety of a given substance. One can optimistically anticipate the time when responsible scientists in academia, government, and industry will place this area of toxicology in its proper perspective. Before a chemical is released on the market, it should be evaluated by a battery of mutagenicity tests performed on intact animals, as is the practice in other areas dealing with important toxicological effects. For those chemicals already on the market, priorities for testing reflecting usage patterns, the persistence of the chemical in the environment, and structure-activity relationships will allow an orderly evaluation of these chemicals starting with the high-priority chemicals. The limitations of epidemiological studies should be clearly understood. It would be difficult to attach any importance to negative findings in epidemiological studies that, by inadequate design, could not detect a minimal but statistically real increase over the background rate if such were the case. One can expect that benefit-risk assessments primarily will rely on data derived from animal testing, as is the practice in other fields of toxicology.

The value of cytogenetic analysis in animals and, more importantly, in man for detecting mutagenic agents has been realized for some time. It is now possible to envisage a comprehensive protocol for surveying human subjects by several procedures in addition to cytogenetics. At least two more procedures can now be considered for population monitoring. The analysis of urine and blood by a variety of indicator systems, after suitable extraction and concentration procedures, now makes it possible to detect indirectly chemicals that might induce gene and chromosomal mutations in the host. One of the most significant advances in population monitoring should be the alkylation of macromolecules in individuals exposed to potential mutagens. This technique should detect the presence of harmful chemicals at a low concentration, and exact dosimetry may be possible. This type of monitoring system, when applied to high risk population such as in industry, should give us an advanced warning of hazardous compounds before actual toxicity occurs, which is in contrast to the after-the-fact information provided by epidemiological studies.

Our greatest opportunity to prevent genetic disease and cancer is our ability to detect those chemicals in our environment that are believed to be responsible for these afflictions. The continued use of animal studies and an expanding human surveillance program utilizing cytogenetics, body-fluid analysis, and the alkylation of macromolecules will help in achieving this goal.

VI. SUMMARY

Mutagenicity studies represent an important area of toxicology and should not be viewed solely as short-term screens for identifying carcinogens. Tier approaches for the identification of mutagens that rely on *in vitro* systems as the primary screen are not acceptable as a result of their inability to standardize the enzyme preparations that supposedly mimick mammalian metabolism or to detect promutagens activated by routes other than the liver (*e.g.* intestinal flora). These problems are overcome by combined *in vitro* and *in vivo* testing. Available procedures for detecting mutagens in animals are as good as methods routinely used to characterize non-genetic, toxological hazards and are economical and short-term relative to carcinogenity studies. Body-fluid analysis and the alkylation of macromolecules (histidine) are two new procedures to be considered along with cytogenetic and epidemiological studies in the context of monitoring human populations for mutagens.

VII. REFERENCES

Armstrong, B. and Doll, R. (1975). *Brit. J. Prev. Soc. Med.*
    29, 73.
Bora, K.C. (1976). *Mutation Res. 41*, 73.
Carr, D.H. (1971). *In* "Advances in Human Genetics". Vol 2. 201.
Carr, D.H. (1972). *Res. in Reprod. 4*, 3.
Connor, T. and Legator, M.S. (1977). *In* "8th Annual Meeting of
    the Environmental Mutagen Society", Colorado Springs.
Connor, T.H., Stoeckel, M, Evrard, J. and Legator, M.S. (1977).
    Cancer Res. 37, 629.
DHEW (1974a). *Cancer Rates and Risks.* DHEW Pub. No. (NIH)
    75-691. GPO, Washington, D.C., p.3.
DHEW (1974b). *What are the Facts about Genetic Disease.* DHEW
    Pub. No. 75-370. GPO, Washington, D.C., p.6.
DHEW (1974c) *Ibid*, p.8.
Dowty, B., Carlisle, D., Laseter, J.L. and Storer, J. (1975).
    Science 187, 75.
Environmental Defense Fund (1974). *The Implications of Cancer-
    Causing Substances in Mississippi River Waters*, Washing-
    ton, D.C.
Epstein, S.S. (1976). *Sci. Total Environ. 6*, 103.
Flamm, W.G. (1974). *Mutation Res. 26*, 329.
Friedrich, U. and Nielsen, J. (1973). *Clin. Genet. 4*, 333.
Gabridge, M.G., DeNunzio, A. and Legator, M.S. (1969).
    Nature (Lond.) 221, 68.

Hartman, P.E., Levine, K., Hartman, Z. and Berger, H. (1971). *Science 172*, 1058.

IARC (1972). *International Agency for Research on Cancer Monographs on the Evaluation of Carcinogenic Risk of Chemicals to Man*, Lyon, Vol. 1: Carbon Tetrachloride. pp. 53-60; Chloroform. pp. 61-65.

IARC (1974a). *International Agency for Research on Cancer Monographs on the Evaluation of Carcinogenic Risk of Chemicals to Man*, Lyon, Vol. 5: Aldrin. pp.25-38; DDT. pp. 83-124; Dieldrin. pp. 125-156; Heptachlor. pp. 173-191.

IARC (1974b). *International Agency for Research on Cancer Monographs on the Evaluation of Carcinogenic Risk of Chemicals to Man*, Lyon, Vol. 7: Vinyl Chloride. pp. 291-318.

Ings, R.M.J., McFadzean, J.A. and Ormerod, W.E. (1975). *Xenobiotica 5*, 223.

Kilian, D.J., Pullin, T.G., Connor, T.H. and Legator, M.S. (1977). *In* "8th Annual Meeting of the Environmental Mutagen Society", Colorado Springs.

Kilian, D.J., Picciano, D.J. and Jacobson, C.B. (1975). *Ann. N.Y. Acad. Sci. 269*, 4.

Kolata, G.B. (1976). *Science 192*, 1215.

Lederberg, J. (1971). *In* Foreword to: Epstein, S.S. and Legator, M.S, authors. "The Mutagenicity of Pesticides: Concepts and Evaluation", M.I.T. Press, USA, p.x.

Legator, M.S., Connor, T. and Stoeckel, M. (1975). *Science 188*, 1118.

Legator, M.S. and Zimmering, S. (1975). *Ann. Rev. Pharmacol. 15*, 387.

Murray, J.L. (1974). *J. Natl. Cancer Inst. 52*, 3.

Odashima, S. (1976). *In* "Screening Tests in Chemical Carcinogenesis", Lyon, IARC Scientific Publications No. 10. pp. 61-75.

Osterman-Golkar, S., Ehrenburg, L., Segerbäck, D. and Hällström, I. (1976). *Mutation Res. 34*, 1.

Purchase, I.F.H., Longstaff, E., Ashby, J., Styles, J.A., Anderson, D., Lefevre, P.A. and Westwood, F.R. (1976). *Nature (Lond.) 264*, 624.

Selikoff, I. and Hammond, E.C. (1972). *In* "7th National Cancer Conference", Los Angeles.

Siebert, D. and Simon, U. (1973a). *Mutation Res. 21*, 257.

Siebert, D. and Simon, U. (1973b). *Mutation Res. 21*, 202. Abstr.

Sram, R.J., Cerna, M. and Kucerova, M. (1976). *Biol. Zbl. 95*, 451.

Stambaugh, J.E., Feo, L.G. and Manthei, R.W. (1969). *J. Pharmacol. Exptl. Therap. 161*, 373.

Temin, R.G., Meyer, H.U., Dawson, P.S. and Crow, J.F. (1969). *Genetics 61*, 497.

Train, R.E. (1976). Speech delivered by the former Administra-
    tor of the U.S. Environmental Protection Agency before
    the National Press Club in Feb. 1976 (as reported in The
    American City and Country. May, 1976, p. 34).
Truong, B.L.T. and Legator, M.S. (1977). *In* "8th Annual Meet-
    ing of the Environmental Mutagen Society", Colorado
    Springs.
U.S. Council on Environmental Quality (1975a). *Environmental
    Quality. 6th Annual Report of the Council on Environmen-
    tal Quality.* GPO, Washington, D.C., p. 28.
U.S. Council on Environmental Quality (1975b). *Ibid,* p. 375;
    pp. 379-381.
U.S. Environmental Protections Agency (1975). *Preliminary
    Assessment of Suspected Carcinogens in Drinking Water.*
    Office of Toxic Substances. A report to Congress, p.3.

INDICATORS OF DNA DAMAGE DETERMINED BY A COMBINED ANALYSIS OF
CHROMOSOME BREAKAGE, SISTER CHROMATID EXCHANGE, ARYL HYDROCAR-
BON HYDROXYLASE, DNA REPAIR SYNTHESIS AND CARCINOGEN BINDING

Felix Mitelman

Department of Clinical Genetics
University of Lund
Lund, Sweden

Ronald W. Pero

Wallenberg Laboratory
University of Lund
Lund, Sweden

It is well established that chromosome aberrations in
somatic cells may be regarded as morphologic visible evidence
of damage to the genetic material. As yet, no satisfactory
general theory of chromosomal aberration production has emerged. However, chromosomal aberrations in eukaryotyic cells
induced by either ionizing radiation, viruses or chemical
agents are often accepted as an indicator of a mutagenic, and
thereby, a potential malignant event. This is supported by
the following facts:

a) Most agents that cause chromosome damage are carcinogenic. This has been shown to be true for radiation, viruses
and chemical substances (Harnden 1977).

b) Individuals with a spontaneous increased frequency of
chromosome breakage, *e.g.* ataxia telangiectasia, Fanconi´s
anemia, Bloom's syndrome and xeroderma pigmentosum, have an
increased risk to develop malignant disorders (Hecht and McCaw
1977).

c) Most malignant tumors have chromosome aberrations and
there seems to be an intimate relationship between the initial
chromosome damage in normal somatic cells and the chromosome
aberrations in the tumors. Thus, intravenous injections of
7,12-dimethylbenz(α)anthracene (DMBA) in rats produce an in-

creased frequency of chromosome breakage in bone marrow cells. The breaks are preferentially located on one particular chromosome of the complement - chromosome No. A2 (Sugiyama 1971). When sarcomas, carcinomas and leukemias are induced in rats by this carcinogen, trisomy for chromosome A2 is a characteristic aberration in the malignant cells (Mitelman 1974).

d) Chromosome aberrations in experimental and human neoplasma are clearly nonrandom and there are strong indications that the karyotypic pattern is determined by the etiologic agent (Mitelman *et al.*, 1972, Mitelman and Leyan, 1976).

Because of the apparent relationship between chromosome abnormalities, mutagenesis and carcinogenesis, analyses of chromosome aberrations have become a useful tool in evaluating potential hazard caused by environmental mutagens and carcinogens. This is, in fact, the only method which so far is practically useful in detecting genetic damage directly in man. It goes without saying, however, that only gross genetic damage is detectable by this method and the absence of chromosome aberrations does not exclude other types of DNA damage caused by environmental carcinogens. Therefore studying chromosomal aberrations in relation to other parameters for evaluating DNA damage may prove a more accurate and useful tool in assessment of the environmental impact on carcinogenesis.

Chemical carcinogens predominate as the oncogenic agents in environmental carcinogenesis which, in turn, make up about 80-90% of the total incidence of human cancer (Heidelberger 1975). The list of known carcinogens to which humans are exposed can be divided into two groups; viz., those that need metabolic activation, such as the polycyclic hydrocarbons, and those that do not, such as the nitrosamides. It is not clear, and often not considered, whether metabolically activated carcinogens or those that need no activation are of major etiologic importance in environmental carcinogenesis. Certainly both proximate and non-proximate carcinogens play a role, but it is their relative importance to overall cancer induction that is the scientist's dilemma. No matter whether the carcinogen needs activation or not, it is still metabolized by the mixed function oxidase system (Fouts 1970, Heidelberger 1975). On the one hand, then, high mixed function oxidase activity would increase cancer incidence by altering a carcinogen to its active form (*i.e.* Aryl hydrocarbon hydroxylase or AHH activation of polycyclic hydrocarbons) and thereby increase carcinogen reactivity with host target cells. On the other hand, if the carcinogen was already active, then increasing mixed function oxidase activity would reduse cancer incidence by removing the carcinogen through metabolic degradation.

The problem of differential carcinogen reactivity in exposed individuals is complicated by the fact that mixed function oxidase activities associated with drug (Conney, 1967, Vessel *et al.*, 1971) and carcinogen (Gelboin, 1967, Fouts, 1970, Heidelberger, 1975) metabolism are also inducible to higher levels of function. Interindividual variation following AHH induction has been demonstrated in human lymphocytes to be under genetic control (Kellermann *et al.*, 1973a, Atlas *et al.*, 1977) and the AHH inductive capacity correlated to overall drug metabolism (Kellermann *et al.* 1973a, Atlas *et al.* 1977) and to lung cancer (Kellermann *et al.* 1973b). However, other evidence in mice does not always support the relation of high AHH induction to high levels of polycyclic hydrocarbon induced cancer (Nebert *et al.* 1972, Benedict *et al.* 1973). Nevertheless, an important role for AHH induction in chemical carcinogenesis cannot be denied, but its importance must be viewed in terms of the type of carcinogen causing the induction.

Still another important point concerning enzyme induction in carcinogen metabolism is whether the constitutive levels of mixed function oxidase activities are sufficient to cope with the environmental exposure. This point has essentially not been studied. However, if this was the case, then mixed function oxidase induction would be of little consequence compared to the interindividual variation in constitutive level of mixed function oxidases. In this regard, Harris *et al.*, 1976 have recently studied the binding of ^{3}H-benzopyrene to DNA of cultured human bronchus epithelial cells after 24 h incubation in 1.5 μM ^{3}H-benzopyrene but without preincubation in the precense of benzopyrene to induce AHH activity. Since benzopyrene needs metabolic activation by AHH to bind to DNA, then their results have suggested a 70-fold interindividual variation in constitutive levels of AHH. The AHH induction in mouse T3 fibroblasts is barely detectable after 24 h incubation in 1.5 μM benzopyrene (Bittner and Ruddon, 1976). Large interindividual variations in constitutive AHH activity have also been demonstrated by Atlas *et al.*, 1977.

Our laboratory (Pero, 1976, Pero *et al.*, 1976) has also investigated ^{3}H-DMBA binding to DNA of resting human lymphocytes where constitutive AHH activity is extremely low but detectable (Whitlock *et al.*, 1972). Our results have shown that interindividual variation occurred according to sex, age, blood pressure and mortality rates in much the same way as was found previously in human drug metabolism (O'Malley *et al.*, 1971). In addition, we also achieved similar results with the proximate carcinogen, N-acetoxy-2-acetylaminofluorene (NA-AAF). These data suggest that high constitutive mixed function oxidase activity is associated with low and not high level DMBA-DNA binding values, since high mixed function oxidase would have removed more of the already "activated" NA-AAF, and thus would have provided fewer opportunities for NA-AAF induced DNA damage. This would be true since our results have shown that high DMBA binding to DNA was correlated to high NA-AAF induced DNA damage values.

It can be concluded from the results discussed above that both the constitutive and induced levels of mixed function oxidase and/or AHH activity can vary from individual to individual. Whether constitutive or induced levels of AHH activity are of primary importance in cancer induction probably depends on whether the environmental exposure of the carcinogen is sufficiently high to cause AHH induction. In our results (Pero, 1976, Pero *et al.*, 1976) where we compared a proximate (NA-AAF) and a non-proximate (DMBA) carcinogen under conditions where mixed function oxidase induction was minimal, then both types of carcinogens gave similar interindividual variations to DNA damage induction in resting lymphocytes. These results suggest that either:

a) constitutive mixed function oxidase activity is quantitatively different from induced mixed function oxidase activity, and thus carcinogen activation and degradation are equalized with the constitutive levels, but carcinogen activation is selectively amplified when induction takes place, or

b) mixed function oxidase induction does not result in proportionally more bound "metabolically activated" carcinogen to DNA; *i.e.*, because of accelerated enzymatic rates limiting the half-life of DNA bound activated carcinogen.

Our approach so far has been based on human lymphocytes' interindividual responsiveness to *in vitro* standardized doses of carcinogens. We have assumed an important role for the mutagenic event in cancer and other diseases, especially those with an immunological component. In other words, if an individual's lymphocytes have a high potential to accumulate DNA damage, then they would be more susceptible to mutation. To test the effectiveness of our model, we have compared individuals from medical risk groups with normal individuals. As we have stated previously, individuals with high blood pressure and older individuals have increased potentials for DNA damage, following an *in vitro* standardized exposure of carcinogen (Pero, 1976, Pero *et al.*, 1976). This was true when we estimated DNA damage by the three parameters discussed above in the same individuals at the same time; *i.e.*,

a) differential stimulation of unscheduled DNA synthesis (an estimate of the repair process) after 1 h induction of DNA damage by 10 μM NA-AFF,

b) the level of NA-AAF induced chromosome aberrations remaining after 8 h of DNA repair synthesis, and

c) the level of ^{3}H-DMBA bound to DNA after 18 h incubation of resting lymphocytes in 5 μM DMBA.

These results have encouraged us as to the effectiveness of our model because we have correlated increased DNA damage and the consequences thereof to well established risk conditions in humans according to sex and mortality rate differences.

Recently we have extended our studies to include not only the interrelationships between DNA repair, carcinogen-DNA binding and chromosomal aberrations in the regulation of overall DNA damage potentials in cultured lymphocytes, but also two other potentially important and possibly interrelated factors governing DNA damage induction; *i.e.*, sister chromatid exchange and carcinogen metabolism via AHH induction.

The recent introduction of techniques for detection of sister chromatid exchanges (SCE) has attracted much interest as a potentially more sensitive and rapid indicator of DNA damage than conventional chromosome breakage. This subject has been reviewed by Wolff, 1977. SCE involves a symmetrical exchange between sister chromatids which does not result in an alteration of overall chromosome morphology; thus previously undetectable damage to the genetic material might in this way be revealed. The biological significance of SCE is, however, so far mostly unknown. The demonstration of SCE depends on the usage of bromodeoxyuridine (BrdU), and it has recently been shown that this agent iself may give rise to SCE (Lambert *et al.*, 1976). There are also strong indications that SCE and conventional chromosome aberrations arise by partly different mechanisms and reflect different primary DNA damages. Furthermore, almost nothing is so far known concerning the relationship between SCE after chronic *in vivo* exposure of agents causing either conventional chromosome breakage or an increased frequency of SCE after application *in vitro*.

It is now generally accepted that AHH activity in resting lymphocytes is present but barely detectable by conventional assay procedures for AHH activity (Whitlock *et al.*, 1972, Atlas *et al.*, 1977, Burke *et al.*, 1977). The most commonly used AHH assay measures the formation of benzopyrene hydroxide, via proposed epoxide intermediate, by extracting benzopyrene exposed cell cultures with an organic solvent (usually hexane: acetone, 3:1) and then back extracting the organic phase with a basic solution (NaOH). The fluorescence emission of benzopyrene hydroxide can then be estimated fluorometrically in the NaOH solution following excitation. In order to measure AHH-activity quantitatively in lymphocytes, they must first be stimulated to divide by mitogens. However, most individuals exposed to environmental agents have non-dividing, resting lymphocytes. Therefore, we have tried to develop an AHH assay in resting lymphocytes, so that the data we collect will

more clearly parallel the natural situation in humans. This
assay is based on the binding of ^{3}H-carcinogen to the DNA
of resting lymphcytes, following a preincubation for 18 h
in either unlabelled benzopyrene or the solvent for benzopyrene. Induction of AHH is measured by dividing the level of
^{3}H-carcinogen bound to the DNA of benzopyrene pretreated
lymphocytes by the level of ^{3}H-carcinogen bound constitutive-
ly to the DNA of the solvent treated cells. In light of the
fact that we have already measured polycyclic hydrocarbon
binding in resting lymphocytes (Pero *et al.*, 1976), we hope
to increase the sensitivity of the AHH assay by using carcinogen-DNA binding measurements with and without benzopyrene pre-
treatment, so as to quantitate AHH in resting lymphocytes.
Our preliminary results suggest the feasability of this approach, and therefore should allow us the opportunity for
more accurate assessment of the influence AHH has on carcinogen metabolism in relation to overall DNA damage induction.

In summary, to provide a more solid basis for understanding the role of carcinogen metabolism and its relationship
to genetically important events such as induction of DNA damage, we plan to study carcinogen binding to DNA, AHH activity,
carcinogen induced DNA repair, chromosome aberrations, and
sister chromatid exchanges in the same individual at the same
time. At least, these five processes are in some way involved in the expression of DNA damage, and therefore signal potential mutagenic changes, expressed through interindividual
differences in sensitivity to genetic damage. We hope that
this approach will eventually establish the true relationship between these parameters.

ACKNOWLEDGEMENTS

This work was supported by grants from the Swedish Cancer
Society and the Unit for Community Care Sciences in Dalby,
Sweden.

REFERENCES

Atlas, S.A., Vesell, E.S. and Nebert, D.W. (1977). *Cancer Res.*
   36, 4619.
Benedict, W.F., Considine, N. and Nebert, D.W. (1973). *Mol.*
   Pharmacol. 9, 266.
Bittner, M.A. and Ruddon, R.W. (1976). *Mol. Pharmacol. 12*, 966.
Burke, M.D., Mayer, R.T. and Kouri, R.E. (1977). *Cancer Res.*
   37, 460.
Conney, A.H. (1967). *Pharmacol. Res. 19*, 317.

Fouts, J.R. (1970). *Toxicol. Appl. Pharmacol. 17*, 804.

Gelboin, H.V. (1967). *Advan. Cancer Res., 10*, 1.

Harnden, D.G. (1977). *In* "Genetics and Human Cancer" (J.J. Mulvihill, R.W. Miller and J.F. Fraumeni, Jr., eds.). p. 87. Raven Press, New York.

Harris, C.C., Autrup, H., Connor, R., Barrett, L.A., McDowell, E.M., and Trump, B.F. (1976). *Science 194*, 1067.

Hecht, F., and Kaiser McCaw, B. (1977). *In* "Genetics and Human Cancer" (J.J. Mulvihill, R.W. Miller, and J. F. Fraumeni, Jr., eds.) p. 87. Raven Press, New York.

Heidelberger, C. (1975). *Ann. Rev. Biochem. 44*, 79.

Kellermann, G., Luyten-Kellermann, M. and Shaw, C.R. (1973a). *Amer. J. Hum. Genet. 25*, 327.

Kellermann, G., Shaw, C.R. and Luyten-Kellermann, M. (1973b). *N. Engl. J. Med. 289*, 934.

Kellermann, G., Luyten-Kellermann, M., Horning, M.G. and Stafford, M. (1976). *Clin. Pharmacol. Therapeut. 20*, 72.

Lambert, B., Hansson, K., Lindsten, J., Sten, M. and Werelius, B. (1976). *Hereditas 83*, 163.

Mitelman, F. (1974). *In* "Chromosomes and Cancer" (J. German, ed.). p. 675. Wiley & sons, New York.

Mitelman, F. and Levan, G. (1976). *Hereditas 82*, 167.

Mitelman, F., Mark, J., Levan, G. and Levan, A. (1972). *Science 176*, 1340.

Nebert, D.W., Benedict, W.F. and Gielen, J.E. (1972). *Mol. Pharmacol. 8*, 374.

O'Malley, K., Crooks, J., Duke, E. and Stevenson, I.H. (1971). *Br. Med. J. 3*, 607.

Pero, R.W. (1976). *Hereditas 84*, 130.

Pero, R.W., Bryngelsson, C., Mitelman, F., Thulin, T. and Nordén, Å. (1976). *Proc. Natl. Acad. Sci. (USA) 73*, 2496.

Sugiyama, T. (1971). *J. Natl. Cancer Inst. 47*, 1267.

Vesell, E.S., Passananti, G.T., Greene, F.E. and Page, J.G. (1971). *Ann. N.Y. Acad. Sci. 179*, 752.

Whitlock, J.P., Cooper, H.L. and Gelboin, H.V. (1972). *Science 177*, 618.

Wolff, S. (1977). *This volume.*

EVALUATION OF DNA REPAIR AS A TEST FOR MUTAGENIC AND CARCINO-
GENIC ACTION OF CHEMICALS

Bo Lambert [1]
Elaine Harper

Department of Clinical Genetics
Karolinska Sjukhuset
Stockholm
Sweden

INTRODUCTION

Convincing evidence suggests that DNA repair in general
constitutes important biological defence mechanisms which help
the cells to overcome genetic damage induced by environmental
agents. The idea that DNA repair may be used as a test for the
evaluation of potentitally mutagenic and carcinogenic agents
is mainly derived from the growing concept that many of these
agents exert their effects by interacting with DNA. Many
different types of DNA damage are recognized by DNA repair
mechanisms. The demonstration of DNA repair as a result of
treating cells with a specific agent should therefore indicate
the property of this agent to induce DNA lesions. Some of these
lesions may be correctly repaired and cause no harmful effect.
Mutation and neoplasm may originate if the cellular repair
mechanisms are insufficient or defective, or if they are not
able to restore the original DNA molecule without causing
errors in the DNA sequence. Thus, DNA repair tests do not
measure mutation or cell transformation *per se*, but rather the
property of an agent to induce repairable DNA damage, the outcome of which may be mutation, malignancy or error-free repair
depending on a number of conditions.

[1]*This work was supported by grants to B.L. from the Swedish
Work Environmental Fund and Karolinska Institutet*

In this paper we will discuss some of these conditions
which may be of importance to the evaluation of DNA repair as
a test for the mutagenic and carcinogenic action of chemicals.
Data on which this paper is based have recently been presented
in several exhaustive reviews (Cleaver; 1974, Cleaver *et al.*,
1976; Howard-Flanders, 1973; Lehmann, 1975; Strauss, 1975) and
proceedings (Beers *et al.*, 1972; Hanawalt and Setlow, 1975).

The Relationship between DNA Repair and Mutagenesis and Carcinogenesis

The general assumption of a relationship between DNA
repair and mutagenesis and carcinogenesis is based on two
sets of data:

I. Increased mutability of certain repair-deficient
   bacterial strains and increased incidence of cancer in
   humans suffering from certain autosomal recessive disorders associated with defective DNA repair (xeroderma
   pigmentosum and ataxia telangiectasia).

These data support the idea that DNA repair in general
protects the organisms against mutation and cellular transformation which may be caused by chemically and physically
induced DNA damage.

II. The stimulation of DNA repair by treatment of cells
   in vitro, or animals and humans *in vivo* with agents
   which are known from other test systems to be mutagenic and/or carcinogenic.

This set of data indicates that a number of mutagenic and
carcinogenic agents are able to induce DNA lesions which
initiate DNA repair mechanisms. It is conceivable, however,
that some types of DNA damage are not recognized by DNA repair enzymes. This implies the possibility of a qualitative
difference between the ability to stimulate DNA repair and
mutagenic and/or carcinogenic potency. In fact some experimental evidence suggests that damage which is less effectively
repaired could be the main cause of cell transformation
(Kleihues and Margison, 1976; Rajewsky and Goth, 1976). The
possibility that chemicals may exert their mutagenic and/or
carcinogenic effect by other means than inducing structural
DNA damage should also be taken into account.

Principles of DNA Repair Mechanisms

Many different types of damage to DNA are caused by physical and chemical agents. The types of damage may differ
between and within the two classes of agents, and may be repaired by different mechanisms, and also at different rates.

In principle, the chances of successful repair will very much depend on the actual time of damage in relation to DNA replication and whether the damage affects one or both DNA strands.

Damage which affects only one DNA strand at a particular region of the DNA molecule and which is introduced during the non-replicative stages of the cell cycle, may be repaired by a *pre-replication* or *excision repair* mechanism. The damaged bases are removed and the original sequence may be correctly restored by repair synthesis and ligation, using the complementary strand as a template. In theory, pre-replication repair of lesions involving one strand only has the qualifications to be error-free, and there is much evidence that this is normally the case.

Damage which affects both DNA strands at close sites could not be repaired by a mechanism similar to that of excision repair, because there would be no intact strand which could serve as a template for repair synthesis and ligation. A similar situation may arise if not all single strand damage is excised before the DNA undergoes replication. The lesion may be by-passed during replication, leaving gaps in the newly synthesized strand. Such gaps may be filled in later (Lehmann, 1975) by a process usually referred to as *post-replication repair*.

In bacteria there is evidence suggesting that this repair of double strand damage and single strand gaps may involve recombinational events, in which homologous DNA provides the correct template for an error-free repair. On the other hand certain bacterial strains which are unable to complete post-replication repair are also less mutable than the corresponding normal strains. This could mean that the normal ability to carry out post-replication repair would result in a higher frequency of mutation (see Howard-Flanders, 1973). This raises the question whether post-replication repair mechanisms in some situations may be error-prone, and lead to mutation.

It is not known whether mammalian cells possess mechanisms for recombinational repair. However, some evidence suggests that cross-links, as well as post-replicative single strand gaps, are acted upon by some type(s) of repair mechanism(s) which can restore the continuity of the DNA molecule (Strauss, 1976).

Comments on DNA Repair as a Test System

There are several different techniques which allow the study of pre-replicative DNA repair mechanisms in animal and human cells *in vitro*. These techniques have been reviewed elsewhere by others (Cleaver, 1975; Regan and Setlow, 1973) and will not be dealt with in detail here. No simple methods for the assay of post-replication repair, nor for the estimation of persistent, post-replicative DNA damage, have been described.

The need for adequate activation systems is a fundamental problem in all *in vitro* assays, and DNA repair is no exception in this regard. Attempts have been made to combine systems for the metabolic conversion of chemicals such as the measurement of unscheduled DNA systhesis in human fibroblasts, with metabolic activation by rat liver microsomes (Stich *et al.*, 1975). Other approaches make use of cells which are capable of both activation and DNA repair, *e.g.* isolated rat liver cells. As with other short term assays, the relevance of the results obtained from DNA repair assays in combination with metabolic activation *in vitro* will largely depend on our knowledge of metabolic mechanisms in the test systems used in relation to man. Without such knowledge results could be interpreted in a false positive or false negative way.

Other simple and rapid techniques take advantage of the possibility to measure chemically induced non-replicative DNA synthesis in human lymphocytes *in vitro* (in which DNA replication is inhibited by hydroxy-urea) (Lieberman *et al.*, 1971), or single strand breaks arising during the course of DNA repair (Ahnström and Erixon, 1973).

All these methods provide only an estimate of one or a few of the many enzymatic steps that are likely to be involved in the complete excision repair process, and are not adequate for interpreting the biological effectiveness of repair. This may be achieved, however, by host cell reactivation of impaired viral genomes (Takebe *et al.*, 1974). DNA damage is introduced into the virus genome by physical or chemical treatment, and repair is measured as the recovery of virus infectivity. Apart from indicating some aspects of functional repair, this technique has the obvious advantage that the types of DNA damage introduced (either repaired or non-repaired) can be more easily studied in the relatively simple virus genome.

The Predictive Value of DNA Repair as a Test for Mutagenesis and Carcinogenesis

The considerations presented in the previous sections lead to the general conclusion that pre-replicative excision repair, by correction of primary DNA damage in a template directed process, protects the cells from secondary events which may lead to mutation and cell transformation. Lack of repair functions for a specific type of DNA damage, or insufficient capacity for repair, would give rise to persistent DNA damage at the time of DNA replication, and therefore could increase

the risk of mutation due to errors in replication or possibly
error-prone post-replication repair. Therefore, mainly damage
which remains unrepaired or which is restricted to post-replication repair mechanisms would be of consequence, in terms of
mutagenesis and carcinogenesis. If this is so, the predictive
value of DNA repair tests could be questioned, because most
of these tests only measure the property of an agent to stimulate DNA repair, and not the amount of unrepaired damage nor
the frequency of errors of repair. However, when human tissues
are exposed *in vivo* there are a number of conditions which may
influence the efficiency of DNA repair mechanisms, the possibilities for successful repair, the amount of unrepaired
damage, and the frequency of errors in repair.

Variability in the Capacity for and Efficiency of Repair:
Some evidence indicates that the mode and capacity for repair
may differ between species (Hart and Setlow, 1974) and cell
types (Connor and Norman, 1971), as well as between cells in
various stages of differentiation (Frey-Wittstein *et al.*,
1969). The existing data do not allow conclusions to be formed
about the influences of these differences on the consequences
of DNA damage and its repair. It is obvious, however, that
false negative results could be obtained if the cells used in
DNA repair tests were incapable of recognizing or repairing
the type of damage induced by the treatment, or if the damage
was repaired by another mechanism than that assayed for.
There may also be age and sex differences as well as individual variability in the total capacity for DNA repair. Individuals extremely sensitive because of a defective DNA repair
system have already been identified. These are subjects with
the rare, autosomal disorders, xeroderma pigmentosum (see
Robbins *et al.*, 1974) and ataxia telangiectasia (Paterson *et
al.*, 1976). Suggestions of disturbances in DNA repair in a
number of other inherited disorders have been brought forward
(see Lambert and Ringborg, 1976). However, there are very few
studies   on   the variability in repair capacity between subjects in the normal population. Such studies should be encouraged, because it is obvious that subjects with reduced
repair capacity could constitute important risk groups.

*Competition between Different DNA Damaging Agents for the
Same Repair Mechanisms:* Different repair processes may have
certain enzyme steps in common. Xeroderma pigmentosum fibroblasts of the same complementation group display reduced repair, not only to UV-induced DNA lesions, but also to damage
induced by a variety of chemicals (see Cleaver, 1974). In
normal cells these agents evoke excision repair much like UV-

irradiation, but clearly this is mainly due to other types of
DNA damage than thymine dimers. It is therefore likely that
the enzyme which is defective or lacking in xeroderma pigmentosum cells would be involved in the recognition and repair of
a variety of DNA lesions in normal cells. It is also conceivable that other repair steps, *e.g.* repair replication involving polymerase(s) and sealing of single strand interruption
involving ligase(s), are common to the repair of various types
of DNA lesions. Thus various DNA lesions may interact in a
competetive manner with regard to certain repair functions.

Repair functions between different types of DNA damage may
well be of importance *in vivo*, *e.g.* during working conditions
involving heterogeneous exposure. There is a need for more
studies of DNA repair after simultaneous exposure of cells to
different agents. Besides giving important information on
additive, competitive, and inhibiting interactions of chemicals on DNA repair, such studies could provide information for
the classification of different repair mechanisms and help to
find the rate limiting steps in the various repair processes.

Dependence of Repair on Cell Cycle Stage: Induction of DNA
damage in proliferating cells is likely to delay DNA replication or cause reproductive death. The former may lead to post-
replication repair and possibly mutagenic errors of repair.
The latter may bring more cells into proliferation in compen-
sation for the cell loss, and this may result in an even
greater fraction of cells that must undergo post-replication
repair. Thus the amount of persisting DNA damage and possibly
error-prone repair is likely to depend on the number of proliferating cells during the period of exposure. It is possible
that tissues with a high rate of cell turnover will be more
susceptible to the effects of chemical exposure than non-
dividing tissues, due to insufficient time for repair and a
higher frequency of persistent DNA damage, which by secondary
events may lead to mutation and cell transformation. These
circumstances are seldom encountered in DNA repair tests *in
vitro*.

Influence of Dose and Duration of Exposure: It seems likely that the amount of persistent DNA damage at the time of DNA
replication will depend not only on the rate of repair and the
total capacity for repair, but also on the total dose and
duration of exposure. Furthermore, prolonged and repeated exposure to a certain DNA damaging agent may bring more repair
enzymes into action and thereby increase the repair capacity,
i.e. DNA repair may be inducible. Accordingly, short exposure
to high doses may have more serious effects than long exposure

to low doses.

Inhibition of DNA Repair: Clearly, inhibitors of DNA repair could have serious effects in terms of cell survival, mutagenesis, and carcinogenesis if they were allowed to interact with cells simultaneously with DNA damaging agents. There are several drugs which appear to have a general inhibitory effect on both DNA replication and DNA repair in mammalian cells (Cleaver, 1969), but so far no specific inhibitors of pre-replication repair have been described. However, caffeine inhibits post-replication repair in some rodent cell lines and the variant form of xeroderma pigmentosum (Lehmann *et al.*, 1975), and its possible effects in normal human cells should be given further attention.

CONCLUDING REMARKS

The consequences of DNA damage and DNA repair in terms of mutagenesis and carcinogenesis for human tissues exposed *in vivo* are likely to depend on a number of conditions, *e.g.* the exposed cell types, cell cycle stage, individual sensitivity, other environmental influences, the total dose and duration of exposure, etc. In theory, DNA repair should impose a threshold in the dose-effect relationship for a given agent, *i.e.* provide an absolute defense mechanism, only if all DNA damage induced by that agent was repaired by an error-free process. Obviously, this is not likely to be the situation in most human tissues exposed *in vivo*. Therefore the property of an agent to stimulate DNA repair in a test system should be taken to indicate its ability to induce DNA damage, which in the *in vivo* situation may lead to mutation and/or cell transformation. However, before more far-reaching interpretations in terms of human mutagenesis and carcinogenesis can be made from results of DNA repair assays, further attention should be given to sources of variability and false responses due to *e.g.* differences in repair and metabolism between cell types and individuals, interaction and competition between chemicals, the efficiency and consequence of DNA repair in acute *versus* chronic exposure, and the possibility of induction and inhibition of DNA repair mechanisms.

There are several possibilities by which DNA repair tests could produce false negative results with regard to chemical mutability and/or carcinogenicity. Therefore, judgements of the mutagenic and carcinogenic potency of chemicals should not rely exclusively on DNA repair tests. Nevertheless, these tests still offer a valuable method of monitoring the action of chemicals in a relatively simple system at a cellular level. They provide a rapid technique of high sensitivity involving a low number of cells. DNA repair assays may also be useful in the evaluation of DNA damage and genotoxic effects caused by chemicals *in vivo*, where different tissues as well as organisms may respond differently. Although the exact nature or consequence of the damage incurred by the DNA cannot be assessed by these methods at present, the results at least give a measurement of the basic interaction of the damaging agent with the DNA.

SUMMARY

Many chemicals induce various types of genetic damage which may give rise to mutation or cell transformation. Mammalian cells possess DNA repair mechanisms which serve to correct such damage. Considerable evidence suggests that DNA repair may protect organisms from mutagenesis and carcinogenesis by correction of primary DNA damage in a template directed pre-replication repair process. Damage which remains unrepaired or which is restricted to the possibly more error-prone post-replication repair mechanism is most likely to be of consequence in terms of mutagenesis and carcinogenesis. Successful repair after *in vivo* exposure to DNA damaging agents will depend on a number of factors such as total dosage and duration of exposure, the capacity for and efficiency of repair, cell type and cell cycle stage, other environmental influences, and individual sensitivity. It is not likely that all these and other variables will combine in such a way that DNA repair could provide an absolute defense against mutation and cell transformation in human tissues exposed *in vivo*. Therefore the measurement of DNA repair as a result of chemical treatment of cells *in vitro* demonstrates the induction of potentially mutagenic or cancerogenic DNA damage. However, before more far-reaching interpretations in terms of human mutagenesis and carcinogenesis can be made from results of DNA repair assays, further attention should be given to sources of variability due to factors such as those mentioned above.

REFERENCES

Ahnström, G., and Erixon, K. (1973). *Int. J. Radiation Biol.* *23*, 285.
Beers, Jr., R.F., Merriott, R.M., and Tilghman, R.C., eds. (1972). "Molecular and Cellular Repair Processes", |Johns

Hopkins University Press, Baltimore.

Cleaver, J.E. (1969). *Radiation Res. 37*, 334.

Cleaver, J.E. (1974). *In* "Advances in Radiation Biology", Vol. 4 (J.T. Lett, M. Adler and A. Zelle, eds.), p. 1. Academic Press, New York.

Cleaver, J.E. (1975). *In* "Methods in Cancer Research", Vol. 11 (M. Busch, ed.), p. 123. Academic Press, New York.

Cleaver, J.E., Goth, R., and Freidberg, E.C. (1976). *In* "Screening Tests in Chemical Carcinogenesis" (R. Montesano, M. Bartsch, L. Tomatis, and W. Davis, eds.), p. 639. International Agency for Research on Cancer, Lyon.

Connor, W.G., and Norman, A. (1971). *Mutation Res. 13*, 393.

Frey-Wettstein, M., Longmire, R., and Craddock, C.G. (1969). *J. Lab. Clin. Med. 74*, 109.

Hanawalt, P.C., and Setlow, R.B., eds. (1975). "Basic Life Sciences", 5A&B, Plenum Press, New York.

Hart, R.W., and Setlow, R.B. (1974). *Proc. Natl. Acad. Sci. (USA) 71*, 2169.

Howard-Flanders, P. (1973). *Brit. Med. Bull. 29*, 226.

Kleihues, P., and Margison, G.P. (1976). *Nature. 259*, 153.

Lambert, B., and Ringborg, U. (1976). *Acta Med. Scand. 200*, 433.

Lehmann, A.R. (1975). *Life Sciences. 15*, 2005.

Lehmann, A.R., Krik-Bell, S., Arlett, C.F., Paterson, M.C., Lohman, P.H.M., de Weard Kastelein, E.A., and Bootsma, D. (1975). *Proc. Natl. Acad. Sci. (USA). 72*, 219.

Liebermann, M.W., Sell, S., and Farber, E. (1971). *Cancer Res. 31*, 1307.

Paterson, M.C., Smith, B.P., Lohman, P.H.M., Anderson, A.K., and Fishman, L. (1976). *Nature. 260*, 444.

Rajewsky, M.F., and Goth, R. (1976). *In* "Screening Tests in Chemical Carcinogenesis" (R. Montesano, M. Bartsch, L. Tomatis, and W. Davis, eds.), p. 593. International Agency for Research on Cancer, Lyon.

Regan, J.D., and Setlow, R.B. (1973). *In* "Chemical Mutagens. Principles and Methods for their Detection", Vol. 3 (A. Hollaender, ed.), p. 151. Plenum Press, New York.

Robbins, J.H., Kraemer, K.H., Lutzner, M.A., Festoff, B.W., and Coon, H.G. (1974). *Ann. Int. Med. 80*, 221.

Stich, H.F., Kieser, D., Laishes, B.A., San, R.H.C., and Warren, P. (1975). *GANN Monograph on Cancer Research. 17*, 3.

Strauss, B.S. (1975). *Life Sciences. 15*, 1685.

Strauss, B.S. (1976). *In* "Aging, Carcinogenesis and Radiation Biology" (K.C. Smith, ed.), p. 287. Plenum Press, New York.

Takebe, H., Nii, S., Ishii, M.I., and Utsumi, H. (1974). *Mutation Res. 25*, 383.

NONHUMAN TEST SYSTEMS

SHORT TERM TESTS FOR DETECTING CHEMICAL MUTAGENS
AND RECOMMENDATIONS FOR THEIR USE IN MONITORING
EXPOSED POPULATIONS

Diana Anderson

Central Toxicology Laboratory
Imperial Chemical Industries Ltd.
Cheshire
England

INTRODUCTION

The problem of the possible induction of genetic damage
after chemical exposure is worrying both to the scientific
community and to the population at large. The increase in
number of different chemical substances and biological synthesis products in man's environment is due to the rapid strides
made by scientific and technical progress. Each year new
chemical substances are introduced in the form of medical preparations, pesticides, food additives and industrial compounds and some of these could induce mutations. There is a
risk that such mutations may accumulate in the population and
affect future generations. In recent years many substances
shown to be mutagens have also been shown to be carcinogenic,
with 80-90% correlations between mutagenicity and carcinogenicity with certain organic chemicals (Ames *et al.*, 1973a and b,
Ames *et al.*, 1975, McCann *et al.*, 1975, 1976, Coombs *et al.*,
1976, Purchase *et al.*, 1976b). These data are in conflict with
similar studies made in the early 1950's on the mutagenicity
of chemical carcinogens which showed little correlation between carcinogenic and mutagenic activity. These early experiments led to the premature demise of the somatic mutation
theory of cancer; metabolic activation systems were not used
and it was not realised that certain strains of bacteria might
detect only specific types of genetic alterations (de Serres,
1975). Nowadays with the newfound interest in the somatic

mutation theory of cancer, a study of the genetic effect of chemical substances may not only be useful for protecting future generations but also for preventing cancer in present generations.

Little is known of the practical use of results obtained from mutagenic studies and how these results might apply in the field of toxicology in general or how relevant they are to man. Whilst cancer in man has a definitive end-point in the production of a malignant tumour and is very emotive, the end-point of genetic damage in man is more obscure. With current techniques we are able to distinguish between molecular changes in the gene and small chromosome aberrations. Most of human traits are inherited as Mendelian units which can be divided into autosomal dominant, autosomal recessive and X-linked recessive units. There are also X-linked dominants and there may be a few Y-linked traits, but these are numerically insignificant (Crow, 1973). McKusick (1971) lists such traits. The dominant expression of a mutant gene is sufficient to cause a recognisable abnormality or disease and there are also many well established recessive traits. However, whilst a change in a dominant trait is immediately evident in the next generation, recessive traits may take very many generations before they are expressed. Therefore, the effect of chemical mutagenesis at the level of the gene may not show immediate results, and data generated by mutagenic tests lose their impact. Chromosome abnormalities may arise by errors in the distribution of chromosomes leading to abnormalities of chromosome numbers (as in non-disjunction), where the effect is seen in the next generation, or by the consequence of chromosome breakage. Consequences of chromosome damage are physical or mental abnormalities, sterility or embryonic death. With regard to the last condition epidemiological evidence suggests that there is an increased incidence in foetal wastage in wives of workers exposed to vinyl chloride (Infante *et al.*, 1976a, b and c). However, there is a conflict concerning the statistical interpretation of the data (Paddle, 1976), and as yet there is thus no unequivocal epidemiological evidence that vinyl chloride can cause germ cell mutations in man. Married women anaesthetists were claimed to have a higher incidence of spontaneous abortion when they worked than when they did not work and the incidence was also higher by comparison with non-anaesthetist, married women doctors (Knill-Jones *et al.*, 1972). However, it is difficult to determine if the abortions were due to genetic factors or embryotoxicity because of systemic involvement after exposure of the females. Many constitutional and degenerative diseases such as epilepsy and schizophrenia may be caused by other irregularities of gene expression or arise from multiple genes (Crow, 1973). The effect of chemical mutagenesis both at the level of the chromosome and on multiple gene effects may also not be immediately obvious. However, since the world population is rapidly increasing, the chance for natural selection in civilised communities is being sharply reduced; whilst it is recognised that not all mutations are harmful and some are necessary for evolution, any increase in mutagenic factors in the environment could cause a potential risk to the population. It is estimated that more than 4% of the population is affected by genetic anomalies and about 1% of children born each year have chromosomal mutations (Bochkov *et al.*, 1975). Mutations that cause sterility or early embryonic loss are detrimental in the Darwinian sense but have little impact on society. Mutations that are more fit biologically may be a heavy burden to society if the affected persons require medical or institutional care (Sutton 1975).

Thus it is necessary to monitor potential chemical mutagens which could increase the percentage of genetic aberrations in the population, and although a great number of relatively simple and practical methods are available the evaluation of mutagenic effects is a complex and extremely difficult task.

FACTORS DETERMINING A MUTAGENIC RESPONSE

The mutagenic activity of chemical substances has been studied in a number of organisms using different methods for calculating both gene and chromosome mutations in somatic and germ cells. Systems from micro-organisms, plants, insects, mammalian and human cells *in vivo* and *in vitro* have been used, and whilst the mutagenic effect of a chemical may be detectable in one system, it may not be in another or even in different organs of the same system. There may be many interactions between a chemical and an organism, all of which may determine whether genetic damage is expressed, unlike ionising radiation which is immediately active in terms of its mutagenic potential. With radiation, extremely short-lived free radicals are generated randomly in cells and tissues very close to or within the genetic target molecules. There are many factors that determine whether a chemical compound reaches and reacts with the critical genetic targets. In mammals among these factors are chemical structure of the compound, duration of treatment, route of administration, absorption, distribution, excretion, drug-protein binding, metabolic transformation, pharmacogenetic make-up (species, strain, sex), membrane barriers, numbers of SH groups, etc. (Matter, 1976). The following examples will illustrate such factors:

Vinyl chloride, a known carcinogen (Maltoni, 1974), requires metabolic activation to exert its effect (Green and

Hathway, 1975). Vinyl chloride exposure causes mutations in
micro-organisms (Bartsch et al., 1975) and in some test systems involving higher organisms. It causes chromosome damage in
human peripheral lymphocytes (Purchase et al., 1976a, and
references therein) but no effect is seen in the dominant
lethal assay in mice (Anderson et al., 1976). This suggests
that various tissues metabolise the substance to differing
extents and perhaps the testes metabolise the substance least.
This is certainly the case with dimethylnitrosamine, another
indirect mutagen, where there is less alkylation in the testes
than in other organs (Swann and Magee, 1968). This compound
also gives a negative result in the dominant lethal assay
(Propping et al., 1972) and in a host-mediated assay with
yeast cells; mutagenic activity is high in the liver, moderate
in the lungs, but only very weak in the testes (Fahrig, 1974,
1975). Other compounds such as methylnitrosoguanidine, a direct
mutagen, is metabolised rapidly in mammals to non-mutagenic
metabolites (Matter, 1976), and 4-nitroquinoline-I-oxide has
high mutagenic activity in yeast cells recovered from the
lungs, slight activity in the liver, and it is not active in
the testes (Fahrig, 1975). Alternatively, such compounds may
be detoxified by organs where little mutagenic activity is
detected.

Other factors determine whether the damage in the target
molecule is expressed as genetic damage. Such factors are the
innate susceptibility of the cell (e.g.: repair capacity),
numbers of susceptible cells, type of genetic target (DNA,
chromosomal proteins) and mutation produced, significance and
importance of the genetic target locus, germinal and somatic
selection processes involved, mode of inheritance etc. (Matter,
1976). It has been found with ethyl methanesulphonate (EMS)
for example, a compound which can alkylate all tissues directly (Jones, 1973), that at below 100 mg/kg doses a dominant
lethal effect is not observed (Anderson unpublished, Generoso
et al., 1974) but at higher doses an effect is clearly manifest
in CD-1 mice at 150 mg/kg body weight (Ray and Hyneck, 1973) as
confirmed by Anderson et al. (1976a and b, 1977). Sperm proteins and sperm DNA are alkylated by EMS at doses as low as
3 mg/kg body weight (Sega et al., 1974). Such discrepancies may
occur because the numbers of animals screened in a dominant
lethal test may be statistically too small to detect a weak
effect at low doses, or unknown factors may affect the expression of damage in germ-cell development which will prevent the
production of mutations (Matter, 1976).

The factors described above which influence a mutagenic
response complicate mutagenicity testing. Many compounds which
would require testing will not behave in the same way as model
mutagens which bind covalently with biologically important

molecules. Many will react reversibly with receptors by mechanisms such as hydrogen bonding and electrostatic binding.
Such compounds may not react with the genetic material unless
they become metabolised to highly reactive forms. Other factors such as temperature, ageing, pH, hypoxia and viral infections can also produce genetic damage (Matter, 1976). Chemicals
may, therefore, disturb normal bodily functions producing a
change in one of these parameters to produce a genetic effect
indirectly.

TEST METHODS AND RECOMMENDATIONS FOR THEIR USE

Geneticists have a large number of experimental methods at
their disposal to assess the mutagenic action of chemicals. No
single method, however, gives any conclusive information about
the genetic risk to a person who has been exposed to a mutagenic substance. Thus it would seem necessary for a number of
test systems to be used. Bochkov *et al.* (1975) have listed the
effects, and have given a useful précis of the results of
different types of chemicals such as cytostatics, antibiotics,
pharmaceutical preparations, food additives and admixtures,
chemosterilants, fungicides, insecticides, herbicides and industrial chemicals in the different test systems of micro-
organisms, plants, fruit fly (*Drosophila*), mammalian cell
cultures, bone marrow and dominant lethal systems of mammals
and human cells in culture and human lymphocytes.
Various recommendations for combinations of test systems
have been put forward, e.g.: Legator and Malling (1969), Vogel
and Röhrborn (1970), Advisory Committee on protocols for safety evaluation (1970), WHO (1971, 1974 and 1975), Ambio special
report No. 3 (1973), Canadian Ministry of Health (1973),
Bridges (1973 and 1974), Flamm (1974), Committee 17 report
(1975), Bochkov *et al.* (1975). All approaches have their advantages and disadvantages. In the more recent reports
Bridges (1973) proposed that assessment at each stage of testing should take into account the population exposed to the substance, its economic value, and the possibility of substituting
a non-mutagenic grouping for the mutagenic grouping in the
structure of a substance. However, non-mutagenic substitutions
are not always possible without altering the desired biological
activity of a product. The testing scheme proposed by Flamm
(1974) includes tests for the detection of heritable translocations and specific locus mutations. Both of these tests
require large numbers of animals, are not always economically
feasible, and do not necessarily generate sufficient information to assess the genetic risk to man.
Various governmental agencies, such as those of America,

Japan, Britain, and other EEC countries, are now preparing
guidelines for testing methods. Some of the more recent ideas
indicate the need for flexibility. This should permit changing
to new and improved methods as soon as their usefulness has
been substantiated. However, there is anxiety that any significant deviation from generally accepted guidelines may be
questioned by regulatory agencies and that the guidelines
which will satisfy the most demanding health authority may
become the generally accepted procedure (Zbinden, 1973).

With all the recommendations in the communications already
in the literature (listed above), those put forward in this
present communication will be *brief* and it must be remembered
that testing organisations if allowed flexibility will have
their own ideas regarding scientific protocols, etc.

The problem with the present testing methods is that not
all systems are equally reliable and reproducible. Even the
systems in which most compounds have been tested, such as the
Salmonella typhimurium plate incorporation mutagenicity assay
with metabolic activation (McCann *et al.*, 1975) gives different results for some compounds in different laboratories
(Purchase *et al.*, in press).

Ideally, a system for assessing chemicals for mutagenic
activity should:

a) detect the potential of the test substance to induce
gene and chromosome mutations in somatic and germ cells;

b) provide quantitative data for extrapolation to man;

c) be capable of detecting metabolic products of the compound which have mutagenic potential;

d) be reliable and reproducible;

e) be economically viable and quick;

No one test system at present meets all these demands.

References will be given for the following test systems
only where systems are not generally discussed later.

Gene mutations may currently be detected by:

a) micro-organisms with metabolic activation *in vitro* and
in vivo;

b) *Drosophila melanogaster* tests;

c) cultures of somatic mammalian and human cells;

d) specific locus tests in mammals (Russel, 1951, Searle,
1975).

Cytogenetic effects in somatic cells can be detected by:

a) the micronucleus test;

b) conventional metaphase analysis or sister chromatid
exchange (Perry and Evans, 1975) analysis of bone marrow cells
from mammals and mammalian and human lymphocyte cultures.

For detecting chromosomal damage in spermatogenesis and
oogenesis, the dominant lethal or heritable translocation test
(Generoso *et al.*, 1974, Léonard, 1975) or the X-chromosome
loss in mice test (Russel, 1976) can be used. *Drosophila* can

also be used for detecting chromosomal damage and since *Dro-sophila* can also detect gene mutation it is a useful "catch-all" system (Sobels, 1974, Vogel and Sobels, 1976), but extra-polation to man is difficult from such an organism.

In addition to methods which detect gene mutations and chromosome aberrations, the potential of a chemical to induce primary DNA damage can be detected by measuring stimulation or inhibition of repair (Stich *et al.*, 1970, 1973, and Stoltz, 1974) or recombinational or gene conversion events (Zimmerman, 1971, 1975).

It is not feasible for all chemicals to be tested by all methods. Therefore, a system of priorities should be arranged. Not all substances are used by man in the same way. There are some substances to which man is chronically exposed and other substances to which man is only subjected by acute exposure under exceptional circumstances. Thus the genetic risk of a substance will depend on its mutagenic potency, where it is known, the extent to which people come into contact with it, and on the individual susceptibility of a person. We have to accept the fact that there is variation in individual suscep-tibility but we can differentiate our test system according to levels of exposure of the population to a chemical. Expo-sure will depend on a combination of two parameters:
  a) the number of people exposed;
  b) the dosage to which the people are exposed.

If the product of the parameters is low then the chemical is a low exposure chemical, if the product is high then the chemical is a high exposure chemical.

Low exposure chemicals such as very low tonnage industrial chemicals including intermediates, non-ingested substances and substances known not to accumulate in the environment or body may be subjected to a simpler screening programme which should, however, cover the induction of both gene mutations and chro-mosome aberrations, e.g. a test on micro-organisms with meta-bolic activation *in vitro* and a cytogenetic examination of bone marrow cells.

High exposure chemicals such as high tonnage industrial chemicals, pesticides, widely used medicines, food additives, ingested products and substances known to accumulate should be subjected to more rigorous testing with at least the two tests mentioned above and a relevant whole mammal test in germ cells. Other tests may be added to this group if the chemical warrants, such as cytogenetic analysis of human lymphocytes of exposed workers, or any other tests which may be considered relevant.

Substances found positive in the first screening, depend-ing on their economic or medical importance, could be subject-ed to more rigorous testing. According to Matter (1976), how-ever, very few compounds (0.1%) originally produced in pharma-

ceutical research ever have a chance of being marketed. He
suggests that extensive early mutagenicity testing of drugs is
of little help since benefit—risk decisions cannot be made at
that stage and that such testing should be concentrated on
those compounds scheduled for clinical trials or commercial
introduction. However, it would seem better to avoid mutagenic
products at an early stage where possible.

Before a substance is tested, consideration should be giv-
en to its chemical structure, to determine if it is related to
compounds which already have known mutagenic, carcinogenic,
teratogenic or general toxic effects. Such a consideration
might give an indication of its mutagenic potential and this
may be useful in setting testing priorities. If it is suspected of having high mutagenic potency, then this factor may
override the other two parameters which determine exposure.

BRIEF COMMENTS ON RECOMMENDED SCHEMES FOR SOME OF THE METHODS

A great deal has already been published about recommended
protocols (e.g. Bochkov *et al.*, 1975, Report of the *Ad Hoc*
Committee for Determining Chromosome Methodologies in Mutation
Testing, 1972). However, some general rules apply for all
testing systems. All assays should be run with concurrent
negative and positive controls. Positive controls should,
where possible, be structurally or mechanistically related to
the compound under test.

Microbial Methods

Much work in different laboratories has been carried out
on the bacterial *Salmonella typhimurium* mutagenicity system
with rat liver microsomal activation, and as such it is well
validated both for the detection of some mammalian and human
carcinogens and bacterial mutagens (however, it is not well
validated for human germ cell mutagens since none are unequivocally known). This does not mean that other microbial systems would not be equally valid for detecting mutagens/carcino-
gens if as much work had been undertaken with them. The
Salmonella typhimurium strains of Ames (Ames *et al.*, 1975) are
able to detect base pair substitution and frameshift gene mutations. A drawback to the system is that it is not quantitative and the potency of mutagens cannot be ranked. However, the
system is useful for a qualitative answer. The system of
Salmonella typhimurium and other systems such as that of the
bacterium *Escherichia coli* (Bridges, 1972) and yeasts (Parry,
1972, and Zimmerman, 1971, 1975) can provide a quantitative

answer if used in liquid cultures in combination with viability studies. Yeasts, however, are used to measure not only gene mutation but more commonly gene conversion and mitotic recombinational events. A very sensitive test for measuring gene mutation is claimed when *E. coli* is used in a fluctuation assay (Green *et al.*, 1976).

Generally in a screening programme compounds should be tested both with and without microsomal activation. Strains to detect both basepair and frameshift mutations should be used as well as at least five concentrations of the test compound, over a wide dose range in order to maximise the chance of obtaining a response, with the highest dose if possible inhibiting the growth of the microbes. This ensures that the compound has entered the microbes. Replicate experiments should be undertaken to determine a reproducible response. Simpler "spot" tests measuring the degree of differential killing in repair deficient microbial strains by comparison with wild type strains (Slater *et al.*, 1971, Kada *et al.*, 1974) provide an indirect answer to the problem of mutation testing since they detect damage to DNA but do not measure gene mutation directly. Such tests at the present time are not considered particularly sensitive, however,

Mammalian Cell Systems
Cell "Transformation"/Mutation Assay

In this laboratory, concurrent with the Ames test, the "cell transformation" assay system of Styles is used (Purchase *et al.*, 1976, and Styles, 1977). It is based on the ability of mammalian cells transformed by a carcinogen to form colonies in soft agar, and this is only one of the accepted criteria for cell transformation. A recent paper of Huberman *et al.* (1976) suggests that the ratio between transformation and mutagenesis for ouabain resistance in normal diploid cells is about 20:1 (after treatment with benzo(a)pyrene and its 7.8 dihydrodiol) which suggests that any one of 20 genes may be involved in transformation as opposed to 1 for mutation. This ratio is also apparently substantiated for normal hamster embryo cells treated with benzo(a)pyrene. If transformation is, therefore, a genetically based system, it may be useful for preliminary genetic monitoring. We have found it to be equally as predictive for carcinogenicity as the Ames test and by inference, therefore, it is equally predictive for mutagenicity. Baby hamster kidney (BHK 21/C13) and either human diploid lung fibroblasts (WI 38) or human liver cells (Chang) were treated as described (Purchase *et al.*, 1977). Five doses of compound were used with the S-9 Mix of the Ames test. Survival was assessed independently and a transformation

frequency calculated. The test is not as rapid as the Ames
test, but in our "blind" study of 120 compounds with these two
tests in combination, the tests only missed detecting as carci-
nogens/mutagens the two compounds diethylstilboestrol and
vinyl chloride. This latter compound was later detected when
tested in the gaseous phase. This, therefore, seems a very
promising system for a preliminary mutation screen.

The limitations of using these two test systems for a
simple screening programme have been discussed elsewhere
(Purchase *et al.*, 1976b).

Lymphoma Cells in Culture

Whilst human cells or normal diploid cells are obviously
desirable for mutation assays in culture, they are more difficult to handle from a screening viewpoint than malignant cells
in culture, due to their low plating efficiencies and lack of
perpetual proliferation. Lymphoma cells which grow in suspension are even easier to handle than cells which grow as a
monolayer. They do not require trypsinisation, are very easily
subcultured and they are not subject to metabolic co-operation
which can cause a loss of mutants. Also, they can be used in a
host—mediated type assay (Fischer *et al.*, 1974) or in a
fluctuation test (Cole *et al.*, 1976).

Lymphoma cells in culture can readily detect direct acting
mutagens and carcinogens (Anderson and Fox, 1974, Anderson,
1975a and b, Clive *et al.*, 1972, Clive, 1973, Clive and
Spector, 1975, Cole and Arlett, 1976, Fox and Anderson, 1974,
Knaap and Simons, 1975) and may also be used in combination
with S-9 Mix to detect indirect acting mutagens and carcinogens (Anderson unpublished). The lymphoma cell mutation system
is manageable in that induced mutation frequencies are readily
detectable and vast numbers of plates do not have to be used
to detect a spontaneous frequency (e.g. 10^5 cells per petri
dish in soft (0.3%) agar and selective medium give a background frequency of about 2-20 colonies in P388F cells with
selective media containing 5-iodo-2-deoxyuridine and excess
thymidine respectively). Induced frequencies can increase
from 10-100 fold above these levels. Absolute increases in
induced colony numbers are observed over some of the dose
range above control values. This is not always the case for
human cell systems where mutants may be very sensitive to the
inducing agent and are being killed more rapidly than nonmutant cells (Cox and Masson, 1974, Anderson and Fox, 1974).

Bone Marrow in Mammals

This is a useful system in that no auxiliary metabolising system is required, animals can be dosed with compounds directly and bone marrow cells do not need stimulation to divide since they have a high proliferative cell activity (Nichols *et al.*, 1972, and Schmid *et al.*, 1971, Schmid, 1973, Tijo and Whang, 1962). Rats are a suitable species to use but others will do. With rodents, groups of 5-8 animals of 8-10 weeks of age should be used per concentration of the test substance and positive control substance. A larger number of animals should be used in the negative control group to provide a better data base for comparison. A minimum of 50 cells, preferably 100, from each animal should be analysed from coded slides to avoid observer bias when scoring chromosome damage. The higher the background frequency, the higher the number of cells need to be examined in order to detect a statistically significant effect (Buffler, personal communication).

A maximum tolerated dose of compound should be used together with a more realistic dose related to human exposure and an intermediate value. In a preliminary screen only the maximum tolerated dose need be used (or even a micronucleus test may be used (Heddle, 1973, Schmid, 1976)).

Animals can be sacrificed at various time intervals after exposure. From our own studies and those of others (Dean, 1969, Schmid *et al.*, 1971, *Ad Hoc* Committee for Chromosome Methodologies in Mutagenicity Testing, 1972) 6 hours seems to be a suitable time after multiple chemical exposures and 24 hours after single exposures. However, these times may not be suitable for all chemicals.

The most relevant route of administration of the compound should be used. If man is exposed to the chemical by inhalation then an inhalation route should be used. If he is likely to ingest a pesticide sprayed on crops then an oral route is recommended either by gavage or in the diet. Rarely does man receive intraperitoneal or intraveneous injections of compound except in the case of some medicines.

Positive control animals should be housed under identical conditions to those dosed with the test compound even if identical routes of administration of test compound and positive control substances cannot be achieved.

Cells should be scored for all types of chromosome aberrations including chromosome gaps. We have shown with several positive control mutagens such as ethyl methanesulphonate, mitomycin C, benzene and vinyl chloride that gaps are at least as sensitive an indicator of damage as other types of damage. They may indicate a toxic event as opposed to or in addition to a genetic event (Anderson and Richardson, unpublished).

Dominant Lethal Mutation Assay

This is one of the whole mammal tests on germ cells (Bateman and Epstein, 1971). The tests should be carried out on random-bred rodents of 8-10 weeks of age. The comments which apply to dosing regimes in bone marrow cells also apply to dominant lethal tests, i.e. a maximum tolerated dose should be used together with a dose related to human exposure levels and an intermediate value. Again the most relevant route of administration should be used and animals treated with the positive control substance should be housed under identical conditions to animals exposed to the test compound.

Each test group should consist of at least 10-20 males if males are exposed (dominant lethal effects in females are more difficult to separate from systemic effects). Each male is bred with 2-4 females once a week for 8 weeks in the case of mice or 10 weeks in rats to sample all stages of the spermato-genic cycle. Negative control animal numbers should be larger, again to provide a better data base for comparison (the size of the negative control group can be determined by the square root of the number of treatment groups, i.e. if there are 9 treatment groups, each of 10 animals, then there should be 30 in the control group (Fieller, 1947)).

A pre-treatment fertility test should be undertaken to check fertility of the animals used and to determine a back-ground dominant lethal frequency.

The statistics used in the dominant lethal test are com-plicated. Various statistical methods can be used each with their own bias. The greater the number of statistical methods used the greater the chance of producing a false positive result (Anderson, unpublished).

Different evaluation methods are used for dead implant data. The US Food and Drug Administration has analysed all data on dead implants in two ways (both on dead implants/total implants/female basis and on dead implants/female) and infer that equally significant data are provided by both (Green and Springer, 1973). If late deaths are eliminated from the analysis of the above parameters a more sensitive index of dominant lethality is obtained. Preimplantation losses can be indicated by comparing values of total implants in females mated with treated males and those mated with control males, as suggested by Epstein (1973), instead of counting corpora lutea.

Decreases in total implants which represent increases in preimplantation losses without a corresponding increase in early deaths may not represent a mutagenic event. Factors other than genetic can explain a preimplantation egg loss. Fertility effects can also be determined in the dominant lethal test if only fertile animals are used for the study.

The dominant lethal test is best conducted with at least 3
concentrations of test compound in order to try to obtain a
dose response relationship from which extrapolation of risk
might be made. Clear cut genetic effects are best claimed
when a dose dependent increase in post-implantational foetal
deaths is evident.

Other methods such as the heritable translocation test
detect transmitted damage and are useful in this respect.
However, such a test as well as the specific locus gene
mutation test are extremely expensive, requiring very large
numbers of animals, extensive housing and maintenance.

A negative result in any one of the systems so far discussed may mean that the wrong species of animals or microsomes has been used or a result may be negative for any one
of the reasons discussed earlier, or it may be a true negative.

Human Peripheral Lymphocytes from Exposed Workers

This sort of study is usually carried out retrospectively
after a compound has been identified as a hazard, e.g. the
many publications on workers exposed to vinyl chloride
(Purchase *et al.*, 1975 and refs. therein). Before such a study
can be undertaken many ethical problems have to be taken into
consideration, e.g. if the exposed work-force need to be told
the results of the study and what the results mean in the
light of current knowledge. Such a study and decisions relating to it involve negotiations with medical officers, workers,
unions etc. Negotiation may be more difficult on a prospective basis where compounds present an unknown hazard. If a
positive response is obtained, while it is possible to show a
correlation between exposure and level of abnormal cells on a
group basis, for exposed people the range of individual values
within groups is wide and overlaps the ranges of adjacent or
control groups. In retrospective studies where exposure has
occurred for up to 20 to 30 years it is often difficult to
know exact exposure levels and exposure can often be only
estimated roughly from occupation.

We are investigating whether such a study is useful in
trying to determine safe exposure levels. If the exposed
population is not significantly different in terms of chromosome damage from the control population, then it might be
assumed that safe exposure levels have been achieved because
a negative result suggests that the chemical concerned is not
a mutagen. However, there are limitations to this approach
since the exposure level may be too low to produce the

chromosome damaging effect, but may still cause sister chromatid exchanges or gene mutations which are not detected by
conventional chromosomal analysis.

Controls should be taken from both on and off site and
where possible should be age and sex matched. Cells are generally cultured for 48 and 72 hours and we have found no significant differences in data at these two times after vinyl
chloride exposure (Purchase *et al.*, 1975), although after irradiation this is not the case (Abbat *et al.*, 1974); 48 hour
cultures are then desirable. At least 100 cells per individual should be analysed from coded slides to avoid observer
bias.

Whilst there is a correlation between carcinogenesis and
mutagenesis (earlier refs.), a review paper by Harnden (1976)
puts the correlation for clastogenic and carcinogenic effects
into perspective, as does the book edited by German (1974).
There appears to be a fairly good but non-quantitative correlation.

Other techniques such as sister chromatid exchange may be
useful on exposed workers but effects are much more short-
lived (Stetka and Wolf, 1976a, b) and may to some extent have
disappeared before culturing is possible.

Once it is established that a chemical is clastogenic,
regular population monitoring of the work-force may be initiated where all workers exposed to the chemical are monitored
both pre-exposure and during employment. The results of the
monitoring will be useful for checking plant hygiene. An increase in abnormal cells in an individual could be used as an
indication that the worker should not be exposed to that
chemical. If an individual is found to have a chromosomal
abnormality linked to a certain disease, he can be advised
through appropriate medical channels of the risk of inheritance of the disease in any children he may have.

If all workers in chemical plants were monitored as part
of a routine medical surveillance service, then many of the
difficulties involved in initiating prospective studies
would disappear.

Computerised microscopic techniques are available which
can reduce by about 40% the time spent by a technician. At
present, however, metaphase spreads are merely located and
the amount of damage still has to be assessed visually. Initial cost of purchasing a computer microscope is very high.

Other techniques for direct application to man include
the use of urine or blood plasma from exposed workers in combination with a microbial assay (Legator *et al.*, 1976),
electrophoretic monitoring of enzymatic markers in man (Neel,
1977), detection of variants in haemoglobin molecules (Nute
et al., 1976, Popp *et al.*, 1977), investigation of sperm

morphology (Wyrobek and Bruce, 1975), and an increase in the presence of YY bodies (see Approaches to Determining the Mutagenic Properties of Chemicals (1977), D.H.E.W. Sub-Committee on Environmental Mutagenesis).

INTERPRETATION OF DATA

Assuming that results are reproducible in well conducted experiments using adequate testing protocols, interpretation of data may still be difficult especially with results within the control range or on the border line of statistical significance. Interpretations may differ depending on whether results from treatment groups are compared with historical (accumulated) or concurrent controls, the type of statistical analysis or whether results are compared at equitoxic doses. If the experiments are repeated, similar equivocal results may be obtained, and if not, there is the problem of whether the first or repeat experiment is correct. The difficulty is in proving and accepting negative data. By comparison, the handling of positive data is much more clearcut.

Systems giving negative results are often considered insensitive, e.g. the dominant lethal test. However, such a test might truly reflect the situation in the germ cells and the testes may really be incapable of metabolising many compounds to their active intermediates. Microbial systems need large concentrations of compounds to detect a positive result by comparison with concentrations expected to be present in whole mammals and as such could equally be considered insensitive.

ASSESSMENT OF GENETIC EFFECT

Making a quantitative risk evaluation for man from data obtained in testing procedures is an even more difficult task. First it must be determined if the data are biologically and/or statistically significant. Chemicals with unknown mutagenic potential are probably more difficult to assess than data for radiation or known mutagens (many of which are "radio-mimetic" agents) unless there are clear-cut dose responses, since chemicals behave differently with different cells, organs, and organisms.

One of the recommended ways of assessing risk is by comparing the values obtained from chemical data with corresponding radiation data. Crow (1973) recommended that the population genetic effects of chemical mutagens be assessed, taking

radiation as an equivalent and equating the population dose of
chemical mutagens to the radiation dose admissable for that
population. Bridges (1973, 1974) in his papers describing the
3-tier system also recommends the principle of a radiation
equivalent dose. With this system, only those substances
which have passed through the first two 2 tiers and are of
great social, medical and economical importance are subjected
to a quantitative assessment. Chemicals which have shown a
positive effect in the first 2 tiers if not widely used should
be prohibited or used with a non-mutagenic derivative substituted for the mutagenic radical. Those which are positive in
the third tier should be expressed as the equivalent of a radiation dose which produced the same effect and this makes it
possible to standardise the level of chemical mutagens to the
limit of the level of ionising radiation. With extrapolation
back to very small doses it would be difficult to decide which
is the line of best fit and with some chemicals there may be
shoulders of varying size on dose response curves due to permeability problems with a chemical or other unknown factors.

Yet another "radiation-equivalent concept", the ABCW
model (Abrahamson *et al.*, 1973), is based on the hypothesis
that the radiation induced mutation rate per radiation unit
and gene locus in a variety of organisms from microbes to
mammals is proportional to their genome size. This model has
been extended to chemical compounds (Wolff, 1975). As a result it would seem possible to estimate risks for man on the
basis of an upward extrapolation. However, this model has been
criticised by Sankaranarayanan (1976).

The problem with these radiation concepts is that chemicals may behave differently from radiation induced free radicals and often have species type specificity. This problem
has been highlighted by Auerbach (1975) and Sobels (1976).

Bochkov *et al.* (1975) recommend that the assessment should
be based on individual and population prognoses. The individual prognosis should be determined by the quantity of the
chemical and its mutagenic activity. The population prognosis
should be determined by the number of persons of reproductive
age who are in contact with the chemical mutagen and the average quantity of this substance for each of them. To make such
an assessment various factors need to be considered:

a) data are required on the test specimen for which the
highest and clearest quantitative dependencies were obtained;

b) the quantity of substance with which an individual
comes into contact over a period of a year;

c) the fraction of the population up to the age of 30 who
are subject to the action of a mutagen;

d) the mean population dose of substance which can be
calculated from the above data;

e) the limit for the admissible level of genetic changes.

They suggest prohibiting a substance with mutagenic activity, replacing it with a non-mutagenic compound, or restricting its use to persons of non-reproductive age in cases where its mean population dose causes an increase of 0.1% above the spontaneous level and a doubling of the spontaneous level on an individual basis.

Again, this approach depends on the reliability of the experimental data obtained and being able to determine information concerning a-e. Auerbach (1975) and Sobels (1976) also criticise the concept of the doubling dose.

Considering the difficulties of interpreting results obtained in a test system any single test can hardly provide absolute answers. Negative results are often underweighted. Results from several practical test methods should be processed and decisions based on the biological and statistical significance of all the results observed, having regard for the normal range of control values in the test systems used. If a socially and economically useful compound is found to be "hazardous" for man, then a detailed examination can be made of levels to which workers are exposed and attempts made to improve plant hygiene where the product is manufactured. Further investigations can be made to determine if and by how much other groups of people or the general population are exposed e.g. in the case of vinyl chloride, manufacturing plant exposure levels have been reduced and attention has been focused on whether any free monomer occurs in plastic food wrappings, etc. Auerbach (1975) feels that in the benefit-risk calculations of a compound, on the benefit side, the calculation should carry a correction factor for the special economic, social or medical situation of the country concerned, e.g. in the country where millions suffer from malaria, the benefits of an efficient preventive or curative drug should be weighted accordingly, or in countries where there is a famine problem, pesticides should be considered similarly.

CONCLUSIONS

Both academic and industrial scientists are well aware of the need for safety evaluation in general toxicological testing and this is certainly true in the field of genetic toxicology. We still do not understand if positive or negative results in a laboratory model test system are really relevant to man because of man's unique metabolism and because of the absence of any convincing "no effect" level data for animals or man. Epidemiological evidence for germ cell mutations after chemical exposure (or in fact, any agent) is sadly lacking. In industrial areas it is often hard to pinpoint the exact chemical or agent which may be causing the problem. Unbiased abortion rates are difficult to obtain in interview by comparison with control or unexposed populations. Even if they are taken from hospital and medical practitioners' records not all abortions are recorded. Auerbach (1975) does not believe that we shall ever be able to identify potential human mutagens with complete confidence and even less shall we be able to feel confident about such quantitative features as thresholds, dose-effect curves and comparisons between mutagens in the human environment. She thinks this is unavoidable when we as a species are both the subject and the object of such investigation.

However, we do have thousands of untested chemicals in our environment and some attempts must be made to identify those potentially hazardous to man. The limitations of the simple short-term tests which are more concerned with the concept of somatic mutation are becoming better understood, and the problem of false positives and negatives has been discussed earlier (Purchase *et al.*, 1976b). At present, we can only do our best with the test systems available and hope that as research progresses our understanding and techniques will improve so that results generated in our model systems will become unequivocal in terms of hazard to man. To achieve this goal, attention will have to be given to studies aimed at assessing the significance to man of positive mutagenic responses given by a test system for a given chemical in addition to the search for better assay procedures.

ACKNOWLEDGEMENTS

The author wishes to thank Professor P. Oftedal, Department of General Genetics, University of Oslo, and Drs. J.A. Styles, I.F.H. Purchase, J. Ashby, L.L. Smith and E.A. Lock of this laboratory for reading the manuscript and offering helpful comments.

REFERENCES

Abbatt, J.D., Bora, K.C., Quastel, M.R., and Lefkovitch, L.P. (1974). *Bull. Wld. Hlth. Org. 50*, 373.
Abrahamson, S., Bender, M.A., Conger, A.D. and Wolff, S. (1973). *Nature. 245*, 460.
Advisory Committee on the Biological Effects of Ionising Radiation (BEIR) (1972). "The Effects on Population of

Exposure to Low Levels of Ionising Radiations". *Natl.*
Acad. Sci. - Natl. Res. Council, Washington D.C.

Advisory Committee on Protocols for Safety Evaluation (1970).
Toxicol. Appl. Pharmacol. 16, 264.

Anderson, D. (1975a). *Mutation Res. 33*, 399.

Anderson, D. (1975b). *Mutation Res. 33*, 407.

Anderson, D., and Fox, M. (1974). *Mutation Res. 25*, 107.

Anderson, D., McGregor, D.B., and Purchase, I.F.H. (1976).
Mutation Res. 40, 349.

Anderson, D., Hodge, M.C.E., and Purchase, I.F.H. (1976).
Mutation Res. 40, 359.

Anderson, D., McGregor, D.B., Purchase, I.F.H., Hodge, M.C.E.,
and Cuthbert, J.A. (1977). *Mutation Res. 43*, 231.

Ambio Special Report No. 31 (1973). "Evaluation of Genetic
Risks of Environmental Chemicals". Royal Swedish Academy
of Sciences, Stockholm.

Ames, B.N., Lee, F.D., and Durston, W.E. (1973). *Proc. Natl.*
Acad. Sci. USA. 70, 782.

Ames, B.N., Durston, W.E., Yamasaki, E., and Lee, F.D. (1973).
Proc. Natl. Acad. Sci, USA. 70, 2281.

Ames, B.N., McCann, J., and Yamasaki, E. (1975). *Mutation Res.*
31, 347.

Approaches to Determining the Mutagenic Properties of Chemicals: Risk to Future Generations (1977). Prepared for the
Department of Health, Education and Welfare Committee to
co-ordinate Toxicology and Related Programmes by the working group of the sub-committee on Environmental Mutagenesis.

Auerbach, C. (1975). *Mutation Res. 33*, 3.

Bartsch, H., Malaveille, C., and Montesano, R. (1975). *Int. J.*
Cancer. 15, 429.

Bateman, A.J., and Epstein, S.S. (1971). *In* "Chemical Mutagens", Vol. II (A. Hollaender, ed.), p. 541. Plenum Press,
New York and London.

Bochkov, N.P., Šrám, R.J., Kuleshov, N.P., and Zhurkov, V.S.
(1975). *Genetika. 11*, 156.

Bridges, B.A. (1972). *Lab. Practice. 21*, 413.

Bridges, B.A. (1973). *Environ. Hlth. Perspect. No. 6*, 221.

Bridges, B.A. (1974). *Mutation Res. 26*, 335.

Clive, D.W. (1973). *Environ. Hlth. Perspect. No. 6*, 119.

Clive, D.W., Flamm, W.G., Machesko, M.R., and Bernheim, N.J.
(1972). *Mutation Res. 16*, 77.

Clive, D.W., and Spector, J.F.S. (1975). *Mutation Res. 31*, 17.

Cole, J., and Arlett, C.F. (1976). *Mutation Res. 34*, 507.

Cole, J., Arlett, C.F., and Green, M.H.L. (1976). *Mutation*
Res. 41, 377.

Coombs, M.M., Dixon, C., and Kissonerghis, A.M. (1976). *Cancer*
Res. 36, 4525.

Committee 17 (1975). *Science. 187,* 503.

Cox, R., and Masson, W.K. (1974). *Intern. J. Biol. 26,* 493.

Crow, J.F. (1973). *Environ. Hlth. Perspect. No. 6,* 1.

Dean, B.J. (1969). *Lab. Anim. 3,* 159.

Epstein, S.S. (1973). *Environ. Hlth. Perspect. No. 6,* 23.

Fahrig, R. (1974). *Mutation Res. 26,* 29.

Fahrig, R. (1975). *Mutation Res. 31,* 381.

Fieller, E.C. (1947). *The Analyst. 72,* 37.

Fischer, G.A., Lee, S.Y., and Calabresi, P. (1974). *Mutation Res. 26,* 501.

Flamm, W.G. (1974). *Mutation Res. 26,* 329.

Fox, M., and Anderson, D. (1974). *Mutation Res. 25,* 89.

Generoso, W.M., Russel, W.L., Huff, S.W., Stout, S.K., and Gossee, D.E. (1974). *Genetics. 77,* 741.

Green, M.H.L., Muriel, W.J., and Bridges, B.A. (1976). *Mutation Res. 38,* 33.

Green, S., and Springer, J.A. (1973). *Environ. Hlth Perspect. No. 6,* 37.

Green, T., and Hathway, D.E. (1975). *Chem. Biol. Interact. 11,* 545.

German, J. (1974). "Chromosomes and Cancer". John Wiley and Sons, New York.

Harnden, D.G. (1976). *Proc. Roy. Soc. Med. 69,* 41.

Heddle, J.A. (1973). *Mutation Res. 18,* 187.

Huberman, E., Mager, R., and Sachs, L. (1976). *Nature. 264,* 360.

Infante, P.F., Wagoner, J.K., and McMichael, A.J. (1976a). *Lancet. i,* 734.

Infante, P.F., Wagoner, J.K., and McMichael, A.J. (1976b). *Lancet. i,* 1289.

Infante, P.F., Wagoner, J.K., and Waxweiler, R.J. (1976c). *Mutation Res. 41,* 131.

Jones, A.R. (1973). *Drug Metabolism Reviews. 2,* 71.

Kada, T., Moriya, M., and Shirasu, Y. (1974). *Mutation Res. 26,* 243.

Knaap, A.G.A.C., and Simons, J.W.I.M. (1975). *Mutation Res. 30,* 97.

Knill-Jones, R.P., Moir, D.D., Rodrigues, L.V., and Spence, A.A. (1972). *Lancet. i,* 1326.

Legator, M.S., and Malling, H.V. (1969). *Genetics. 61,* S5.

Legator, M.S., Zimmering, S., and Connor, T.H. (1976). *In* "Chemical Mutagens", Vol. 4 (A. Hollaender, ed.), p. 171. Plenum Press, New York and London.

Léonard, A. (1975). *Mutation Res. 31,* 291.

Maltoni, C., Lefemine, G., Chieco, P., and Carreti, D. (1974). *Rivista "Gli Ospedali della Vita" 1,* 7.

Matter, B.E. (1976). *Mutation Res. 38,* 243.

McCann, J., Choi, E., Yamasaki, E., and Ames, B.N. (1975). *Proc. Natl. Acad. Sci. USA. 72,* 5135.

McCann, J., and Ames, B.N. (1976). *Proc. Natl. Acad. Sci. USA.* *73*, 950.

McCann, J., Spingarn, N.E., Kobori, J., and Ames, B.N. (1975). *Proc. Natl. Acad. Sci. USA.* *72*, 979.

McKusick, V. (1971). "Mendelian Inheritance in Man", 3rd ed. Johns Hopkins Press, Baltimore.

Neel, J.V. (1977). *Genetics.* In press.

Nichols, W.W., Moorhead, P., Brewen, G. (1972). *Toxicol. Appl. Pharmacol. 22*, 269.

Nute, P.E., Wood, W.G., Stammatoyannopoulos, G., Olivery, C., and Fialkow, P.J. (1976). *Brit. J. Haemat. 32*, 55.

Paddle, G.M. (1976). *Lancet. i*, 1079.

Parry, J. (1972). *Lab. Practice. 21*, 417.

Perry, P., and Evans, H.J. (1975). *Nature. 258*, 121.

Popp. R.A., Hrisch, G.P., and Bradshaw, B.S. (1977). *Genetics.* In press.

Propping, P., Röhrborn, G., and Buselmaier, W. (1972). *Molec. Gen. Genet. 117*, 197.

Purchase, I.F.H., Richardson, C.R., and Anderson, D. (1975). *Lancet. ii*, 410.

Purchase, I.F.H., Richardson, C.R., and Anderson, D. (1976a). *Proc. Roy. Soc. Med. 69*, 290.

Purchase, I.F.H., Longstaff, E., Ashby, J., Styles, J.A., Anderson, D., Lefevre, P.A., and Westwood, F.R. (1976b). *Nature. 264*, 624.

Ray, V.A., and Hyneck, M.L. (1973). *Environ. Hlth. Perspect. No. 6*, 27.

Report of the Ad Hoc Committee of the Environmental Mutagen Society and the Institute for Medical Research (1972). *Toxicol. Appl. Pharmac. 22*, 269.

Russel, W.L. (1951). *Cold. Spring Harbor. Quant. Bio. 16*, 327.

Russel, L.B. (1976). *In* "Chemical Mutagens", Vol. 4. (A. Hollaender, ed.), p. 55. Plenum Press, New York and London.

Sankaranarayanan, K. (1976). *Mutation Res. 35*, 341.

Searle, A.G. (1975). *Mutation Res. 31*, 277.

Sega, G.A., Cumming, R.B., and Walton, M.F. (1974). *Mutation Res. 24*, 317.

Schmid, W. (1973). *Agents and Actions. 3*, 77.

Schmid, W., Arakaki, D.T., Breslau, N.A., and Culbertson, J.C. (1971). *Humangenetik. 11*, 103.

Schmid, W. (1976). *In* "Chemical Mutagens", Vol. 4 (A. Hollaender, ed.), p. 31. Plenum Press, New York and London.

de Serres, F.J. (1975). *Mutation Res. 33*, 11.

Slater, E., Anderson, M.D., Rosenkranz, H.S. (1971). *Cancer Res. 31*, 970.

Sobels, F.H. (1974). *Mutation Res. 26*, 277.

Sobels, F.H. (1976). Some thoughts on the evaluation of environmental mutagens. From the address presented at the

7th Annual EMS Meeting in Atlanta, Georgia, USA.

Stetka, D.G., and Wolff, S. (1976a). *Mutation Res.* *41*, 333.

Stetka, D.G., and Wolff, S. (1976b). *Mutation Res.* *41*, 343.

Stich, H.F., and San, R.H.C. (1970). *Mutation Res.* *10*, 389.

Stich, H.F., and San, R.H.C. (1973). *Proc. Soc. Exptl. Biol. Med. 142*, 155.

Stoltz, D.R., Poirier, L.A., Irving, C.C., Stich, H.F., Weisburger, J.H. and Grice, H.C. (1974). *Toxicol. Appl. Pharmac.* *29*, 157.

Styles, J.A. (1977). *Brit. J. Cancer.* In press.

Sutton, H.E. (1975). *Mutation Res.* *33*, 17.

Swann, P.F., and Magee, P.N. (1968). *Biochem. J. 110*, 39.

Testing of Chemicals for Carcinogenicity, Mutagenicity and Teratogenicity (1973). *Health and Welfare*, Ottawa, Canada.

Tijo, J.H., and Whang, J. (1962). *Stain Technol. 37*, 17.

Vogel, F., and Röhrborn, G. (1970). *In* "Chemical Mutagenesis in Mammals and Man" (F. Vogel, and G. Röhrborn, eds.), p. 453. Springer-Verlag, Berlin, Heidelberg and New York.

Vogel, E., and Sobels, F.H. (1976). *In* "Chemical Mutagens", Vol. 4 (A. Hollaender, ed.), p. 93, Plenum Press, New York and London.

WHO Report (1971). *Wld. Hlth. Org. Techn. Rep. Ser. No. 482.*

WHO Report (1974). *Wld. Hlth. Org. Techn. Rep. Ser. No. 546.*

WHO Report (1975). *Wld. Hlth. Org. Techn. Rep. Ser. No. 563.*

Wyrobek, A.J., and Bruce, W.R. (1975). *Proc. Natl. Acad. Sci. USA. 72*, 4425.

Zbinden, G. (1973). "Progress in Toxicology", Vol. 1. Springer Verlag, Heidelberg.

Zimmerman, F.K. (1971). *Mutation Res. 11*, 327.

Zimmerman, F.K. (1975). *Mutation Res. 31*, 71.

THE USEFULNESS OF BACTERIAL TEST SYSTEMS FOR MONITORING
MUTAGENIC AND CARCINOGENIC AGENTS

Claes Ramel
Ulf Rannug

Environmental Toxicology Unit
Wallenberg Laboratory
University of Stockholm
Stockholm
Sweden

INTRODUCTION

The way of life of industrialized countries has become
totally dependent on countless numbers of chemicals manufactured for industrial purposes. Every year hundreds of
thousands of new organic compounds are synthesized, although
the number put into industrial use of course is much smaller -
of the order of a thousand. These compounds are added to perhaps 20,000 or 30,000 compounds already in use. This large
output of chemicals obviously has serious consequences from a
toxicological point of view. It is far easier to synthesize
new chemicals than to evaluate their risk to man and other
organisms. It has therefore become an impossible task for
toxicologists and biologists to keep pace with chemical technology. In the race between the chemists and the toxicologists,
the toxicologists have been hopelessly left behind. As a result we have a very poor knowledge of adverse effects and
health risks for most of the chemicals used - to say noting
of possible synergistic actions. This problem is bad enough
for acute toxic effects, but in such cases one can at least
hope for reasonably early warning signals from epidemiological
studies of exposed human populations. The situation concerning
long term effects like heritable changes and cancer induction
is more difficult in that respect. Even if a highly exposed
human population can be identified the long latent period
usually makes it impossible to observe any effects for decades.

It is evident that mutagenic and carcinogenic chemicals in the environment must be identified long before they give rise to observable effects on humans and have created a problem at a population level. This identification of mutagens and carcinogens has to be done primarily through tests on experimental organisms. The testing for mutagenicity and carcinogenicity implies one advantage as compared to corresponding tests for acute toxic effects. A wealth of data indicates that changes in the genetic material is a common denominator both for heritable changes and cancer induction. This conclusion can be drawn from several lines of evidence. Thus there is a chemical accordance between known carcinogens and mutagens, since both exhibit electrophilic properties (Miller and Miller, 1871). When metabolic conversion of procarcinogens is taken into account, present data indicate that all carcinogens are also mutagens (Ames *et al.*, 1973; McCann *et al.*, 1975). Mutations in the repair mechanisms of DNA lead to tumor induction, as in Xeroderma pigmentosum in man (Cleaver, 1968). Setlow and Hart (1975) showed on fishes that transplantation of tissue, irradiated with UV, gives rise to a high frequency of tumors, but that this effect could be drastically counteracted by illuminating the transplant after UV with visible light. Evidently the visible light acted on the photoreactive repair of thymin dimers. Finally, data strongly suggest a monoclonal origin of tumors, which is in accordance with a mutational event (Möller and Möller, 1975).

Although available data indicate that tumors originate from somatic mutations, it is not possible to eliminate an epigenetic origin in all cases. However, at least the vast majority of tumors seems to be of a mutational origin and from an operational point of view cancer can be dealt with as a mutation problem. Consequently, Druckrey has suggested the term genotoxic as a common expression for mutagenic and carcinogenic effects (Ramel, 1973).

The fact that both mutagenic and carcinogenic effects can be traced back to lesions in DNA makes it possible to extrapolate from one organism to the other.

As a principle, a compound acting on DNA in a microorganism can be assumed to act also on DNA in man. Whether it does that or not is a matter of metabolic conversion of the compound and its ability to reach DNA in the human cells. The mutagenicity testing therefore comprises two problems – whether the compound can interact with DNA and what happens to it in the mammalian body. This poses an experimental dilemma. On the one hand the aim should be to approach the human metabolism and physiological conditions as closely as possible, on the other hand to apply an experimental test

system with sufficient resolution power to discover also
weaker mutagenic effects. Unfortunately there is no test
system which can meet these requirements satisfactorily.
Therefore one has to deal with the problem of mutagenicity
and metabolic conversion to some extent separately.

The problem of combining a suitable test system with the
mammalian metabolism has been successfully tackled in the last
few years by combined test methods. Microorganisms, particularly bacteria, are used as indicator organisms for mutagenic effects but a mammalian metabolizing system has been
added to the bacterial culture in one way or another. This
procedure has vastly increased the relevance of tests on
microorganisms for the extrapolation to man.

THE *SALMONELLA* TEST SYSTEM

Many prokaryotic test systems have turned out to be useful
for mutagenicity testing – transforming DNA as well as various
viruses and bacteria. It is not possible to deal with these
various test systems here. Hence, we will concentrate on the
one which has become the dominating test for practical purposes; namely, that for reverse mutations at the histidine locus of *Salmonella typhimurium*. This genetic test system has
been developed by Bruce Ames (Ames, 1971; Ames *et al.*, 1975).

The test is performed in the standard way for reverse
mutations. It is based on mutants in a gene which is involved
in the biosynthesis of the amino acid histidine. Such mutants
can not grow on a substrate lacking histidine unless the
mutant gene reverts to its original state and the ability to
synthesize histidine is restored. If a large number of such
histidine mutant bacteria are plated on a substrate lacking
histidine, a few colonies will be formed, representing
spontaneous reverse mutations. The addition of a mutagenic
agent increases this number of revertants. The histidine
reversion test system has acquired a high degree of sensitivity and sophistication by various genetic manipulations.
Different histidine mutants have been selected which respond
specifically to base substitution or frame shift. The test
therefore also reveals the mechanism for the mutagenic action
of the test substance.

The sensitivity of the test system has been increased
by addition of various mutations to the tester strains. One
such mutation eliminates the excision repair of DNA. Mutant
genes taking away parts of the lipopolysaccarid bacterial
wall has increased the permeability of the bacterial cells,
making them more like the cells of higher organisms. Further-

more, the addition of an episome carrying ampicillin resistance has increased the sensitivity to some chemicals rather
strikingly, apparently through the induction of an error prone
repair (McCann *et al.*, 1975).

IN VITRO METABOLIC ACTIVATION

The potentiality of combining a genetic test system on
microorganisms with a mammalian metabolism was shown by
Garbridge's and Legator's (1969) invention of the host-mediated assay. By this method microorganisms serving as test
organisms were injected into the peritoneal cavity of a mouse
or a rat to which the test substance had been given. The
bacteria were left in the animal for about three hours, taken
out and tested for induced mutations. Thus the test organisms
- usually *Salmonella* - were exposed to metabolites formed in
the mammalian body. Although the host mediated assay implied
an important step in mutagenicity testing, there are some
disadvantages with this method. Negative results do not constitute any reasonable safety against mutagenic effects,
particularly as the test system poses a strong restriction as
to the doses which can be applied. There is often an undesir-
able interaction between the test organism and the animal,
causing, for instance, mutations when *Neurospora* has been
used (Legator and Malling, 1971). Because of the interaction
with the host animal, the application of some of the most
useful *Salmonella* strains is also excluded.
A simplified solution to the problem of adding a metabolizing system to microorganisms was introduced by Malling
(1971) and further developed by Ames *et al.* (1973). The procedure is based on the fact that the essential metabolic
activation occurs in the liver, particularly through the
oxygenase system with cytochrom P 450. The liver fractions,
which contain this enzyme system, operate also *in vitro*.
When such a microsomal fraction from rat liver (S 9-Mix) is
introduced to the bacterial test system, Ames could show that
the bacteria also responded very efficiently to indirect
carcinogens (Ames *et al.*, 1973). In fact surveys of hundreds
of chemicals have revealed that over 90% of carcinogens can
be traced by this method (McCann *et al.*, 1975).

APPLICATIONS OF BACTERIAL TESTS

The Ames test system with liver metabolic activation is clearly an extremely useful tool to screen chemicals for mutagenicity and to draw attention to compounds causing genotoxic risks to man. Of particular advantage is, of course, the fact that the test is cheap and can be performed in a few days. Environmental chemicals can be subjected to testing for mutagenicity and potential carcinogenicity at a scale which would be entirely out of question with tumor tests on whole animals. Since Ames's test was introduced in 1973, it has become a widespread tool, particularly to trace carcinogenic substances in the environment.

Beside these routine tests, there are, however, several other applications of bacterial systems. The can serve as an efficient bioassay to identify genetically active molecules in many other situations. *Salmonella* seems to lack the metabolism present in mammals with exception of nitro reductase (McCann *et al.* 1975); this is an advantage of using *Salmonella* in an analytical bioassay. It means that the metabolic activation can be switched on and off by means of liver microsomes and the mutagenic effect and metabolic activation can be studied separately. The metabolic system can furthermore be affected experimentally by enzyme induction and by other means. Also the metabolizing effect of other organs than the liver can be investigated, as has been demonstrated by Brusick *et al.* (1976).

The *Salmonella* test system can thus be regarded as a biological analytical instrument, corresponding to the analytical devices used in chemistry. As a matter of fact the bacterial test is particularly suitable to use along with chemical analyses to characterize mutagenic steps in metabolic processes or to identify mutagenic components in complex chemical mixtures. We would like to illustrate this with a series of investigations which we have done, starting with vinyl chloride.

After the carcinogenic property of vinyl chloride was discovered, tests in our laboratory (Rannug *et al.*, 1974) as well as in other laboratories (Bartsch *et al.*, 1975; Malaville *et al.*, 1975; Loprieno *et al.*, 1976) showed that vinyl chloride exhibits a mutagenic effect on *Salmonella* after metabolic conversion with liver microsomes. The effect was restricted to a bacterial strain responding to base substitutions, which indicated that the reactive metabolite is an alkylating agent.

In order to investigate the mutagenic pathway of vinyl chloride in the liver, a system for trapping the metabolites was worked out in the organic chemical division of the laboratory by Göthe *et al.* (1974). Vinyl chloride-containing air was passed over a rat liver microsomal fraction (Fig. 1) in the presence of a molecule specifically constructed for trapping alkylating metabolites. This molecule was 3,4-dichlorobenzenethiol, which reacts rapidly with alkylating species. The dichlorophenyl group makes it easy to detect the reaction products by gas chromatography - mass spectrometry technique. In this way, 3,4-dichlorobenzenethioacetaldehyde was identified indicating the formation of chloroacetaldehyde as a metabolite. This compound can presumably be formed in two days - either via the formation of an epoxide: chloroethylene oxide, or 2-chloroethanol. The final product should in both cases be chloroacetic acid.

According to that schedule, four possible metabolites are involved in the metabolism of vinyl chloride. They were all tested for mutagenicity in TA 1535 without metabolic activation. It turned out, not unexpectedly, that chloroethylene oxide was the most powerful mutagen (Rannug *et al.*, 1976). In fact it was 10,000 - 15,000 times more effective than ethylene oxide, used as a positive control. Chloroacetaldehyde, however, also showed a positive effect, although considerably weaker than the effect of chloroethylene oxide; the difference was about 450 times. A very weak mutagenic effect was also obtained with 2-chloroethanol, which is in accordance with date on *E.coli* (Rosenkranz and Wlodkowski, 1974).

From this data it can be concluded that one or both of the two metabolites chloroethylene oxide and chloroacetaldehyde are responsible for the carcinogenic effect of vinyl chloride. The most likely compound giving rise to angiosarcoma in the liver is chloroethylene oxide, considering its extremely powerful mutagenicity. The interpretation of chloroethylene oxide as the most active carcinogenic metabolite is supported by the data of Malaville *et al.* (1975). This compound has a very short life length and one can speculate that tumors formed outside the liver by exposure to vinyl chloride might be induced by chloroacetaldehyde, which is more persistent than chloroethylene oxide.

The occupational risk with vinyl chloride focused attention also on the waste products which are formed in the vinyl chloride manufactoring process. When vinyl chloride is synthesized, a tar-like product is formed, called ethylenedichloride-tar or EDC-tar from one of its main components 1,2-dichloroethane. EDC-tar is of importance from an environmental point of view, because it has been dumped in the sea in very large quantities - several hundred thousand tons per year in the North Sea (Jensen *et al.*, 1970).

The tar has been analyzed in our laboratory by Jensen, who

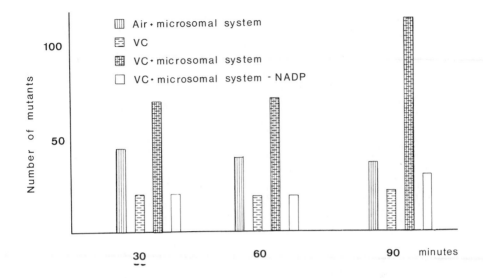

FIGURE 1. *The mutagenic effect of treatment for 30-90 minutes of 20% (v/v) gaseous vinyl chloride monomere on TA 1535 with and without microsomal system and the cofactor NADP.*

could identify over 80 components (Jensen *et al.*, 1975). EDC-tar was tested for mutagenicity and was found to be mutagenic. The mutagenicity was greatly enhanced by microsomal activation (Rannug and Ramel, 1977).

Such complex mixtures of chemicals represent the actual situation in many cases in industries and therefore it seemed of interest to use the EDC-tar as a model substance for analyses of genotoxic properties alongside with chemical analyses. This work is being done at our laboratory in collaboration with Jensen's group in analytical chemistry. We will report some of the results available so far (Fig. 2).

The EDC-tar has been fractionated by distillation in four fractions, depending on boiling temperature. The first fraction contains almost only 1,2-dichloroethane and the second 1,1,2-trichloroethane. These two substances, which comprise a large proportion of the tar, were tested as pure substances. 1,1,2-trichloroethane did not give any indication of a mutagenic effect in any connection.

1,2-dichloroethane, on the other hand, showed a mutagenic effect, which is in accordance with previous findings in *Drosophila* (Rapoport, 1960) and on *Salmonella* (McCann *et al.*,

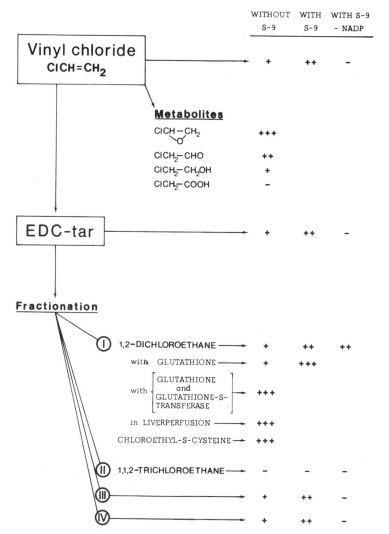

FIGURE 2. *The mutagenicity testing scheme of vinyl*
chloride, its metabolites and its waste pro-
ducts (EDC-tar). Four distillation fractions
of the EDC-tar have been tested. The diagram
shows the analysis of the mutagenic property
of 1,2-dichloroethane, constituting the first
fraction. The relative strength of the muta-
genic effects is arbitrarily indicated by the
number of + signs or by a - sign. All data
refer to experiments with Salmonella typhi-
murium, strain TA 1535, responding to base
substitution.

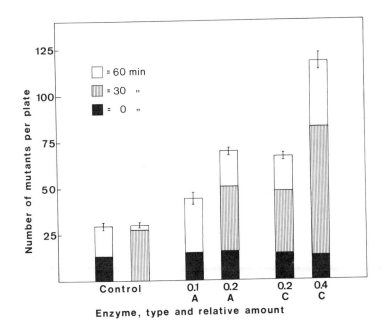

FIGURE 3. *The mutagenic effect on TA 1535 of 1,2-di-
chloroethane (10 mM) after preincubation with
glutathione (5 mM) and glutathione-S-trans-
ferases A and C for 0, 30 or 60 minutes.*

1975). The results with dichloroethane were, however, of some
theoretical interest. Although there was a direct effect without metabolic activation, the addition of liver microsomes
caused an enhancement of the effect. Attempts by McCann *et al.*
(1975) to obtain such an activation of dichloroethane were unsuccessful. We have therefore performed a special investigation on the activation of 1,2-dichloroethane (Rannug and
Ramel, in manuscript).

It was found that the activation of dichloroethane is not
dependent on NADP, which excludes the possibility that the
mixed function oxygenase system is involved. It furthermore
turned out that the metabolic activation is not connected with
membrane bound enzymes as the activation disappeared when pure
microsomes were used. However, the supernatant after separa-
ting the microsomes again showed the activation. The fact that
the soluble fraction was responsible for the metabolic activation pointed to a conjugation mechanism and this was tested
for glutathion. Tests with only glutathion and 1,2-dichloro-

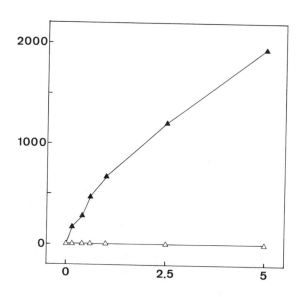

FIGURE 4.   *The mutagenic effect on TA 1535 of the syn-*
thetic conjugation product chloroethyl-S-cys-
tein (filled triangles) as compared to 1,2-
dichloroethane (open triangles).

ethane were negative, but if glutathion-S-transferase A or C
was added to the system, a very strong mutagenic effect was
obtained (Fig. 3). Glutathion-S-transferases are responsible
for the enzymatic conjugation with glutathion and also apparently responsible for the mutagenicity in this case. A synthetic conjugate chloroethyl-S-cystein was made and tested on
Salmonella; it also showed a strong mutagenic effect (Fig. 4).

It is obvious that activation of compounds of mutagens is
not restricted to the microsomal enzymes, but can occur by
other means. An effect like the one found with 1,2-dichloro-
ethane can easily be missed, unless a sufficient quantity of
the liver fraction (S 9-Mix) is added to the bacterial culture.

In order to overcome the limitation with the liver fraction from this point of view, experiments with 1,2-dichloro-
ethane were performed with liver perfusion. The perfusion
system used was worked out by Brita Beije in our laboratory

(Beije, 1977), and the work was done in cooperation with her (Rannug and Beije, in preparation). The perfusion is performed *in vitro* and under the conditions employed the liver can be kept for at least five hours without any essential change in the activity of the microsomal detoxication enzymes (Beije, 1977).

1,2-dichloroethane was introduced in the perfusion system and samples were taken from the blood and the bile and tested for base substitution mutagenicity on *Salmonella* TA 1535. The mutagenicity of the blood samples was very small and at the border of significance. Samples from the bile, however, showed a drastic induction of mutations with samples taken 15-30 minutes after the introduction of 1,2-dichloroethane in the perfusion system. The mutagenic effect of the bile after treatment with 1,2-dichloroethane also occurs *in vivo*. This was shown by tests on bile taken from the gall bladder of mice treated with the substance *in vivo*.

The mutagenic effect of 1,2-dichloroethane in the bile could be suspected, as an excretion of the conjugate through the bile can be assumed.

What mutagenic and possible carcinogenic effect this glutathione conjugate could imply from the point of view of health risk is unknown and depends on factors like reabsorption in the intestine and possible action during the process of excretion.

Considering the last two fractions of the EDC-tar, both of them showed a clear mutagenic effect which was enhanced by metabolic activation. As these fractions are free from 1,2-dichloroethane, the mutagenicity of these two fractions show that 1,2-dichloroethane does not contribute in a major way to the mutagenicity of EDC-tar in the direct or the indirect mutagenicity test. It should finally be mentioned that chemical analysis has also shown that the vinyl chloride monomere is not present in sufficient concentration to make any noticeable contribution to the mutagenicity of EDC-tar.

In conclusion we would like to emphasize that the bacterial tests with metabolic activation through mammalian tissues are extremely useful both for routine screening of chemicals and the assessment of genotoxic effects in various analytical connections. However, it should also be kept in mind that the available test systems have limitations. Although the empirical data from testing with Ames's method is encouraging, cancer induction is a complicated process where interaction between procarcinogens and promotors plays an important role. The bacterial tests cannot be expected to reflect this complicated process in a satisfactory way.

It must be emphasized that mutagenicity tests with prokaryotes indicate damage at the nucleotide level, but they

don't take into consideration the interaction of chromosomal
aberration and nondisjunction. For such effects other test
methods are indispensible.

Presumably the weakest point in bacterial tests with *in
vitro* metabolic activation is the fact that detoxication and
activation of substances in the mammalian body are far more
complicated processes than the *in vitro* systems can account
for. It also seems likely that the most important improvements
of these test methods are to be expected within this area.
New approaches to account for the mammalian metabolism, like
the use of body fluids, liver perfusion, the use of genetic
metabolic variants, the application of tissue fractions other
than from the liver and so forth should help to increase the
reliability of the *in vitro* test systems.

The bacterial test systems as they stand today are not
sufficient to make final predictions about mutagenic and
carcinogenic risks in man, and they must be used in combination with other test methods. As regards the prescreening
of the thousands of chemicals which are let loose into the
environments, the bacterial tests, however, are necessary in
order to raise a warning flag against chemicals which may
constitute risks. This is particularly evident for chemicals
used in industries where the massive number of chemicals
excludes an elaborate set of tests.

SUMMARY

The rapid increase in the number of chemicals used for
industrial purposes makes it necessary to apply rapid and
sensitive tests to trace chemicals which may constitute
genetic and carcinogenic risks to man. Bacterial tests with
in vitro activation by mammalian liver or liver preparation
constitute a valuable tool in this respect. Beside the function of such tests for routine screening, they are of great
value as a bioassay to identify genetically active compounds
in complex chemical mixtures or reactive metabolites with indirectly acting carcinogens and mutagens. These applications
are illustrated by work done on vinyl chloride, where the
metabolite most likely involved in the carcinogenic effect is
chloroethylene oxide. The waste product from the vinyl chloride industry, EDC-tar, has been analyzed chemically by fractionation along with bacterial test for mutagenicity of the
fractions. 1,2-dichloroethane was shown to be activated by
a liver microsomal fraction. The investigation indicates,

however, that this activation is performed by the conjugation
to glutathione by means of the soluble glutathione-S-trans-
ferase enzymes. This conclusion was supported by mutagenicity
tests on bacteria with liver perfusion.

ACKNOWLEDGMENTS

The authors are greatly indebted to Dr. Erik Arrhenius,
Professor Lars Ehrenberg, Professor Sören Jensen, Dr. Lars
Renberg and Dr. Carl Axel Wachtmeister for valuable discussion
and advice. The skillful assistance of Mrs. Agneta Hedenstedt,
Miss Annica Sundvall and Miss Britt Marie Lidesten is also
gratefully acknowledged. This work has been supported by the
Swedish Board for Technical Development and the National
Swedish Environmental Protection Board.

REFERENCES

Ames, B.N. (1971). *In* "Chemical Mutagens. Principles and
    Methods for Their Detection" (A. Hollaender, ed.),
    pp. 267-282, Vol. 1. Plenum Press, New York.
Ames, B.N., Durston, W.E., Yamasaki, E., and Lee, F.D. (1973).
    Proc. Natl.Acad. Sci. (USA). 70, 2281
Ames, B., McCann, J., and Yamasaki, E. (1975). *Mutation Res.
    31,* 347.
Bartsch, H., Malaveille, C., and Montesano, R. (1975). *Int.
    J. Cancer. 15,* 429.
Beije, B. (1977). *Statens Naturvårdsverk PM. 822,* 1.
Brem, H., Stein, A.B., and Rosenkranz, H. (1974). *Cancer Res.
    34,* 2576.
Brusick, D., Bakshi, K., and Jagannath, D.R. (1976). *In*
    "*In Vitro* Metabolic Activation in Mutagenesis Testing"
    (F.J. de Serres, J.R. Fouts, J.R. Bend, and R.M. Philpot,
    eds.), pp. 125-141. North Holland Publ. Co.
Cleaver, J.E. (1968). *Nature. 218,* 652.
Garbridge, M.G., and Legator, M.S. (1969). *Proc. Soc. Expl.
    Biol. 130,* 831.
Göthe, R., Calleman, C.J., Ehrenberg, L., and Wachtmeister,
    C.A. (1974). *Ambio. 3,* 234.
Jensen, S., Lange, R., Berge, G., Palmork, K.H., and
    Renberg, L. (1975). *Proc. R. Soc. Lond. (Biol). 189,* 333.
Legator, M.S., and Malling, H.V. (1971). *In* "Chemical Muta-
    gens. Principle and Methods for Their Detection" (A. Hol-
    laender, ed.), pp. 569-589, Vol. 2. Plenum Press, New York.

Loprieno, N., Barale, R., Baroncelli, S., Bauer, C.,
    Bronzetti, G., Cammellini, A., Cercignani, G., Corsi, C.,
    Gervasi, G., Leporini, C., Nieri, R., Rossi, A.M.,
    Stretti, G., and Turchi, G. (1976). *Mutation Res. 40*, 85.
Malaveille, C., Bartsch, H., Barbin, A., Camus, A.M.,
    Montesano, R., Croisy, A., and Jacquignon, P. (1975).
    Biochem. Biophys. Res. Commun. 63, 363.
Malling, H.V. (1971). *Mutation Res. 13*, 425.
McCann, J., Choi, E., Yamasaki, E., and Ames, B. (1975).
    Proc. Natl. Acad. Sci. (USA). 72, 5135.
McCann, J., Springarn, N.E., Kobori, J., and Ames, B.N. (1975).
    Proc. Natl. Acad. Sci. (USA). 72, 979.
Miller, E.C., and Miller, J.A. (1971). *In* "Chemical Mutagens.
    Principles and Methods for Their Detection" (A. Hollaender,
    ed.), pp. 83-113, Vol. 1. Plenum Press, New York.
Möller, G., and Möller, E. (1975). *J. Nat. Cancer Inst. 55*,
    755.
Ramel, C. (1973). *In* "Report of a symposium held at Skokloster,
    Sweden", Ambio Spec. Report, pp. 1-27.
Rannug, U., Göthe, R., and Wachtmeister, C.A. (1976). *Chem.-
    Biol. Interactions. 12*, 251.
Rannug, U., Johansson, A., Ramel, C., and Wachtmeister, C.A.
    (1974). *Ambio. 3*, 194.
Rannug, U., and Ramel, C. (1977). *J. Toxic and Envir. Health.
    2*, in press.
Rapoport, I.A. (1960). *Akad. Nauk. SSSR, Dokl. Biol. Sci. 134*,
    745.
Setlow, R.B., and Hart, R.W. (1975). *Proc. Fifth. Int. Congr.
    Rad. Res. Radiation Res.*, 879.

MOLECULAR EVENTS IN RESPONSE TO GENE DAMAGE IN THE EUKARYOTIC ORGANISM *TETRAHYMENA PYRIFORMIS*[1]

Ole Westergaard
Johan Chr. Leer
Kjeld A. Marcker
Ole F. Nielsen

Department of Molecular Biology
University of Aarhus
Aarhus
Denmark

INTRODUCTION

Evaluation of the risk of genetic damage in man requires a knowledge of the enzymology of DNA metabolism as well as a proper understanding of the chemistry of the mutational event itself. Many of the enzymes of DNA replication and repair processes contribute to the diversity of the mutational response. In order to obtain a better understanding of some of the molecular events which occur in the eukaryotic chromatin we have studied the DNA metabolism in the lower eukaryotic organism *Tetrahymena pyriformis*. In the following we describe how various types of damage to DNA cause accumulation of both a particular DNA polymerase and intermediates in the DNA synthesis. In addition we have developed an *in vitro* system which allows us to study the effect of gene damage on the transcriptional processes of a specific gene in its chromatin stage.

[1]*Part of this investigation was supported under contract No. 122-74-1 Bio-DK with EURATOM, CEC, Brussels.*

MATERIALS AND METHODS

Culture of Cells

    Cultures of *Tetrahymena pyriformis*, strain GL (amicronu-
cleate) were grown at 28°C in defined or complex medium (Johnson and Westergaard, 1976; Leer *et al.*, 1976). Cells were harvested and washed by centrifugation at 250 x g. The cell number was determined using a Celloscope electronic cell counter
(Ljungberg, Stockholm) equipped with a 125 μ aperture.

Gradient Centrifugation

    Cellular DNA and chromatin were analyzed on velocity or
density gradients as described by Nymann and Westergaard
(1976), Johnson and Westergaard (1976) and Leer *et al.* (1976).

Enzyme Assays

    DNA polymerase activity was measured according to the
method of Westergaard (1970) and specific activity defined as
pmole deoxyribonucleotides incorporated into DNA in 20 min per
0.3 mg protein. The results were corrected for overlapping
nuclear DNA polymerase activity (see Westergaard, 1970). RNA
polymerase activity was measured as pmole nucleotide incorporated into RNA in 15 min.

DNA Electrophoresis

    Electrophoresis was conducted in slab gels of 0.6% agaro-
se. The buffer was 40 mM Tris and 1 mM EDTA adjusted to pH 7.9
with phosphoric acid. Samples of chromatin for electrophoresis
were dissolved in the buffer supplemented with 1% SDS and 1 mM
EDTA.

Other Manipulations

    Electron irradiation was performed by a 10 Mev linear
electron accelerator (10^9/sec) at the Danish Atomic Energy
Commission Research Establishment. Exponentially growing cells
were irradiated in a 1 litre beaker covered by aluminium foil.
The radiation time was about 4 min. Thymine starvation was performed by the method of Zeuthen (1968). Fifteen μM methotrexate
plus 1.5 mM uridine were used for intensive thymine starvation
in a defined medium, and starvation was overcome by addition of
1 mM thymidine or bromodeoxyuridine. Cells starved in a complex medium were treated with 0.1 mM methotrexate plus 20 mM
uridine.

RESULTS AND DISCUSSION

Deoxyribonucleotide starvation has in several organisms been found to cause chromosome damage consisting of gaps and open breaks. A similar effect has been observed for the eukaryotic organism *Tetrahymena*, where intense thymine starvation causes excision-reparable damage (Westergaard and Pearlman, 1969). Figure 1 demonstrates that the damage causes a drastic increase in the specific activity of a particular DNA polymerase activity during starvation (Westergaard, 1970). The activity increases up to 35 fold during the treatment, but levels off and returns to normal level if thymidine is added to overcome the starvation (Westergaard and Pearlman, 1969).

Radiation of *Tetrahymena* with ultraviolet light or electrons also causes a considerable increase in the specific activity of the DNA polymerase. Table I demonstrates that the specific activity of the polymerase fraction is induced up to 17 fold by electron irradiation. It is evident from the table that the higher the dose of radiation (<400 Krad), the greater the induction of the polymerase. Time course studies demonstrate that the polymerase rises until the cells have recovered from radiation and are able to undergo normal mitosis. Thereafter the level of the enzyme declines towards the normal level in untreated cells (Keiding and Westergaard, 1971).

Subcellular fractionation of treated cells demonstrates that the induced polymerase is located in the mitochondria. However, the biosynthesis of the enzyme is probably under nuclear control since the induction is blocked by cycloheximide (Westergaard *et al.*, 1970). The enzyme can also be induced by treatment with low concentrations of ethidium bromide or chloramphenicol (Westergaard *et al.*, 1970; Tabak and Borst, personal communication; Keiding, personal communication), which indicates that the activity is controlled by a repressor encoded in the mitochondrial DNA (Barath and Küntzel, 1972). The synthesis or the regulation of the putative repressor may thus be directly linked to: (1) a step in the mitochondrial DNA synthesis or (2) a change in the DNA structure (*i.e.* relaxation of a supercoiled form). These hypotheses are further supported by observations that large amounts of replicative intermediates accumulate in the mitochondrial DNA under conditions where the polymerase is induced (Upholt and Borst, 1974).

In addition to these phenomena, exposure of *Tetrahymena* to excision-reparable damage of DNA also results in accumulation of replicative intermediates in the nuclear DNA.

Figure 2 shows that gene damage caused by thymine starvation leads to accumulation of a distinct class of "interdensity" nuclear DNA (region II) (Johnson and Westergaard, 1976).

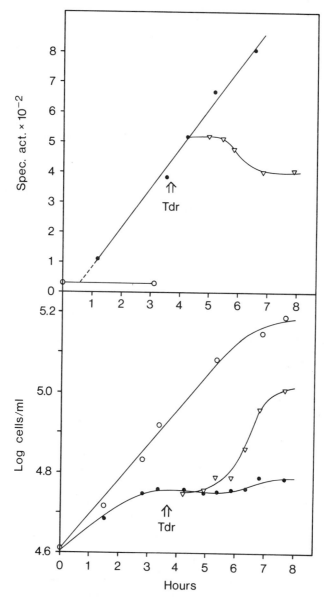

FIGURE 1. *Effect of thymine starvation on cell number and specific activity of DNA polymerase.*

The cells were grown in a complex medium and starved for thymine at time 0 as described in the methods section.

The culture was after 3.6 hrs of treatment divided into two parts. One mM thymidine was added to one part to overcome the starvation, while the other part was starved.

(o——o) *untreated, exponentially growing cells;*
(•——•) *thymine starved cells;* (∇——∇) *thymine starved cells*
with thymidine added.

In this experiment, cells prelabelled with (^{3}H)-thymidine
have been exposed to damage by thymine starvation and then exposed to a 45 min long pulse of bromodeoxyuridine and (^{32}P)-
phosphate. The interdensity DNA is precursor to the fully hybrid DNA in region I and contains newly synthesized strands,
which are smaller than the parental strand. Furthermore, all
interdensity DNA contains single stranded regions in contrast
to bulk DNA (region I) where only a few single stranded regions are found. These observations have led us to believe
that the DNA of region II represents replicating structures
accumulated in front of damage within the individual *replicons.*
Identical observations have been made (Kato and Strauss, 1974;
Rommelaere *et al.*, 1974) for UV irradiated cells, while it has
not been possible to observe any significant accumulation of
intermediates in exponentially growing cells. Thus, the system
described here might allow us to study the interaction between
replication and repair. In particular it might be possible to
solve the long debated question as to how much repair synthesis occurs at the growing point.

In order to study the conditions around the replication
fork further, we have recently developed methods which allow
us to isolate *replicons* of a distinct size (Nymann and Westergaard, 1976). Figure 3 demonstrates that all newly synthe-

TABLE I. Induction of DNA Polymerase in Response to
Electron Irradiation.

Dose of radiation Krad.	*Specific activity of DNA polymerase x 10^{-2}*	
	6½ hours after irradiation	*9 hours after irradiation*
0	3.1	2.9
5	8.0	4.9
20	12	4.3
40	22	8.1
80	31	8.8
120	35	14
150	41	23
300	53	29

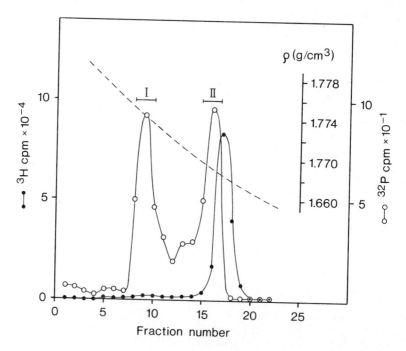

FIGURE 2. Neutral CsCl buoyant density gradient of DNA synthesized in the presence of bromodeoxyuridine after thymine starvation.

Cells grown in defined medium were prelabelled with (^{3}H)-thymidine (10 μC/ml) and starved for thymine as described in the methods.

After 18 hrs of starvation the cells were incubated with bromodeoxyuridine and (^{32}p)-phosphate (100 μC/ml) for 45 min. The nucleic acid was isolated and centrifuged to equilibrium in a neutral CsCl density gradient (see Johnson and Westergaard, 1976). The hydrolyzed fractions were assayed for acid precipitable radioactivity. The gravity field is directed to the left. (●——●) ^{3}H; (○——○) ^{32}p.

sized DNA can be accumulated as *replicons* if the cells have been treated with low concentrations of cycloheximide. In this experiment, cells prelabelled with (^{14}C)-thymidine have been exposed to a 30 min pulse of (^{3}H)-thymidine in the presence or absence of cycloheximide followed by a 60 min chase with cold thymidine. It is evident from the figure that all newly synthesized DNA accumulates as *replicons* (~41S, MW~17 x 10^6 dalton) in the presence of the drug, while the ^{3}H-labelled DNA

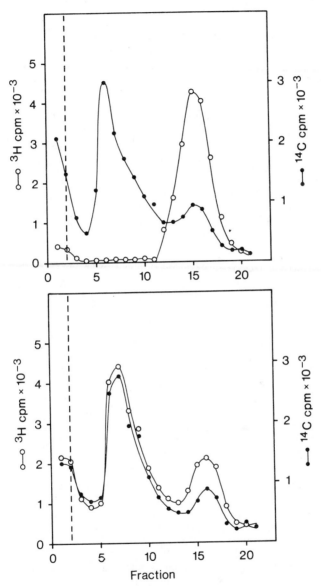

FIGURE 3. *Effect of cycloheximide on the maturation of*
replicons *into high molecular weight DNA.*
Cells prelabelled with (14C)*-thymidine were exposed to a*
30 min pulse of (3H)*-thymidine* (5 µC/ml) *followed by a 60 min*
chase with cold thymidine (100 µg/ml) *in* (top) *presence and*
(bottom) *absence of 1.6 µM cycloheximide. The cellular DNA was*

then sedimented on alkaline sucrose gradients as described by Nymann and Westergaard (1976). The gravity field is directed to the left. The fractions were assayed for acid precipitable radioactivity (o——o) ^3H; (●——●) ^{14}C.

in the control experiment is converted into high molecular weight material ($\geq$86S; MW$\geq$80 x 10^6 dalton). When the inhibitor is removed from the cells the accumulated *replicons* are joined into high molecular weight DNA. It has not been possible to detect any intermediates between the *replicons* and the high molecular weight DNA. The low concentration of cycloheximide has under these conditions only a slight effect (<10%) on the rate of the DNA synthesis, although the rate of the total protein synthesis is nearly completely blocked.

Investigations of the effect of mutagens on the structure and function of the eukaryotic chromosome have for a long time been complicated by the extreme complexity of the genetic material and the problems associated with its fractionation. In order to study some of these effects on the DNA metabolism and the transcription of specific gene sequences in chromatin, we have recently developed techniques which allow isolation of the ribosomal RNA gene from *Tetrahymena* in the state of transcriptionally active chromatin (r-chromatin) (Piper *et al.*, 1976; Leer *et al.*, 1976). The isolated chromatin is more than 95% pure with respect to DNA as seen from Figure 4 where DNA from the two final purification steps has been exposed to electrophoresis on an agarose gel. The fastest moving DNA band is r-DNA according to the following criteria: (1) It hybridizes to r-RNA in the gel using the gel hybridization technique. (2) It gives DNA fragments on cleavage by EcoRI restriction endonuclease identical to those yielded by digestion of r-DNA purified by other procedures (Engberg *et al.*, 1976; Karrer and Gall, 1976). (3) It has the buoyant density and size of r-DNA.

The r-chromatin is transcriptionally active, and estimations show an average of 10 active RNA polymerase molecules per gene. Furthermore, data indicate that (1) the transcription is faithful; (2) the *in vitro* transcription proceeds with a rate of 3-4 nucleotides/sec/RNA polymerase, which is about 20-25% of the *in vivo* rate, and finally (3) the transcript increases from a size of 2S up to approximately 17S during the transcription. A detailed study of the transcription process of the isolated chromatin using hybridization to c-DNA of 17S and 25S RNA, is now in progress.

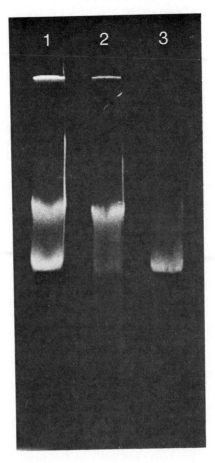

"Bulk" DNA

r DNA

FIGURE 4. *Agarose gel electrophoresis of DNA from isolated r-chromatin.*
The r-chromatin was isolated according to Leer et al.
*(1976). Slot 1 and 3 demonstrate the purity of r-DNA in the
two final purification steps. The positions of r-DNA and the
"bulk" DNA are indicated.*

In conclusion, we have developed conditions which allow
accumulation of intermediates in response to gene damage in
the eukaryotic organism *Tetrahymena pyriformis.* A detailed
study of these intermediates might allow us to obtain a more
detailed picture of some of the molecular events which occur

on the chromosome. Finally, our studies on specific chromatin
fragments might be of importance for the understanding of how
various chemicals and radiation effect the cellular differentiation on the transcriptional level.

SUMMARY

   Studies on the eukaryotic organism *Tetrahymena* have demonstrated a clear correlation between gene damage and molecular
events on chromatin. In this paper we have described how:
(1) Radiation or starvation for deoxyribonucleotides causes
induction of a specific DNA polymerase up to 35 fold. The
results show that there is a clear correlation between gene
damage and induction of polymerase. Furthermore, time course
studies demonstrated that the polymerase activity declines towards the normal low level at the time the cells have recovered from the damage and are able to undergo mitosis.
(2) Cells after exposure to gene damage accumulate large
amounts of newly synthesized DNA as DNA intermediates. The
isolated intermediates probably represent replication forks
lined up in front of damage within the individual *replicons*.
   Finally, we have described the isolation of a specific
gene in the state of transcriptionally active chromatin, and
how this gene can be used as a probe for studies of the effect
of gene damage on the transcription process.

REFERENCES

Barath, Z., and Küntzel, H. (1972). *Nature New Biol. 240*, 195.
Engberg, J., Anderson, P., Leick, V., and Collins, J. (1976).
    J. Mol. Biol. 104, 455.
Johnson, B., and Westergaard, O. (1975). *Eur. J. Biochem. 62*,
    345.
Karrer, K.M., and Gall, J.G. (1976). *J. Mol. Biol. 104*, 421.
Kato, K., and Strauss, B. (1974). *Proc. Natl. Acad. Sci. (USA)
    71*, 1969.
Keiding, J., and Westergaard, O. (1971). *Exp. Cell Res. 64*,
    317.
Leer, J.C., Nielsen, O.F., Piper, P.W., and Westergaard, O.
    (1976). *Biochem. Biophys. Res. Comm. 72*, 720.
Nymann, O., and Westergaard, O. (1976). *FEBS Lett. 64*, 139.
Piper, P.W., Celis, J., Kaltoft, K., Leer, J.C., Nielsen, O.F.,
    and Westergaard, O. (1976). *Nucl. Acid Res. 3*, 493.
Rommelaere, J., Fauris-Miller, A., and Errera, M. (1974). *J.
    Mol. Biol. 90*, 491.

Westergaard, O., and Pearlman, R. (1969). *Exp. Cell Res.* 54, 309.
Westergaard, O. (1970). *Biochem. Biophys. Acta 213*, 36.
Westergaard, O., Marcker, K.A., and Keiding, J. (1970). *Nature 227*, 708.
Upholt, W.B., and Borst, P. (1974). *J. Cell Biol. 61*, 383.
Zeuthen, E. (1968). *Exp. Cell Res. 50*, 37.

USE OF *DROSOPHILA* IN TESTS
FOR ENVIRONMENTAL MUTAGENS

Marja Sorsa[1]

Department of Genetics
University of Helsinki
Helsinki
Finland

INTRODUCTION

Just as the history of genetics is closely linked with *Drosophila melanogaster*, so has the "noble fly" played an important role in the research on chemical mutagenesis (cf. review by Auerbach, 1975), even though the first attempts of Muller in 1914 to produce mutations in *Drosophila* by alcohol or ether proved to be negative.

The general advantages of using *Drosophila* in mutagenicity tests are many and well-known (Sobels, 1974). The load of genetic information compiled during fifty years of *Drosophila* research is now available as the vast number of tester strains, genetic and chromosomal markers at only a fraction of the cost and time of the *in vivo* mammalian test systems.

At this stage, largely due to the intensive work done at the Sylvius Laboratory in Leiden, *Drosophila* study is gaining high priority among the short-term submammalian tests for mutagenicity.

AVAILABLE PROCEDURES FOR TESTING

Without doubt *Drosophila* is the only indicator animal which enables *in vivo* testing of the various forms of mutational events from the molecular to the chromosomal level,

[1]*Senior research worker, Natural Research Council for Sciences, Academy of Finland*

and from the germ line to somatic cells. In germ line cells, tests for the sensitivity of different cell stages and types can easily be arranged in both sexes by brood matings. The production of delayed genetic effects, characteristic of chemical mutagenesis, can be studied either by storing treated sperm in the receptacle of the female or by scoring delayed recessive lethals in successive generations of the treated animal.

The array of available procedures with *Drosophila* enables one to search for the most relevant tests in the detection of human genetic damage. This question of ultimate concern to the whole test philosophy is complicated by the huge variety of chemical compounds, their possible specificities in the mode of action in the living organism, and the various internal and external factors affecting the biotransformation process. The human genetic load is inevitably increased firstly, by point mutations and small deficiencies, causing deleterious effects. Secondly, chromosome breakage effects are expressed as balanced or unbalanced translocations and inversions, or deletions and duplications mainly causing severe abnormaliti-es or prenatal death. Thirdly, the non-disjunction events leading to aneuploids of various phenotypic expression are almost totally primary events and not hereditary, even though they comprise an important part of acutely expressed congeni-tal abnormalities. All of these mutational events can be tested in *Drosophila*. In addition to these high-risk mutatio-nal types in the germ line (for technical details of the tests, see Abrahamson and Lewis, 1971, Sobels, 1972), also various somatic events can be successfully tested in *Droso-phila*.

The area of somatic mutations, somatic non-disjunction, and mitotic recombination has been rather neglected among *Drosophila* workers as far as the use of somatic recombination tests for mutagenicity studies is concerned. This idea arises from the classical works of Auerbach (1945) on the induction of mitotic recombination with mustard gas. Even though the meaning of somatic recombination events from the point of view of human genetic health is not clear, such tests in *Drosophila* would be quite applicable by using eye colour markers (Becker, 1975) or bristle markers (Martensen and Green, 1977) to detect somatic mosaics.

A new promising application of *Drosophila* tests seems to be the testing of mutagenic substances present in mammalian peripheral blood and urine (Legator *et al.*, 1976) or tissue extracts (Gee *et al.*, 1972). The use of human body fluids and *Drosophila* as indicator could provide direct information on genetic and carcinogenic risks in man.

THE SEX-LINKED RECESSIVE LETHAL TEST

The recessive lethal test has gained special importance among the variety of available tests in *Drosophila* due to the recently published data of Vogel (1976) revealing a high correlation of known indirect carcinogens to their potency to induce recessive lethal mutations (Table I).

TABLE I. Mutagenic Response of Drosophila to 52 Indirect Carcinogens (Data of Vogel, 1976)

Class of chemical	Number of chemicals tested	Activity in the recessive lethal test
Triazenes	15	+
Nitrosamines	4	+
Hydrazo-, azoxyalkanes	3	+
Oxazaphosphorines	3	+
DDT, DDA	2	+
HEMPA	1	+
Aflatoxin B_1	1	+
Pyrrolizidine alkaloids	12	+
Vinyl chloride	1	+
Aromatic amines, polycyclic hydrocarbons	10	−

The test is relatively easy to perform with the *Basc* strain, and the result can always be precisely and unequivocally scored by re-testing. Furthermore, altogether some 1000 loci comprising about 20% of the total DNA of the genome can be tested in the same experiment. Also treated females can be tested, even though male treatment is more usual and safer, because it obviates the problem of pre-existing lethals in the X chromosome of the test animal.

The data of Vogel (1976, see Table I) indisputably reveal that, in addition to the ability to detect efficiently germ line mutations of deleterious type, the recessive lethal test simultaneously gives valid information on the carcinogenic potentials of the chemicals tested. It should be pointed out that the resolving power of the recessive lethal test measured as LEC/LD_{50} ratio seems to be 10 times (*e.g.* MMS) to 250 times (*e.g.* triazines) more efficient as compared to the induction of dominant lethals or chromosome loss in *Drosophila* (Vogel, 1976).

MUTAGENICITY *VS.* CARCINOGENICITY IN *DROSOPHILA* TESTS

As far as the risk of increasing the human genetic load is concerned, transmittable point mutations and small deletions deserve the greatest attention. The valuable achievements of Ames (Ames *et al.*, 1973, McCann and Ames, 1976) have demonstrated a striking overlap of carcinogens in the possession of mutagenic activity in the *Salmonella*/microsome test. These results suggest that the most informative types of mutations, on the basis of human carcinogenic risk, are point mutations (base pair substitutions and frameshift mutations). The recessive lethal test, able to detect small deletions and lethal code substitutions simultaneously in one fifth of the genome, is therefore superior to tests for other mutational types. In fact, Vogel's data (Vogel, 1976) on indirectly acting carcinogens clearly show that over 80% of these are effective in producing recessive lethals, but either do not induce chromosome breakages at all, or do so at much higher concentrations of the test chemical.

Results with *Drosophila* (op. cit.) reveal indirectly that the fruit fly is capable of metabolic activation reactions similar to mammals. Direct evidence of mixed-function oxidases has told us that even though the biotransformation processes are not concentrated in one organ, as they are in the mammalian liver, insect microsomes exhibit the same degree of metabolic versatility and substrate nonspecificity as mammalian microsomal enzymes (Sobels and Vogel, 1971, Vogel and Sobels, 1976).

Doubts about the validity of short-term submammalian mutagenicity tests as a prescreen for carcinogenicity have gradually subsided during the last two years' results with the microbial assay systems and with *Drosophila*. Indirectly, they give new and strong support to the old theory of somatic mutations as the ethiological cause of cancer (Boveri, 1914). At this stage, it is no longer unwarranted to expect that damage to DNA by the human chemical environment may be the main cause of death and disease in industrial societies (Cairns, 1975, McCann and Ames, 1976).

USE OF MUTAGEN-SENSITIVE STRAINS OF *DROSOPHILA*

The assumption about the identity of the processes of meiotic recombination and DNA repair has gained strong support from the results showing the sensitivity of the bacterial strains deficient of recombination (see review of Howard-Flanders, 1975) as well as of eukaryotic organisms (see review of Catchecide, 1974) to both radiation and chemical mutagens. Extensive evidence on the essential relationship of somatic

cell repair and meiotic recombination has recently been
obtained in studies with mutagen-sensitive strains of *Drosophila* (Boyd *et al.*, 1976b, Boyd and Setlow, 1976).

The mutagen-sensitive mutants were primarily selected from
EMS-induced flies on the basis of their sensitivity to methyl
methanesulphonate (MMS) at larval stages (Boyd *et al.*, 1976a).
Thirteen of these MMS sensitive mutants have been assigned to
ten complementation groups in the X chromosome, four of the
mutants being allelic to earlier known meiotic mutants (Baker
and Carpenter, 1972). Mutagen sensitivity is expressed after
treatments with various mutagens - in addition to MMS, nitrogen mustard, 2-acetylaminofluorene, γ-rays and UV. On the
basis of their data Boyd *et al.* (1976a) suggest assignment of
the mutants into three classes of mutagen sensitivity: firstly,
mutants that are sensitive to γ-rays and MMS, secondly,
mutants that are sensitive to nitrogen mustard and acetylaminofluorene, and thirdly, mutants that are sensitive to all of
these agents. The results show a reasonable correlation with
the data of Brendel and Haynes (1973) about mutagen-sensitive
mutants in *Saccharomyces*. When tested for their capacity to
repair damage induced by X-rays and UV in embryonal DNA, the
mutants have been tentatively assigned to four classes of repair capacity (Boyd and Setlow, 1976). Interestingly enough,
several of these MMS-sensitive mutants also show deficiency
in meiotic recombination as estimated from the increased frequency of spontaneous sex chromosome non-disjunctions and
losses (Boyd *et al.*, 1976a).

These data validate the assumption of the involvement of
recombinational mechanisms in somatic repair also in higher
eukaryotes. Furthermore, the future use of sensitive eukaryotic indicators, *e.g.* the repair defective strains of the fruit
fly, would make *Drosophila* even more valuable in mutagenicity
testing.

SUMMARY

A review of the various advantages of *Drosophila* in assay
systems for environmental mutagens is given. In addition to
its capacity to detect the whole spectrum of relevant genetic
damage, recent findings have revealed the capacity of *Drosophila* to carry out in a similar way metabolic activations on
the mammalian liver. In attempts to prescreen for carcinogens
with short-term submammalian tests, the relevance of the recessive lethal test is emphasized, with its ability to detect
forward mutations in 1000 loci simultaneously. In future, pro-

gress may be gained from the use of special indicator strains of *Drosophila*, sensitive to mutagens due to repair defectiveness.

In conclusion, *Drosophila*, especially in the test for recessive lethals, deserves high priority in screening programmes connected with environmental chemicals causing potential genetic risk to man.

REFERENCES

Abrahamson, S., and Lewis, E.B. (1971). *In* "Chemical Mutagens", Vol. 1 (A. Hollaender, ed.), p. 461. Plenum, New York.

Ames, B.N., Durston, W.E., Yamasaki, E., and Lee, F.D. (1973). *Proc. Natl. Acad. Sci. (USA). 70*, 2281

Auerbach, C. (1945). *Proc. Roy. Soc. Edinb. B 62*, 120.

Auerbach, C. (1975). *In* "Chemical Mutagens", Vol. 3 (A. Hollaender, ed.), p. 1. Plenum, New York.

Baker, B.S., and Carpenter, A.T.C. (1972). *Genetics. 71*, 255.

Becker, H.J. (1975). *Molec. gen. Genet. 138*, 11.

Boveri, T. (1914). "Zur Frage der Entstehung maligner Tumoren". Fischer, Jena.

Boyd, J.B., Golino, M.D., Nguyen, T.D., and Green, M.M. (1976a). *Genetics. 84*, 485.

Boyd, J.B., Golino, M.D., and Setlow, R.B. (1976b). *Genetics. 84*, 527.

Boyd, J.B., and Setlow, R.B. (1976). *Genetics. 84*, 507.

Brendel, M., and Hayes, R.H. (1973). *Molec. gen. Genet. 125*, 197.

Cairns, J. (1975). *Sci. Amer. 233*, 64.

Catcheside, D.G. (1974). *Ann. Rev. Genet. 8*, 279.

Gee, P.A., Sega, G., and Lee, W.R. (1972). *Mut. Res. 16*, 215.

Howard-Flanders, P. (1975). *In* "Molecular Mechanisms for Repair of DNA" (P.C. Hanawalt, and R.B. Setlow, eds.), p. 265. Plenum, New York.

Legator, M.S., Zimmering, S., and Connor, T.H. (1976). *In* "Chemical Mutagens", Vol. 4 (A. Hollaender, ed.), p. 171. Plenum, New York.

Martensen, D.V., and Green, M.M. (1977). *Mut. Res.*, in press.

Mc Cann, J. and Ames, B.N. (1976). *Proc. Natl. Acad. Sci. (USA). 73*, 950.

Sobels, F.H. (1972). *Arch. Genet. 45*, 101.

Sobels, F.H. (1974). *Mut. Res. 26*, 277.

Sobels, F.H., and Vogel, E. (1976). *Environ. Health Persp. 15*, 141.

Vogel, E. (1976). *In* "Screening Tests in Chemical Carcinogenesis" (R. Montesano, H. Bartsch, and L. Tomatis, eds.), p. 112, IARC, Lyon.

Vogel, E., and Sobels, F.H. (1976). *In* "Chemical Mutagens", Vol. 4 (A. Hollaender, ed.), p. 93. Plenum, New York.

ANTENATAL DIAGNOSIS

THE ROLE OF ANTENATAL DIAGNOSIS IN THE PREVENTION
OF BIRTH DEFECTS CAUSED BY ENVIRONMENTAL MUTAGENS

Aubrey Milunsky[1]

Harvard Medical School,
Genetics Division
Eunice Kennedy Shriver Center
and
Massachusetts General Hospital
Boston, Massachusetts

Certain considerations and limitations characterize
efforts at antenatal diagnosis. It is necessary to know exactly what type of disease is being sought and, for biochemical
disorders, the actual subtype. Frequently there is only one
opportunity for diagnostic studies, the pressure of time coup-
led with the slow growing amniotic fluid cells being the prime
problems. Only limited approaches exist at present for pursuing antenatal diagnosis. Efforts still depend largely on the
manifestation of genetic disorders in cultured and non-
cultured cells, on the measurements of substances in cell-
free amniotic fluid and upon examination of the fetus by non-
invasive techniques, e.g., ultrasound, or by invasive techniques such as amniography and fetoscopy. In the future, it may
become possible to induce cultured cells to reflect their
gene dysfunction. Studies have already been done demonstrating
the possibility that fetal cells may be of diagnostic use,
when found in the maternal circulation (Schröder and de la
Chapelle, 1972; Schröder and Herzenberg, in press).

[1]*Supported in part by U.S.P.H.S. Grants HD05515, GM07015
and HD09281*

Extent and Impact

There are about 2,400 recognizable genetic disorders with a simple mode of Mendelian inheritance that are primarily caused by point mutations (McKusick, 1975). Serious congenital malformations, mental retardation or genetic disease occur in 3-4% of all births. Available estimates concerning the causes of developmental defects are in reasonably close agreement (Brent, 1976; Wilson, 1976) (Table I). The etiology of developmental defects is at present not recognized in about two-thirds of the cases. No clue to the diagnosis is evident in at least a third of cases with severe mental retardation (Milunsky, 1975). Extremely worrisome is the report by Funderburk *et al.* (1977) confirming earlier papers (Breg *et al.*, 1972; Jacobs, 1974) and demonstrating a significant increase in non-Robertsonian translocations among individuals with mental retardation or genetic disease (Clow *et al.*, 1973).

Surveillance

Considerable difficulties are encountered in the monitoring of birth defects and in the interpretation of the observations made. For example, the incidence of certain defects may change with statistical significance even due to random variation. Such conclusions could be construed as false positive results and are exemplified from incidence data obtained in Atlanta by the Center for Disease Control where the evidence thus far has suggested that Warfarin, Dilantin or even alcohol have *no* teratogenic actions (Oakley, 1976).

TABLE I. Causes of Developmental Defects

	Estimates (%)	
	Wilson (1977)	Brent (1977)
Known genetic transmission	20	20
Chromosomal disorders	5	5-10
Environmental causes		
Radiation	1	1
Infections	2-3	2-3
Drugs and environmental		
agents	2-3	1
Maternal disease	1-2	2-3
Unknown	65-70	62-69

Published data, however, do implicate all these agents as
having teratogenic effects (Kerber *et al.*, 1968; DiSaia, 1966;
Meadow, 1968; Monson *et al.*, 1973; Jones *et al.*, 1974). A
clear and balanced perspective based on extensive carefully
collected and analyzed data has recently been provided by
Heinonen and associates (1977).

False negative reports may also arise because a certain
teratogenic agent may affect only a small number of exposed
embryos or because the exposure is very brief. Oakley (1976)
at the Center for Disease Control calculated that if 5% of the
pregnant women were exposed to an agent that increased the
incidence of cleft lip with or without cleft palate 4-fold,
the incidence would be expected to change only 17% - a level
that might be difficult to discern from the background noise.
Only if the risk for the defect was increased some 20-fold
would the monitoring probably detect a change.

The degree of teratogenicity of a compound will obviously
affect the ability with which it can be detected. Calculations
suggest that to detect a teratogen that doubled the incidence
of a defect which has an incidence of 1 in 1000 (e.g., anencephaly), would require study of some 23,000 newborns whose
mothers had actually been exposed during the first trimester
(Sullivan, 1974). The thalidomide tragedy serves to illustrate
this point rather well. The natural incidence of limb-reduc-
tion deformities was estimated to be between 1 in 100.000 and
1 in 1.000.000, while perhaps half of the embryos exposed during the critical period were affected. Sullivan pointed out
(1974) that the increase was 50.000 - 500.000 times greater
than normal. Even so it took 4.000 - 5.000 cases in Germany
from 1956-1961 to come up with a cause and effect relationship. It would seem that laboratory techniques will probably
be superior to epidemiological approaches.

SELECTED ASPECTS OF ENVIRONMENTAL TERATOGENESIS

The limited role of antenatal diagnosis in the detection
of disorders of environmental origin immediately becomes apparent when the known environmental teratogens are considered.
The frequency with which certain disorders occur may vary with
the season of the year. Seasonal clustering has, for example,
been noted for patent ductus arteriosus (Rutstein *et al.*,1952),
ventricular septal defect (Rotman and Fyler, 1974), congenital
dislocation of the hip (Cohen, 1971), and neural tube defects
(McKeown and Record, 1951; Record, 1961). The evidence concerning the seasonal occurrence of Down's syndrome remains conflicting (Penrose and Smith, 1966). A vital etiology for these
disorders remains most likely.

Drugs as transplacental teratogens may not only cause malformations, but lead to abortion or act as mutagens, the effects of which may be delayed a few decades. Diethylstilbestrol and vaginal carcinoma is one important example (Herbst *et al.*, 1971). The use of nitrosourea compounds in pregnant rats and the subsequent development of brain tumors in the progeny is another example (Druckrey *et al.*, 1972). Drugs with indisputed teratogenic effects are shown in Table II. There are many other compounds in man which may also be teratogenic but where the evidence remains conflicting. The difficulty in elucidating the teratogenic effect of drugs is confounded by a variety of interacting factors. The critical nature, for example, of the particular stage or even day of exposure during gestation is exemplified by experience with thalidomide in humans and nitrosourea compounds in rats. Other important variables include the strength (dosage) of exposure, the effects of synergism or interaction with other compounds (for example aspirin and benzoic acid in rats), the effects of genetic heterogeneity as reflected by slow and fast

TABLE II.   Drugs with Known Teratogenic Effects

Drugs	Effect on fetus
Amphetamines	Congenital heart disease, transposition of great vessels
Aminopterin (methotrexate)	Abortion, multiple malformations
Methyltestosterone	
17-alpha-Ethanyl-19-nor-testosterone (Norlutin)	
Progesterone	Masculinization of female fetus
17-alpha-Ethanyl-testerone (Progestoral)	
Diethylstilbestrol	Masculinization of female fetus, vaginal adenocarcinoma in adolescense
Thalidomide	Phocomelia and other malformations
Trimethadione and paramethadione	Abortion, multiple malformations, mental retardation
Warfarin	Hypoplasia of nasal structures, optic atrophy and mental retardation

drug metabolizers and the genetic predisposition reflected by a certain histocompatability type and the association of particular defects (e.g., H-2 locus and corticosteroid induced cleft palate in the mouse) (Goldman *et al.*, 1977).

The many known *chromosomal clastogens* (Shaw, 1970), whether physical or chemical in nature, are not clearly related to congenital malformations.

Occupational exposure (treated more fully by others at this conference) to teratogenic compounds raises extremely serious questions about the degree of risk, the possibilities of prevention and of antenatal diagnosis. A number of studies have shown that surgical operating room female personnel (including anesthetists, nurses and technicians) have a higher frequency of spontaneous abortion as well as a greater likelihood of having offspring with congenital defects (American Society of Anesthesiologists, 1974; Askrog and Harvald, 1970; Cohen *et al.*, 1971; Corbett *et al.*, 1974; Knill-Jones *et al.*, 1972). No specific type of malformation has been discerned. A particularly disturbing point was that the same observations have been made regarding wives of *male* anesthetists (American Society of Anesthesiologists, 1974).

Recent studies (discussed more fully in this conference) of individuals exposed to polyvinyl chloride have also pointed to a higher frequency of congenital malformations among the offspring of those exposed (Infante *et al.*, 1976a; Infante *et al.*, 1976b) (Table III). Again, a specific pattern of defects among the offspring of exposed individuals has not emerged. However, central nervous system defects featured prominently, the majority of serious malformations having been neural tube defects (personal communication from Dr. P.F. Infante). If these data are borne out in further studies then there would be clear implications in the context of antenatal diagnosis for exposed women. Much, however, has still to be done to define the nature and extent of the risk of polyvinyl chloride exposure in pregnancy before any firm offer of amniocentesis and antenatal studies could be made.

A veritable multitude of environmental factors exist which could singly or together affect the developing sperm or ovum, or embryonic development. Defects may arise from changes in the cell surface at an early embryonic stage leading ultimately to problems in cell recognition, cell migration and organization, as well as changes in the antigenic nature of the cell surface and alterations of programmed cell death. Our present inability to predict or anticipate with any certainty the consequences of so many environmental agents makes antenatal diagnosis a poor choice in the approach to prevention.

Whether certain teratogenic drugs act directly on the

TABLE III. Observed, Expected and Relative Risk for Specific Congenital
Anomalies in Index Areas Including N. Ridgeville, 1970-73[a]

| Defect category | Number of defects | | RR[b] |
	Observed	Expected	
All defects (740-756, 758, 759)[c]	109	56.0	1.95
Central nervous system (740-749)	17	5.6	3.02
Cleft palate and lip (749)	10	6.5	1.53
Genital organs (752)	16	8.4	1.90
Clubfoot (754)	23	8.2	2.79
All other defects	43	27.2	1.58

[a]Excludes skin, hair and nails (757)
[b]RR, relative risk (observed/expected).
[c]International Classification of Disease codes, 8th revision, are
shown in parentheses.

(From Infante, et al. (1976b). Mutation Res. 41, 131)

fetus to cause characteristic defects or whether metabolites of the drug exert the grim effects is unknown. Thalidomide is an important example illustrating the complexity of the necessary considerations. Thalidomide is a lipid soluble drug which easily crosses the placental barrier into the embryo or the fetus. In the embryo, it spontaneously hydrolyzes to produce certain water-soluble highly polar compounds. These compounds are unable to cross the placenta to re-enter the maternal circulation (Sullivan, 1973). Hence it appears that metabolites of thalidomide accumulate in the fetus (Williams *et al.*, 1965) and may well exert the actual teratogenic effects.

A recent new experimental observation is important and concerns the combined effect of *transplacental and postnatal exposures to chemicals*. It appears that fetal exposure to a chemical carcinogen facilitates the effect of postnatal contact with the same or different carcinogen (Fraumeni, 1974). Extrapolating from these findings may mean that a person's lifetime risk of developing cancer might be conditioned by prenatal chemical exposures. The same compound that may cause abortion in very early gestation may produce malformations some weeks later or induce neoplasia which could appear decades after *in utero* exposure.

The general consensus is that *smoking in pregnancy* does not yield offspring with an increased frequency of congenital malformations (Russel *et al.*, 1966; Russell *et al.*, 1968). Most studies appear to concur about the diminished body weight of the offspring of smokers (Russell and Millar, 1969; Mulcahy *et al.*, 1970; Yerushalmy, 1971) as well as the increased perinatal morbidity and mortality (Meyer and Comstock, 1972; Andrews and McGarry, 1972; Rush and Kass, 1972; Lubs, 1873). However, pregnant rats who have been exposed to cigarette smoke have been noted to produce offspring who had lower body and organ weights in contrast to controls, with diminished cell number and size in carcasses and hind brains (Haworth and Ford, 1972). Moreover, nicotine and metabolites administered intravenously to pregnant rats have been found in higher concentrations in fetuses than in maternal plasma from 30 minutes to 20 hours after injection (Mosier and Jansons, 1972). These investigators also found these compounds in rat amniotic fluids, which could constitute a reservoir for recycling through the fetus.

We have studied the amniotic fluid of women who smoked in pregnancy (Vanvunakis *et al.*, 1974). Measurements were made of the amniotic fluid content of nicotine and cotinine during the second trimester. Smokers were easily detected by these amniotic fluid studies. Most disturbing, though perhaps not all that surprising in view of the experimental

evidence just noted, are the results of British studies indicating the increased frequency of learning disorders among the
offspring of smoking mothers (Butler and Goldstein, 1973).

A prelude to disaster was heralded when in the villages
along Minimata Bay in the Kyushu district of Japan in the
middle 1950's cats went mad and died. These incidents retrospectively had signaled a disorder in humans subsequently
called Minimata disease and found to be due to pollution of
water with *methylmercury* (Harada, 1968; Miller, 1974). The
mercury originated from waste flushed into the Bay by a factory that made vinyl plastic. A chemical concentrated in fish
and affected those humans eating the contaminated fish. Adults
and older children suffered a degenerative neurological
disease. Only later was it observed that about 6% of births
(in contrast to about 0.5% elsewhere in Japan) resulted in
children with cerebral palsy. This disorder was the first
evidence that a chemical pollutant could harm the human fetus.
Some of the Minimata Bay children who had previously had no
overt signs of methylmercury poisoning have subsequently displayed deficiencies in athletic ability in their teens.

Years later in the United States, a farmer and his family
were exposed to methylmercury containing fungicide. They ate
contaminated pork and later three of the children developed
severe brain damage and the pregnant mother delivered a child
with cerebral palsy and mental retardation (Roueche, 1970;
Pierce *et al.*, 1972; Snyder, 1971). Inadvertent mercury
poisoning of many thousands of people has also occurred in
Iraq (Bakir *et al.*, 1973).

It is the alkyl mercury compounds of which methylmercury
is most toxic that are most hazardous to man. Methylmercury
may cause chromosomal damage even in minimal concentrations.
There has however, been no clear evidence for the causation
of malformations by mercury compounds. A disturbing aspect is
that mercury does cross the placenta and may cause fetal damage even at concentration levels that produce no symptoms or
signs in the mother. The propensity of mercury for fetal *vs.*
maternal tissues has become evident through observations
showing higher concentrations of mercury in fetal red blood
cells (umbilical cord) as compared to maternal blood (Suzuki
et al., 1971). Moreover, the level of mercury in fetal brain
has been found to be twice that in maternal brain (Clegg,
1971).

Another *chemical pollutant* (polychlorinated biphenyls)
which contaminated foodstuff also affected the fetus. Contamination occurred through the use of a particular brand of
cooking oil. Ten pregnant women who used the cooking oil gave
birth to cola-colored babies (Kuratsune *et al.*, 1972;
Yamaguchi *et al.*, 1971). While the color faded during infancy,

the possibility of delayed effects remains uncertain.

Airborne *lead pollution* has been implicated in the causation of lead poisoning over 40 years ago. More recently,
cord blood levels have been studied in the offspring of women
living in areas with high lead pollution (near expressways).
These blood samples have demonstrated higher than normal lead
concentrations (Scanlon, 1971) - observations which raise
further questions about the known relationship of low chronic
lead exposure to the subsequent development of learning disorders or even mental retardation (de la Burde and Choate,
1972; Pueschel, 1974; David, 1974).

Lead exposure has been implicated in the virtual doubling
in the rate of spontaneous abortion in Japan (84/1000 after
exposure compared to 46/1000 before exposure) (Catz and
Yaffee, 1976). Certainly in animals the passage of lead across
the placenta has been demonstrated (Charlotte *et al.*, 1976).
Lead poisoning during the last trimester may lead to the birth
of infants who are underweight, anemic and who have slow
development (Angle and McIntire, 1964).

Insecticide Exposure

In one U.S. Report, pregnant women in a rural agricultural
area were found to have levels of residues of chlorinated
hydrocarbon insecticides in their serum comparable to those
found in those occupationally exposed (D'Ercole *et al.*, 1976).
Persons living in such areas often have more insecticide exposure than do the general population, and may be exposed by
ingestion of contaminated foodstuffs and drinking water and by
insecticide inhalation through aerial spraying. The cord blood
of offspring of these women were also found to have significant residue levels. There were no signs or symptoms of acute
insecticide poisoning in their newborns. The incidence of
congenital malformations did not appear to differ from that of
the general population (note, however, that only 339 women
were studied). Others have also observed high insecticide
residue in maternal newborn pair studies (Selby *et al.*, 1969).
The mean level of DDT metabolites was 2-5 times higher for
both mothers and newborns than previously noted in the United
States. Concentrations of DDT metabolites were comparable to
the mean levels reported in occupationally exposed chemical
company employees (Edmundson *et al.*, 1969). It is clear that
exposure to DDT probably occurs throughout fetal life since
residues have been detected in tissues of abortuses (O'Leary
et al., 1970). Residues were detectable in cord blood from a
20 week fetus in this study. Even though the study was done
at a time when almost no DDT was in use, it showed that over

90% of the mothers, and 45% of the white newborns were exposed.

Claims have also been made that dioxin exposure (a herbicide) led to almost immediate abortion in 22 of 73 pregnant women (Laporte, 1977). Further claims indicate a high frequency of chromosomal aberrations in those exposed to dioxin as well as a higher stillbirth rate and congenital malformation rate (Laporte, 1977).

Behavioral Teratogens

Since the thalidomide tragedy, much attention has been paid to the possible effects of drugs and chemicals on the developing fetus. It is only relatively recent that this interest has extended to the consideration of non-morphological abnormalities. Biochemical or functional disturbances resulting in behavioral changes may well prove to be equally or more important than gross morphological congenital malformations that have occupied our major attention. Most of the available evidence for these possibilities is still based on animal experimentation.

Methylmercury administered to rats during pregnancy may produce enduring behavioral deficits without any obvious morphological abnormalities (Rosenthall and Sparber, 1972; Spyker and Sparber, 1971; Spyker *et al.*, 1972). There is supporting evidence that differences in drug dosage may cause variable effects. High doses of methylmercury or vitamin A, for example, may cause severe congenital malformations especially of the central nervous system (Kalter, 1968), whereas much lower doses may result in behavioral deficits (Butcher *et al.*, 1972; Hutchings *et al.*, 1973; Hutchings and Gaston, 1974; Rosenthall and Sparber, 1972; Spyker and Sparber, 1971; Spyker *et al.*, 1972). Chlorpromazine may induce behavioral anomalies in the offspring of rats exposed to this drug during pregnancy (Golub and Kornetsky, 1974). The prenatal administration of delta-9-tetrahydrocannabinol may lead to a delay in physical maturation and behavioral development (Borgen *et al.*, 1973). Certainly in the experimental model system, any drug capable of causing physical defects in exposed progeny may well have an affect on behavior in lower doses. Experimental neonatal exposure to halothane may produce enduring learning deficits as well as cerebral synaptic malformation (Quimby *et al.*, 1974). A degree of behavioral teratogenesis might therefore be dependent not only upon the dose, but also upon the stage of gestation when exposure occurs. Finally, the subsequent behavior of rats irradiated at low doses while *in utero* appers to be altered in a reproducible and quantitative fashion (Mullenix *et al.*, 1975).

THE CURRENT AND POTENTIAL POSSIBILITIES FOR ANTENATAL
DIAGNOSIS

The brief review given above of some of the environmental
agents which may damage the human organism *in utero* reflects
the difficulty in predicting specific and/or diagnosable
defects in early pregnancy. Only when an environmental agent
can be associated with specific fetal defects could antenatal
diagnosis have any important role in prevention. The technology and approach to prevention have been fully described
(Milunsky, 1975) and for antenatal diagnosis recently updated
(Milunsky, 1976).

CONCLUSIONS

1. Antenatal diagnosis is not regarded as a screening
tool but rather as a diagnostic tool. Hence antenatal diagnosis will be of the greatest value where it is clear what
disorder should be sought for. In time, at risk groups may be
recognized (e.g., wives of anesthetists; wives of polyvinyl
chloride workers, etc.) who could then benefit from antenatal
diagnostic studies.

2. While test systems are being perfected and epidemiologic surveillance continues, extremely careful monitoring
of pregnancies and births in high risk professions and
industries should be pursued.

3. The role of antenatal diagnosis in the prevention of
defects resulting from environmental agents will be insignificant until such time as new specific risk data emerge.

REFERENCES

American Society of Anesthesiologists (1974). *Anesthesiology.*
  41, 321.
Andrews, J., and McGarry, J.M. (1972). *J. Obstet. Gynecol. Br.
  Commonw. 79*, 1057.
Angle, C.R., and McIntire, M.S. (1964). *Amer. J. Dis. Child.
  108*, 436.
Askrog, V., and Harvald, B. (1970). *Nord. Med. 83*, 498.
Bakir, F., Damluji, S.S., Amin-Zaki, L.,*et al*. (1973).
  Science. 181, 230.
Borgen, L.A., Davis, W.M., and Pace, H.B. (1971). *Toxicol.
  appl. Pharmac. 20*, 480.

Breg, W.R., Miller, D.A., Allerdice, P.W., *et al.* (1972).
    Amer. J. Dis. Child. 123, 561.
Brent, R.L. (1976). *In* "Prevention of Embryonic, Fetal and
    Perinatal Disease" (R.L. Brent, and M.I. Harris, eds.),
    p. 211. Bethesda, U.S. Government Printing Office,
    Washington, D.C.
Butcher, R.E., Brunner, R.L., Roth, T., *et al.* (1972). *Life
    Sci. 11*, 141.
Butler, N.R., and Goldstein, H. (1973). *Brit. Med. J. 4*, 573.
Catz, C.S., and Yaffee, S.J. (1976). *In* "Prevention of
    Embryonic, Fetal and Perinatal Disease" (R.L. Brent, and
    M.I. Harris, eds.), p. 119. U.S. Government Printing
    Office, Washington, D.C.
Clegg, D.J. (1971). "Embryotoxicity of mercury compounds;
    special symposium on mercury in man's environment."
    Monograph, Ottawa, Canada.
Clow, C.L., Fraser, F.C., Laberge, C., *et al.* (1973). *In*
    "Progress in Medical Genetics", Vol. 9 (A.G. Steinberg,
    and A.G. Bearn, eds.), p. 159. Grune and Stratton, Inc.,
    New York.
Cohen, E.N., Bellville, J.W., and Brown, B.W., Jr. (1971).
    Anesthesiology. 35, 343.
Cohen, P. (1971). *J. Interdiscipl. Cycle Res. 2*, 417.
Corbett, T.H., Cornell, R.G., Endres, J.L., *et al.* (1974).
    Anesthesiology. 41, 341.
David, O.J. (1974). *Environ. Health Perspect. 7*, 17.
de la Burde, B., and Choate, M.S., Jr. (1972). *J. Pediat. 81*,
    1088.
D'Ercole, A.J., Arthur, R.D., Cain, J.D., *et al.* (1976).
    Pediatrics. 57, 869.
DiSaia, P.J. (1976). *Obstet. Gynecol. 28*, 469.
Druckrey, H., Ivankovic, S., Preussmann, R., *et al.* (1972).
    In "The Experimental Biology of Brain Tumors" (W.M. Kirsch,
    E. Grossi-Paoletti, and P. Paoletti, eds.), p. 85.
Edmundson, W.F., Davies, J.E., Nachman, G.A., *et al.* (1969).
    Public Health Rep. 84, 53.
Fraumeni, J.F. (1974). *Pediatrics. 53*, 807.
Funderburk, S.J., Spence, M.A., and Sparkes, R.S. (1977).
    Amer. J. Hum. Genet. 29, 136.
Goldman, A.S., Katsumata, M., Yaffee, S.J., *et al.* (1977).
    Pediat. Res. 11, 456.
Golub, M., and Kornetsky, C. (1974). *Devel. Psychobiol. 7*. 79.
Harada, Y. (1968). *In* "Congenital (or fetal) Minamata disease:
    Minamata disease," p. 73. Study Group of Minamata Disease,
    Kumamota University, Japan.
Haworth, J.C., and Ford, J.D. (1972). *Amer. J. Obstet. Gynecol.
    112*, 653.

Heinonen, O.P., Slone, D., and Shapiro, S. (1977). "Birth Defects and Drugs in Pregnancy". Publishing Sciences Group, Inc., Littleton, Mass.

Herbst, A.L., Ulfelder, H., and Poskanzer, D.C. (1971). *New Eng. J. Med. 284*, 878.

Hutchings, D.E., and Gaston, J. (1974). *Devel. Psychobiol. 7*, 222.

Hutchings, D.E., Gibbon, J., and Kaufman, M.A. (1973). *Devel. Psychobiol. 6*, 445.

Infante, P.F., McMichael, A.J., Wagoner, J.K., *et al.* (1976a). *Lancet. 1*, 734.

Infante, P.F., Wagoner, J.K., and Waxweiler, R.J. (1976b). *Mutation Res. 41*, 131.

Jacobs, P.A. (1974). *Nature. 249*, 164.

Jones, K.L., Smith, D.W., Streissguth, A.P., *et al.* (1974). *Lancet. 1*, 1076.

Kalter, H. (1968). "Teratology of the Central Nervous System". The University of Chicago Press, Chicago.

Kerber, U.J., Warr, O.S., and Richardson, C. (1968). *J.A.M.A. 203*, 223.

Knill-Jones, R.P., Rodrigues, L.V., Moir, D.D., *et al.* (1972). *Lancet. 1*, 1326.

Kuratsune, M., Yoshimura, T., Matsuzaka, J., *et al.* (1972). *Environ. Health Perspect. 1*, 119.

Laporte, J.-R. (1977). *Lancet. 1*, 1049.

Lubs, M.E. (1973). *Amer. J. Obstet. Gynecol. 115*, 66.

McKeown, T., and Record, R.G. (1951). *Lancet. 1*, 192.

McKusick, V.A. (1975). "Mendelian Inheritance in Man. Catalogs of Autosomal Dominant, Autosomal Recessive, and X-Linked Phenotypes", 4th ed. The Johns Hopkins University Press, Baltimore.

Meadow, S.R. (1968). *Lancet. 2*, 1296.

Meyer, M.B., and Comstock, G.W. (1972). *Amer. J. Epidemiol. 96*, 1.

Miller, R.W. (1974). *Pediatrics. 53*, 792.

Milunsky, A. (1975). "The Prevention of Genetic Disease and Mental Retardation". W.B. Saunders Company, Philadelphia.

Milunsky, A. (1976). *New Eng. J. Med. 295*, 377.

Monson, R.R., Rosenberg, L., Hartz, S.C., *et al.* (1973). *New Eng. J. Med. 289*, 1049.

Mosier, H.D., Jr., and Jansons, R.A. (1972). *Teratology. 6*, 303.

Mulcahy, R., Murphy, J., and Martin, F. (1970). *Amer. J. Obstet. Gynecol. 106*, 703.

Mullenix, P., Norton, S., and Culver, B. (1975). *Exp. Neurol. 48*, 310.

Oakley, G.P., Jr. (1976). *In* "Cytogenetics, Environment and Malformation Syndromes". Birth Defects: Original Article Series, Vol. XII (D. Bergsma, and R.N. Schimke, eds.), p. 1. The National Foundation-March of Dimes.

O'Leary, J.A., Davies, J.E., and Feldman, M. (1970). *Amer. J. Obstet. Gynecol. 108*, 1281.

Penrose, L.S., and Smith, G.F. (1966). "Down's Anomaly". Churchill, London.

Pierce, P.E., Thompson, J.F., Likosky, W.H., *et al.* (1972). *J.A.M.A. 220*, 1439.

Pueschel, S.M. (1974). *Environ. Health Perspect. 7*, 13.

Quimby, K.L., Aschlenase, L.J., and Bowman, R.E. (1974). *Science. 185*, 625.

Record, R.G. (1961). *Brit. J. Prev. Soc. Med. 15*, 93.

Rosenthall, E., and Sparber, S.B. (1972). *Life Sci. 11*, 883.

Rothman, K.J., and Fyler, F.C. (1974). *Lancet. 1*, 193.

Roueche, B. (1970). *New Yorker,* August 22.

Rush, D., and Kass, E.H. (1973). *Amer. J. Epidemiol. 96*, 183.

Russell, J.K., and Millar, D.G. (1969). *In* "Perinatal Factors Affecting Human Development", No. 185, October. Pan American Health Organization Scientific Publication.

Russell, C.S., Taylor, R., and Law, C.E. (1968). *Brit. J. Prev. Soc. Med. 22*, 119.

Russell, C.S., Taylor, R., and Maddison, R.N. (1966). *J. Obstet. Gynecol. Br. Commonw. 73*, 742.

Rutstein, D.D., Nickerson, R.J., and Heald, F.P. (1952). *Amer. J. Dis. Child. 84*, 199.

Scanlon, J. (1971). *Amer. J. Dis. Child. 121*, 271.

Schröder, J., and de la Chapelle, A. (1974). *Transplant. 17*, 346.

Schröder, J., and Herzenberg, L.A. (in press). *In* "Hereditary Disorders of the Fetus: Diagnosis, Prevention and Treatment" (A. Milunsky, ed.), Plenum Press, New York.

Selby, L.A., Newell, K.W., and Hauser, G.A., *et al.* (1969). *Environ. Res. 2*, 247.

Shaw, M.W. (1970). *Ann. Rev. Med. 21*, 409.

Snyder, R.D. (1971). *New Eng. J. Med. 284*, 1014.

Spyker, J.M., and Sparber, S.B. (1971). *Pharmacologist. 13*, 275.

Spyker, J.M., Sparber, S.B., and Goldberg, A.M. (1972). *Science. 177*, 621.

Sullivan, F.M. (1973). *In* "Transplacental Carcinogenesis" (L. Thomas and U. Mohr, eds.), No. 4, IARC Scientific Publications, Leon.

Sullivan, F.M. (1974). *Pediatrics. 53*, 797.

Suzuki, T., Miyama, T., and Katsunuma, H. (1971). *Bull. Environ. Contam. Toxicol. 5*, 502.

Van Vunakis, H., Langone, J.J., and Milunsky, A. (1974).
    Amer. J. Obstet. Gynecol. 120, 64.
Williams, R.T., Schumacher, H., Fabro, S., *et al.* (1965).
    "Embryopathic Activity of Drugs", Churchill, London.
Wilson, J.G. (1976). *In* "Prevention of Embryonic, Fetal and
    Perinatal Disease" (R.L. Brent and M.I. Harris, eds.),
    Bethesda, U.S. Government Printing Office, Washington,
    D.C.
Yamaguchi, A., Yoshimura, Y., and Kuratsune, M. (1971).
    Fukuoka Acta Med. 62, 117.
Yerushalmy, J. (1971). *Amer. J. Epidemiol. 93*, 443.

RESULTS OF FOUR YEARS OF SYSTEMATIC,
PROPHYLACTIC FETAL CHROMOSOME ANALYSIS

John Philip

Section of Teratology
Department of Obstetrics and Gynecology
Rigshospitalet
University of Copenhagen
Denmark

INTRODUCTION

One of the purposes of this meeting is to investigate the value of diagnostic methods in detection of genetic disease in the *foetus*. One aspect of this problem, *i.e.* the indicence of *chromosomal* disease in the foetus, will be considered here.

Changes in frequencies of genetic diseases may be the result of environmental influences. Recognition of such changes is only possible when the results of studies of large populations are available.

MATERIAL

The material consists of 1086 consecutive foetal chromosome analyses in a hospital which accepts risk-cases as well as normal women from a large area. This area is not a well defined catchment area so the type of bias in referrals cannot be defined.

When referred, all women who have an elevated risk of having a chromosomally abnormal baby, are offered amniocentesis (so far, less than ten have not accepted the offer). Furthermore, the offer of amniocentesis is given to women with risks of other detectable genetic diseases such as inborn errors of metabolism or to women with a risk of anencephaly or

spina bifida. In all these cases a chromosome analysis is also
carried out. A number of women have amniocentesis for other
reasons; among these are women with a specific wish for amniocentesis but without any known elevated risk of a chromosomally abnormal foetus.

METHODS

Amniocentesis and culture of foetal cells were carried out
according to earlier published methods (Bang and Northeved,
1972, Philip *et al.*, 1974).

RESULTS AND DISCUSSION

The material may be divided into two parts:
1) Women *with* elevated risk of having children with
chromosomal disease, and
2) Women *without* known elevated risk of having children
with chromosomal disease.
Table I shows the results of foetal chromosome investiga-

*TABLE I. Results of Studies of Foetal Chromosomes.
Mothers with Elevated Risk of Having Abnormal Foetuses.*

	No. investigated	No. with abnormalities	% abnormal
Known familial translocation	4	1[a]	25
X-linked disease	20	(9 males)	(45)
Mother has chromosome abnormality	3	0	—
Mother 40 years or more	243	7[b]	2.9
Mother 35-39 years of age	422	1[c]	0.2
Previous child with chromosome abnormality	47	0[c]	—
Total	739	9 (+9 males)	1.3 (2.5)

[a] +3 cases were carrying the familial translocation in balanced form
[b] +2 balanced translocations
[c] +1 balanced translocation

TABLE II. *Results of Studies of Foetal Chromosomes.*
Mothers with Average Risk of Having Chromosomally Abnormal
Foetuses

	No. investigated	No. with abnormalities	% abnormal
Previous child with congenital malformation, inborn errors of metabolism, oligophrenia or other disease	68	0[a]	—
Previous child with spina bifida, anencephaly or hydrocephalus	71	1[b]	1.4
Mother has spina bifida or hydrocephalus	2	0	—
Down's syndrome "in family"	75	3[c]	3.8
Patient wishes investigation	70	1[d]	1.4
Other	61	0	—
Total	347	5	1.4

[a] +1 *case of congenital adrenal hyperplasi*
[b] +2 *cases: elevated* α*-foetoprotein − anencephaly*
[c] +1 *case of 46,XX male discovered after delivery*
[d] +1 *balanced translocation*

tion of the series of cases with *elevated* risk of having
children with chromosomal disease. Of 739 cases 9 or 1.3% had
a chromosome abnormality. Furthermore, 9 male foetuses were
found in families with X-linked disease.

   Table II shows that 347 foetuses were investigated chromosomally *without* having known elevated risk of chromosome
anomaly. Five chromosome abnormalities (1.4%) were
found. The numerical risk of a foetus with a chromosome abnormality was thus found to be identical in the two groups
(p >99%).

   The total number of balanced translocations was 7, three of
which occurred in families without known familial translocation.

TABLE III. List of Abnormal Cases, According to Known Risk

Elevated		Not elevated	
Unbalanced translocation	(1 case)	47,XYY	(2 cases)
13 trisomy	(1 case)	47,XXX	(1 case)
18 trisomy	(1 case)	46,XX male	(1 case[a])
21 trisomy	(5 cases)	45,XO/46,XY	(1 case)
47,XXX	(1 case)		
9 males in families with X-linked disease		2 cases of anencephaly 1 case of congenital adrenal hyperplasia	

[a]*Diagnosed after delivery*

In Table III the chromosome abnormalities are listed, divided by groups. The balanced translocations are not included. As the numbers are admittedly small the distribution of abnormalities (sex chromosome abnormalities/autosomal abnormalities) may be due to chance. The chromosome abnormalities causing severe malformations, that is autosomal trisomies, were all found in the group with elevated risk. The abnormal cases in the group without known elevated risk all had sex chromosome abnormalities; assuming that the risk of severe (autosomal) abnormalities is the same in the two groups the probability of the present findings is 3.7%.

*TABLE IV. Results of Studies of Foetal Chromosomes.
Survey of Series with Large No. of Females who Did not Have
Elevated Risk of Having a Foetus with Chromosome Abnormality*

	No. of pregnancies	No. with abnormalities	%
European Collaborative Study (Galjard, 1976)	711	12	1.7
Canadian Medical Research Council (1977)	191	2	1.0
German Medical Research Council (31.12.1976)	843	3(+?)	0.4(+?)
Ferguson-Smith and Ferguson-Smith (1976)	497	3	0.6
Philip and Bang Present material	359	5	1.4
Total	2601	25	0.96

The following figures are from the literature: in 2601 foetuses investigated in many different laboratories (but found in 5 publications) the number of abnormal chromosome constitutions was almost 1% (Table IV). All these foetuses were investigated chromosomally although they did not have any known elevated risk of chromosomal disease. Thus the literature seems to confirm our findings.

The average frequency of chromosome abnormalities in live-born babies is about 0.5% (Jacobs *et al.*, 1974). The frequency of chromosome abnormalities in the two groups of foetuses in the present material is significantly different from 0.5% (p= 1.2% and 3.6%, respectively). It should be remembered that other authors found higher than expected frequencies of abnormalities in risk groups studied prenatally (Ferguson-Smith and Ferguson Smith, 1976).

Still, the lack of difference in frequency of chromosome abnormalities in foetuses from mothers with and without known elevated risk and the significant difference between foetal and neonatal series may be due to chance because of small numbers and sampling errors. One possible explanation may be that the so-called non-risk groups do in fact include a number of patients with the known or unknown risk factors; it should be noted that the three pairs of parents of abnormal children in the group Down's syndrome in the family had normal chromosomes, except one mother who had an unknown balanced translocation involving chromosomes 6 and 8. Furthermore, all mothers were less than 35 years of age. The age of the 347 mothers is seen in Table V.

The small numbers may account for the differences in the frequency of autosomal abnormalities in risk and non-risk groups.

A number of questions may be raised on the basis of the findings:

Is it justified to screen the whole population of foetuses (all pregnant women), considering that a number of not severe abnormalities will be found?

TABLE V. Age of Mothers with not Known Risk of Having Children with Chromosome Abnormalities

<20	20-24	25-29	30-34
5	69	138	146

Is it justified *not* to screen the whole population of foetuses, considering the possibility of a 1% risk of chromosomal abnormalities in non-risk groups and the low risk of amniocentesis now demonstrated?

Is it justified to abort foetuses with chromosome abnormalities which are not really causing very severe congenital malformations and/or mental retardation?

Time will not allow further discussion of the problems, which have to be considered before decisions are made on an enlargement of the profylactic programs for prevention of chromosomal diseases.

The conclusion is that our definition of risk groups may not be correct at present.

REFERENCES

Bang, J., and Northeved, A. (1972). *Am. J. Obst. Gyn. 114*, 559.

Canadian Collaborative Study (1977). *Canadian Medical Research Council. Report No. 5*, Ottawa.

Deutschen Forschungsgemeinschaft (1976). *Informationsblatt 10*, München.

Ferguson-Smith, M.A., and Ferguson-Smith, M.E. (1976). *In* "Prenatal Diagnosis Colloqium" (E. Boué, ed.), p. 81. INSERM, Paris.

Galjaard, H. (1976). *Cytogenet. Cell Genet. 16*, 453.

Jacobs, P.A., Melville, M., Ratcliffe, S., Keay, A.J., and Syme, J. (1974). *Ann. Imm. Genet. 37*, 359.

Philip, J., Bang, J., Hahnemann, N., Mikkelsen, M., Niebuhr, E., Rebbe, H., and Weber, J. (1974). *Acta obstet. gynec. scand. Suppl. 29*, 1.

COST OF MUTATION

THE COST OF MUTATION

J.H. Edwards

Infant Development Unit
Birmingham Maternity Hospital
Edgbaston, Birmingham, England

> *Important deficiencies in this accumulated body of knowledge and theory become strikingly apparent, however, whenever attempts are made to apply it to problems that are of social as well as scientific importance (Newcombe 1971).*

In classical usage mutation refers to a change at a locus which retains the capacity to replicate and excludes changes at the level of the chromosome. Recent usage has been vague, the term casually extending over all elements in the genetic hierarchy, and being widely confused with the consequences of some somatic mutations (carcinogenesis) and with impaired embryonic development (teratogenesis). This confusion is very deep, and has already had a number of unfortunate consequences. The Flowers report (1976) for example, defines genetic effects in its glossary as "Effects produced by radiation in the offspring of the person irradiated, usually malformations". A mutant genetic unit may be transmitted by a gamete and be associated with the development of carcinomata, or, more usually, other tumours, and may also be a cause of malformation. However, genetically transmissable mutants associated with neoplasia are rare, accounting for well below 1% of neoplasms, and genetical causes of malformation are likewise rare.

Since neoplasia seems largely consequent upon some cellular event which involves the genetic apparatus of the cell it is hardly surprising that most mutagens are carcinogens and most cancinogens are mutagens. And since embryogenesis is the most exacting test of the control of the genetic elements which determine the differences in cell types, and their birth rates and death rates, it is hardly surprising that most carcinogens

and mutagens are teratogens. However, since a wide range of
small molecules, and of parasites, are teratogenic it is probable that most teratogens are not mutagenic or carcinogenic.

As neoplasia is usually demonstrable in mammals only after
many months, and mutagenesis may be demonstrable in organisms,
or mammalian cells, which reproduce in minutes or hours, mutagenic tests are now used to define probable carcinogens. However, the major hazard of mutagens is mutation, and the fact
that our generation may suffer from cancers by being exposed to
them is no excuse for ignoring the cumulative consequences we
may be imposing on our remote descendants.

I will use mutagen in the specific sense of causing a
quantitative or qualitative abnormality in the genetic apparatus which is transmitted by a gamete, but will try and restrict
"mutation" to a qualitative change, and refer when possible to
the chromosomal changes of quantity, or arrangement, as "chromosomal mutation", a clumsy and unsatisfactory term now made
unnecessary by the terms "clastogen" and "turbigen" which cover
numerical and structural chromosomal abnormalities respectively (Shaw, 1970; Brøgger, 1977).

Mutation, if unqualified, usually implies point mutation
in classical usage, but even here difficulties arise since a
gene can no longer be considered as having a one-to-one relationship to the smallest unit capable of resolution by optical
microscopy in *Drosophila* chromosomes. A gene, like any other
physical unit, is defined in terms of the most precise techniques available, and is now usually defined as that segment of
DNA which transcribes into one messenger RNA; the problem of
polycistronic messengers does not appear to arise at the level
of translation in mammals, and has only been observed in primitive organisms. The phrase point mutation is usefully retained
as a generic term for mutations of units beyond optical resolution. One of the classical point mutations in man, retinoblastoma, is now known to be frequently associated with an in-
terstitial deficiency, and a steady promotion from genic units
to those higher up the hierarchy may be anticipated through new
techniques. Other dominants, including achondroplasia, Apert's
syndrome and Marfan's syndrome are known to be associated with
paternal age. In Huntington's chorea and myotonic dystrophy,
two common and very unpleasant dominants in which new mutants
have yet to be unequivocally observed, there is a very severe
form transmitted from the mother or the father respectively,
suggesting a unit prone to duplication or other derangment at
meiosis. It would seem convenient to use the word segmental for
that class of mutants related to submicroscopic structures
larger than a genic locus. Such a term would merely be a temporary limbo for phenomena not yet explained in terms of the
structures involved. On this basis we may attempt to partition
the types of mutation as follows (Fig. 1):

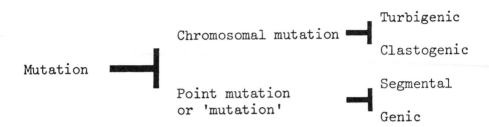

FIGURE 1. *Types of mutational event.*

Mutation imposes a very special responsibility on those involved in advice, or legislation, since it differs essentially from other problems of factory or population health. Most industrial processes, however dirty or dangerous, impose their hazards locally so that those who suffer are usually within the same political unit as those who profit, and these hazards are unlikely to involve those yet unborn. The products, however toxic, can become innocuous purely by dilution, since most can be regarded as harmless when below a certain threshold.

Potentially mutagenic processes, such as nuclear power, not only impose hazards at great distances in space, since the absence of any threshold does not limit the consequences to any neighbourhood, but also impose hazards at great distances in time. The recessive disorders, which are the main hazard in man, will not be evident at all for at least three generations, and will reach their peak in a hundred or more generations under conditions of outbreeding (Haldane, 1947). Further, while a worker exposed to a conventional industrial hazard can but die once, a worker exposed to a mutational hazard could cause many deaths, usually preceded by grossly disturbed lives, among his remote descendants.

The matter is at least serious, and deserves a more precise use of words, and a more accurate accountancy of both our ignorance and our knowledge than it appears to have received. Some twenty years ago a number of committees made various reports, largely against the background of war and experimental war, and there was considerable consistency of expert opinion. Even in retrospect there seems little to criticise (for example, see Weaver *et al.*, 1956 and WHO, 1957).

Since then, the complexity of the chromosomal apparatus and the scale and variety of transmission errors in man has doubled the scope for genetic disability (Fig. 2), while more precise definitions of genetic disease have reduced the incidence. I will attempt a deductive development of the subject, as I attempted in "The Mutation Rate in Man" (Edwards, 1974), in the hope of clarifying the use of the words and thereby defining those areas of optional ignorance which must be eliminated if we are to attempt to audit the various debts which we will be exporting to the future.

The basic problem of mutation may be seen in Fig. 3, which shows an environment (E) and a family of organisms (O) which are serially related by gametes which convey the heredi- tary factors.

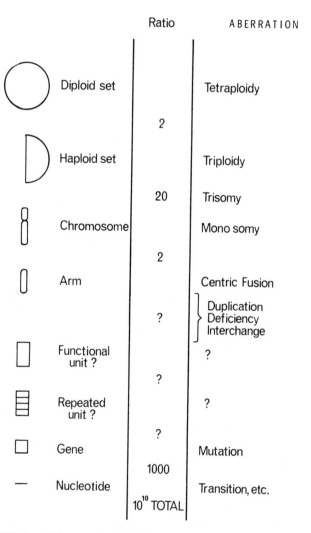

FIGURE 2. The genetic hierarchy and the derangements to which it is subject.

If some of the atoms in (E) are "mutable" so that they either change their elemental form, or emit energy sufficient to displace atoms in molecules, then some of the mutable atoms, either by gaining entry to the organism, or from the outside, will cause a change in the hereditary apparatus; that is, a mutation, or mutated molecule, which may so disturb the development or function of an organism that it is classified as a "mutant". In addition some mutations may be caused by energy acting directly from outer space.

It is clear from such a diagram that, since not all mutations are secondary to mutagens, the proportion in the increase in the rate of production of such mutants must necessarily be less than the proportional increase in the "mutable" or "mutagenic" atoms or waves in the environment. This is basically the argument used in most studies of human mutation. It may be made more complicated by the concept of a "doubling dose", but this merely introduces a factor defined in another species to replace the unknown proportion of mutants due to mutagens. This has the disadvantage that it does not exploit the certainty that the upper limit of this proportion is unity.

The diagram is, of course, a gross simplification, and the argument assumes an approximate linearity of response, a steady exposure, and that the mutants are related to single mutational units.

An audit requires that we should know the number of deaths and disabilities due to simple genetic mechanisms so that this may be multiplied either by the proportionate increase in mutagenic activity to give an estimate of cost, or by the absolute increase divided by the doubling dose, which is defined as the dose which will double the natural mutation rate. That is, if c is the cost, d the doubling dose, p the proportional increase in mutagens, q the absolute increase in mutagens, and a the incidence of mutant disease we may state

$$c \leq ap \qquad \text{and} \qquad c = aq/d$$

Estimates of a in man are difficult, and were originally based on lethal or semi-lethal dominant disorders, which were thought to be the major form of genetic disease in man. Unfortunately these disorders are difficult to ascertain accurately, and there seems a discrepancy between the experience of those who see patients and those who record surveys. The earlier surveys were handicapped by an absence of such precise diagnostic aids as X-rays and slit-lamp microscopy so that mild manifestation in parents was often missed.

In addition, interest tended to be aroused by the presence of cases, so that areas of high incidence were selectively studied. All that can be said of dominant disease in man is that there is a very wide range of mutation rate, from well over 10^{-5} in achondroplasia to under 10^{-6} in most others, that it breeds true with great precision within and between families, that homozygous manifestation has rarely been observed, that the genetic unit involved has rarely been defined, and that most disorders, including the commonest ones, seem restricted to man. At least, they have not been seen in mice or domestic animals. Exceptions include Waardenburg's syndrome and the split-hand deformity. There are probably many dominant dementias and cataracts and forms of deafness which are rarely manifest due to death from other causes. Polycystic kidney disease, for example, is a common autopsy finding, but a relatively rare clinical problem.

Whether we work from mutation rates and doubling doses, or from proportions, we are in the gravest difficulty. It is probably true to place the incidence of severe symptomatic dominant disease at death at below 1%, in my opinion well below this. Retinoblastoma should be classified as a chromosomal mutant. Lethal disorders of infancy of unknown cause should be unclassified.

The common severe X-linked disorders, which behave as dominants in the male, have a fairly accurately defined incidence and, since the commonest form (Duchenne's muscular dystrophy) is lethal, it has a defined mutation rate. In view of this high mutation rate (almost 10^{-4}), and the failure to observe it in mice or domestic animals, it seems likely to be commoner in man and therefore the consequence of some unit other than a gene. The haemophilas (haemophilia and Christmas disease in Europe; hemophilia A and B in America) are jointly almost as common. The other X-linked disorders have a combined frequency which is about equal to that of Duchenne, and certainly not more than double this, so that severe X-linked recessives have an incidence of about 1 in 2000 boys or 1 in 4000 births.

These peculiarities of the dominants and of some X-linked recessives, and especially their apparent restriction to man, make the application of a doubling dose derived from mice difficult to justify. Some seem related to paternal age. No increase in incidence has been described in either populations or individuals exposed to radiation. It seems reasonable to suppose that informal methods of ascertainment, encouraged by the desire to define an external cause, would have lead to any substantial increase in incidence in the children of radiologists becoming recognised.

In principle some of these disorders could be used as "sentinel phenotypes" (Sentinel is a French word adopted in English and now corrupted to "Sentry"). In practice the diagnostic difficulties impose an upper limit of ascertainment of about 50% by the age of ten in most disorders, and as most have an incidence of less than one per paediatrician lifetime, reliable clinical diagnosis on a population scale would seem impractical. Achondroplasia is confounded by the commoner, but often similar, lethal neonatal forms, and most other dominants are both more difficult to define accurately in the child and to exclude reliably in a parent.

The common X-linked disorders might be more suitable as soon as precise methods of definition are available at the amino acid level.

The chromosomal disorders behave as dominants, and are readily induced by mutagens in mice and in cell culture: there is also disturbing evidence in man on the induction of non-disjunction by diagnostic X-rays (Patil *et al.*, 1977, Uchida and Curtis, 1961). These are usually considered as dominants in mutational studies and, although a chromosome is an organ conveying many tens or hundreds of thousands of loci, it is assumed to have the same doubling rate as its constituent loci. While possible, this seems unlikely (Uchida and Freeman, 1977). Even in embryonic material there is a gross inequality in the proportions of the 22 trisomies. It also seems unlikely that any single event of radiation should liberate sufficient energy to disturb so large a unit unless the intensity of the short diagnostic exposure is more important than its total energy.

The recessive disorders are usually severe, and, due to their high recurrence rate in sibships, fairly precise estimates of the incidence of recessive disease can be made from the study of unclassified cases of blindness (Friedman and Fraser, 1967), deafness (Fraser, 1976) and mental deficiency (Penrose, 1938). Where diagnostic standards are high, most cases of these disorders are unclassified. Their burden exceeds that of the dominant disorders, and since they result from defective genic function, it seems reasonable to regard them as peculiarly at risk for mutation. The concept of a "mutation pressure" putting identical mutants into a pool, to be withdrawn by recessive lethals many generations later, seems difficult to accept in view of effective populations usually being far less than the reciprocal of the mutation rate. It is true for the class of recessives, but not for individual recessives. A common ancestor is possible in most recessive disorders, a possibility which should soon be clarified by exact protein analysis.

The basic Mullerian concept of mutation was based on the chromosome as the mere vehicle for a limited number of "genes", whose effects could be regarded as largely independent, so that any mutant with any deleterious effect would, on average, cause the equivalent of a genetic death. This valuable foundation is now somewhat obscured by the sheer complexity of the

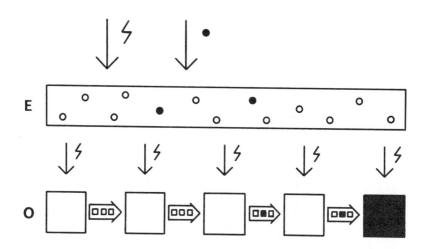

FIGURE 3. *The relationship between the environment (E)
of mutable atoms and ionizing radiation, mutated genetic units
and mutant organisms (O). Successive generations of organisms
are represented by squares and the intervening gametes by
arrows.*

genetic apparatus, by the profligacy of the genic material,
and by its extreme variety, over a third of loci showing allelic forms in most species adequately studied, and by the
strange dialects of the satellite DNA's.

If we consider the hierarchy of genetic elements (Fig.2)
our knowledge is limited to those parts big enough to see and
to those small enough to study by chemical analysis of their
products. Between these is the no-mans land of structural
units, functional units, spacer units, repeated units, fossil
loci, and inert regions whose repetitiveness would be meaning-
less on translation. It is here that complexity might impose
a high risk of mutagenic disturbance, and this disturbance
would not be revealed in non-nuclear organisms. Since the
genetic apparatus is a complex structure a simple audit which
assumes additivity without interaction is doomed to optimism,
and can but suggest a minimum.

Chromosomal Mutants

This category, which was excluded by Muller from the unqualified term mutant, is now known to impose a far larger
fetal mortality than those due to diseases from all other de-

fined genetic units. This mortality is so high, and, in practice, so small a contribution to human misery, as well as so difficult to quantify, that there are advantages in completely ignoring it in this context. Although common, involving at least 4% of conceptions, the numerical abnormalities, which comprise well over 95% of abortions due to chromosomal imbalance, have too small a recurrence rate to make an appreciable contribution to childlessness or to impaired fertility.

A small minority of these fetuses survive and, after birth, about 0.2% of children have a chromosomal defect sufficient to impose a serious handicap, the commonest being trisomy 21 (Down's syndrome or mongolism), and about 0.2% have a relatively minor handicap of a degree which is poorly understood, and, although common, a rare cause of serious disability. Examples include XXX, XYY and XX/XO mosaicism (excluding cases ascertained through disability). The combined incidence of these disorders is commoner than trisomy 21. New structural disorders are far rarer.

Non-disjunction, a "turbigenic event", is easily induced in mice by radiation (Uchida and Freeman, 1977) and there is similar evidence in man (Uchida and Curtis, 1961; Patil *et al.*, 1977). The evidence relating to natural radiation (Kochupillai *et al.*, 1976), although quoted in the Flowers' report, is inconsistent with the doubling doses defined in mice, which seems similar to those related to dominant disorders in mice.

These chromosomal, dominant, and X-linked, disorders have been covered by the International Committee of Radiation Protection and have been the basis of various estimates. These suffer from a precision which is inconsistent with the poor data and rough logic from which they are compounded. A recent estimate gives 130 per million (Ash *et al.*, 1977); between 10 and 1000 would be more realistic. As these estimates relate to events which would be noticed within the lifetime of many of those exposed and some of those advising on exposure, the estimates are capable of validation or modification in the light of experience.

If we descend further down the hierarchy of genetic units to the level of the gene we are again ignorant of both numbers and of mutational rates, and even more ignorant of their product, which is part of the cost of mutational exposure.

There are 3×10^9 nucleotide pairs in a human gamete. This is a large number, and comparable to the population of the earth, and to the duration of human life in seconds. If we conveyed our genetic message in morse code it would take twenty years. A proportion of this message contains the genetic units, and a proportion of these units are necessary, in the sense of catalysing metabolic pathways in which there is little redundancy, so that defects are manifest as recessive disorders. While there is room for three million genic units, some esti-

mates, all very indirect, imply that the greater part of DNA
is otherwise employed in mechanical structures and in inert
fossil loci, for which there is no known mechanism for exclusion or ejection at an acceptable cost in mortality and in re-
petition.

The range of 100,000 ot 1,000,000 would seem a reasonable
estimate of the number of genes. The complexity of the metabolic map would suggest that the great majority of enzyme
pathways were sufficiently buffered by redundancy for homozygous defects to be tolerated, in the same way as normal text
is sufficiently redundant for most spelling errors not to im-
pair the meaning.

The number of metabolic pathways in which a "block" or
"unit" will lead to serious phenotypic effects, that is, the
number of recessive disorders in man, may well be between
10,000 and 100,000. Most of these will be very rare, or even
potential, and, since they are likely to involve those organs
in which speed and precision are more important, in an evolu-
tionary sense, than redundancy, the eye, the ear, and the
brain, are likely to be the main targets. Since difficulties
in biopsy and autopsy make definite diagnosis unusual and enzymatic dissection almost unknown, except in the disorders
consequent on the accumulation of normal substances, enzymatic diagnosis is likely to be difficult. Muller's estimate of
10,000 for *Drosophila* (Muller, 1929, 1955a) would seem consistent with about 100,000 in mammals.

Loci may be inactivated either by a transition or trans-
version causing change in gene product, or by the loci being
excised or deleted. Deletion mutants seem commoner than nucleotide mutants following radiation, and may involve the deletion of many other loci whose products are buffered by redundancy. It is difficult to see how radiation could lead to
"clean" excision. Such defects would be incapable of back-
mutating, or of mutating to useful forms, and might lead to
disturbances to homologues at meiosis due to incomplete pairing. That is, such mutants could be mutagenic, leading to a
cascade phenomenon of mutagenesis in homologous alleles.

A recessive locus, by which we may term one capable of
causing a recessive disorder, can be inactivated in many ways.
It would seem that recessive disorders are the main hazard
from mutagenesis in man. Each mutant will, on average, lead to
the genetic death of an individual if fully recessive, or to
almost two if not (Haldane, 1937).

Recessive disorders have an incidence of about 0.25% at
birth in Nordic Caucasians, in most of whom fibrocystic disease of the pancreas (mucoviscidosis) is the commonest with an
incidence of 1 in 2000 births. In view of its frequency and
its racial restriction it is necessary to postulate heterozygote advantage. In some South Caucasians and Asians thalasse-

mia is the commonest recessive, and in some Negro groups
sickle cell disease: both show heterozygous advantage in the
presence of malaria. If disorders maintained by heterozygous
advantage, at any rate in the past, are excluded, then 0.25
becomes the upper limit of recessive disorders maintained by
mutation. By making assumptions of the number of alleles involved it is possible to estimate the proportion of heterozygotes and derive the number of lethal equivalents in man. The
answer comes to about 0.3, or, even after making the most ge-
nerous assumptions, less than 1.0 (Edwards, 1974). A similar
figure was derived by Arner (1908) using simple arithmetic on
data from cousin marriages. More elaborate methods of estimation have extended this simple method to include the concept
of lethal equivalents (Morton *et al.*, 1956) and led to figures
as high as 4 per gamete or 8 per individual. These figures have
been given a wide currency in various textbooks and circulars,
few of whose authors distinguish between a recessive lethal
and an equivalent. It is difficult to see how the raw data on
recessive disease could be so bad that the upper estimate of
one recessive lethal per gamete could be in error. The basic
principles of estimation, and the wide scatter of the estimates derived, is discussed in Cavalli-Sforza and Bodmer (1971).
These figures are consistent with the frequency of recessive
lethals in *Drosophila*. Data from mammals do not seem to be
available: they would be very difficult to acquire since most
of the recessive disorders of man have only been resolved by
studies which would be difficult on a small mammal in which
detailed examination of the eye, the ear, and the nervous
system, and a detailed chemical study of blood and urine are
difficult, impractical, expensive, or even fatal.

Mutation

The proportion of genic units which are necessary is unknown; the wealth and complexity of the metabolic maps suggest
it is a minority. The remainder will respond to mutation by a
failure of developmental precision, or of the speed and precision with which the organism can defend itself from the ass-
aults of parasites or the counter-attacks of plant toxins, or
can repair the consequences of exposure to mutagens and the
wear and tear of age, or can behave in a way conductive to
posterity.

We may define the consequences of derangements in these
units in terms of "health". That is, health will be impaired
through faulty development, using "fault" to cover any margin
between what is reached and what would otherwise be reached,
or by an impaired resistance to disease or any other impaired
ability to survive and reproduce.

Since only about 1% of disease is due to simple genetic causes, and since all disease is related to the genetic background, the remaining 99% of disease is best regarded in terms of a derangement in health. These disorders are sometimes called multifactorial, but it is difficult to see the justification for such a terminology. As almost all death is secondary to disease, the effect of any genetic damage will be to change the proportions by reducing some ages of onset. No amount of mutation will increase the death rate beyond 100% and, if any cause is increased, another must be reduced. These may seem simple matters, but they are easily over-looked. There seems no justification for the common statement that 6% of deaths are part genetic. All are.

The weight to be given to the mutational damage relating to this 99% of disorders not related to simple genetic mechanisms, but which are probably related to the bulk of the genetic apparatus, is not clear. It would seem larger than that from all the simple genetic mechanisms combined.

We may attempt to enumerate this (Table I) by defining the several units, their number, and their mutation rate. However, this is of little value beyond charting out the depth of our ignorance. Fig. 4 attempts to portray the consequence of a sudden mutational exposure. The chromosomal and dominant disorders respond rapidly, first cases continuing to arise until exposed fetuses and infants have reproduced and some cases continuing to later generations. The recessive disorders will not reveal themselves until children are born with a common exposed ancestor and none will be expected during the lifetime of those exposed, the peak incidence being centuries or millenia later (Haldane, 1947).

TABLE I. Basic procedure for estimating cost from independent mutational events. The common approximation C= UxMxS has little justification

Unit	Rate	Detriment	Product
u_1	m_1	S_1	$u_1 \times m_1 \times S_1$
u_2	m_2	S_2	$u_2 \times m_2 \times S_2$
u_3	m_3	S_3	$u_3 \times m_3 \times S_3$
.	.	.	. . .
.	.	.	. . .
U	M	S	C

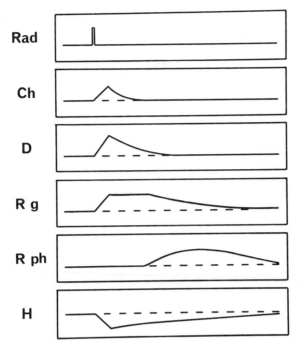

FIGURE 4. The response to a mutational event on genetic units and phenotypes (chromosomes, dominant units, recessive units, recessive phenotypes, and health)

Unfortunately we are ignorant of the relative debts which posterity will incur from these various types of mutational damage, and any costing must involve some exchange rate between small and large effects. All that can be asserted at present is that the total debt must exceed that easily detected through chromosomal and dominant mechanisms, and may exceed it manyfold. At present the artificial increase to our load of mutations would seem to be small, and that part due to radiation cannot exceed the small proportion by which total radiation has increased. Since natural radiation is likely to be only a minor cause of natural mutation, the proportionate increase in mutational disease may be only a small proportion of the increase in mutation activity. Even so, due to the large numbers at risk the number of individual casualties may be very large. Muller (1955b) using conservative estimates based on published data inferred that the deaths from testing atomic weapons in the USA were about equal to the deaths which had resulted from using them against Japan, although, due to dilution among other causes of death, there was no noticeable effect in the USA. In assessing the cost of mutation, whether or not the cost can be detected is, of course, irrelevant to the justification of any policy, a fact frequently overlooked in official reports.

Newcombe (1975) has defined the cost in monetary terms as the engineering costs of additional protection for each mutant saved, reaching a figure of the order of 10 million dollars per mutant, assuming the exposure to a doubling dose would lead to a 6% increase in severe disease, with a range of the estimate of the order of tenfold either way. Since the question is an operational one of reactor design and siting this would seem the simplest approach. The practice of estimating the degree of vagueness is not common in this context, although few estimates of genetic damage can attain much more precision.

The legislation relating to radiation exposure does not adequately distinguish between somatic and genetic effects, the latter, included under gonad dose, being obscurely related to gametic function, and taking no account for age and sex. A more appropriate classification would be to distinguish a new class of dosage as gametic, defined as the dose multiplied by the expectation of effective gametes which will be produced, an effective gamete being defined as one which leads to a birth. This is easily available from published figures on age and reproduction, an early example featuring in Fisher (1930) on the reproductive value of Australian women. This recommendation was made by the Committee on Genetic effects of Atomic Radiation (Weaver *et al.*, 1956).

For practical purposes expectation of further posterity declines rapidly after the mid-thirties in both sexes, reaching zero in women at fifty and almost zero in men. This concept of a gamete dose would to much to clarify the problem, and to draw attention to the serious negligence in the shielding of gonads which is still common. If, as seems clear from the evidence, the genetic risks are more serious than the somatic risks, then the use of a gametic dose could make possible very low levels by deploying older workers in hotter areas. Since most somatic effects seem to have a long incubation period contemporary generations might also benefit.

A further apparent defect in the nature of the legislation, which is based on the concept of a threshold which, rightly, pervades much industrial legislation, is the absence of any limit per unit of production, however measured. It is possible to conform to the law by diluting the exposure over large numbers of workers although this does nothing to reduce the genetic risk. It would seem impractical to give guidance to designers of mutagenic machines, or purveyors of mutagenic substances (and laboratory methods are now available which might be expected to exonerate few organic chemicals) unless

the mutational load per megawatt, kilogram, etc. were to be
defined. While there will be difficulties in such designations,
they will be less than the difficulties from none.

Reducing the Cost of Mutation

Mutational damage from radiation can be reduced by reducing the exposure, and there is now an extensive series of recommendations, largely from the International Commission on
Radiation Protection, which makes the hazards from somatic
effects lower than those from the products of almost any other
industrial process. The risk to workers is of the order of
that from smoking a cigarette every other day. However the
genetic risks are less clearly stated, vague words and adven-
turous arithmetic being used to derive a risk of breathtaking
precision, notwithstanding the lack of adequate data and the
omission of recessive disorders and the impairment in health
of subsequent generations. The recommendations probably provide a safe guideline, but there would seem advantages in
leaving the ICRP to advise on somatic effects, and to leave
the genetic effects to some other body whose responsibilities
would include the majority of our descendants and cover all
mutational consequences, rather than just three varieties
(chromosomal, dominant and X-linked) for a mere two generations.
    The matter can be put in tabular form, where waves and
particles are used to summarize these overlapping mechanisms
(Table II). To a first approximation workers in nuclear power
plants are exposed to waves, and the surrounding population,
and often the distant population, to particles.
    The cost of mutation could, in theory, be reduced if the
consequences could be mitigated by treatment, prevented by
non-conception, or eliminated by abortion. In practice treat-

TABLE II. Reduction in mutational cost

Engineering:	*Waves*	*Shielding*
	Particles	*Effluent control*
Legislation:	*Dose level*	*Person*
		Reactor
		Gamete
Medical:	*Prevention*	*Non-reproduction*
	Elimination	*Fetal diagnosis, abortion*
	Treatment	*Haematology, surgery, etc.*

ment is likely to remain difficult or impossible in most disorders and, if effective, will merely lead to stability at
higher frequencies so that more people will suffer, but each
individual will suffer less. Non-conception is unlikely to be
effective since most mutants will involve only small numbers
of individuals, and the first will be unexpected. From the
point of view of genetic consequences, as well as of that majority who wish to rear children, a non-birth is not a trivial
event. Elimination following fetal diagnosis is likewise not
trivial, and, likewise, cannot be anticipated until an afflic-
ted child has been born. In recessive disorders about 85% of
cases will be the first in the sibship with the present distribution of family size. The dynamics of mutation is such
that each lethal mutant will have an expectation of between
one and two, with a wide scatter, and most mutants will be to
disorders of extreme rarity: in the case of recessives it is
likely to be many years before the cataloguing of disorders is
such that most new mutants will have been catalogued. Mutation
can have little if any contribution to the maintenance of frequencies of the commoner recessives in any racial group. The
impracticality of medical advance having any substantial effect
on the debts with which we are threatening our descendants
might seem too obvious to be worthy of mentioning. However,
notwithstanding this being outside the terms of reference, it
has featured in various discussions in the past, and has even
been used to justify the exclusion of recessive disorders
from consideration.

Potential for research

    If even one per cent of the investment in research on
atomic nuclei had been spent on the nuclei of cells we might
anticipate more solid grounds on which to base our inferences,
and might even anticipate being able to relax some of the more
extravagant practices recommended, on grossly inadequate data,
in radiation workers.

    However, we are at present ill equipped to advise on the
extent to which we are polluting the future, and Table III,
which summarizes some of the data on the genetics of mouse and
man, makes this clear. Extrapolation from any single mammal to
man might seem unwise, especially if selected for small size,
large litter size, and ease of adaption to laboratory conditions. It would seem only prudent to extend this base to at
least one other species, preferably one whose chromosomes are
fewer and more distinct than those of mice, and in which both
laboratory and wild type species can be studied. While any
constraint on what is studied is likely to demoralise many research workers, and might prevent the discovery of some system

TABLE III.   Some of the information lacking in mammals.
The last unknown is purely semantic

Number of necessary genes in mammals
        1,000-100,000

Number of functional genes
        10,000-2,000,000

Genetic nature of dominant
                Unknown in all cases leading to physical
                abnormality

Genetic nature of mutable units studied in mice
                Unknown

Number of recessive lethals in man
        0.3-8 per gamete

Incidence of serious dominant disorders at birth
        1/1,000-10/1,000

Incidence of part genetic or multifactorial disorders
        6% to 99%

---

which would open up new mechanisms of heredity or new modes of thought, at least some research should be directed to the detailed study of those remarkable loci which are well known in either mice and men. In particular the loci related to haemoglobins, immunoglobulins, histocompatibility and those few disorders in which close homologies seem to exist between man and mouse (testicular feminisation, Menkes' syndrome, and albinism), or in which common mutation-maintained disorders had not been found in the mouse, possibly for technical reasons (haemophilia, Duchenne's disease, achondroplasia).

Since there are reasons to regard recessive disorders as a major contribution to our future debts, and since inferences about these are likely to be based in large part on the assumption that the unshielding of recessives is the cause of inbreeding depression, a study of this strange phenomenon would seem indicated. Experienced laboratory workers have difficulties in inbreeding some animals, including guinea pigs and rabbits, which cannot be explained by recessivity alone. The human data are difficult to assess due to incest taboos being so strong that they are preferentially broken by the defective and psychopathic, while the milder form of cousin marriage is often influenced by considerations of comfort or property, and lacks any adequate control.

It seems at least possible that a study of inbreeding under laboratory conditions might reveal some explanation of the discrepancy of tenfold between the number of recessive lethals and equivalents per gamete estimated by direct and indirect methods.

Finally we may attempt to define the present state of acquisition of genetic debt (Table IV). I do not think there is any expert disagreement on the relative magnitude. Natural events are still the dominant cause, and are considerably reduced by protection from stone, slate or cement housing (and considerably increased by granite). This reduction has been somewhat offset by the use of medical radiation for diagnostic purposes, which continues to be a hazard substantially exceeding that from present levels of nuclear power, or even experimental war. However, this serious situation is largely preventable by increasing the quality of those x-rays which need to be taken, and protecting the gonads. In the USA the predatory activities of malpractice lawyers is a substantial cause of unnecessary x-raying, and is likely to continue to be until unnecessary x-rays become included within those actions amenable to malpractice suits. Non-ionising radiation is poorly understood, and due to its use in obstetrics is likely to be used on many millions of fetuses and mothers before its mutagenic effects, if any, have been adequately estimated. In Britain the exacting standards of factory safety do not extend to patients, and requirements for testing diagnostic machines are less exacting than those required for drugs, or even aeroplanes. Since everybody has been, and most of us will be, a patient, some tightening of legislation would seem necessary if hospitals are to lose their leadership as seedbeds of mutational disease.

TABLE IV.   Various mutational hazards in man

	Waves	*Particles*	*Chemicals*	*Ultrasound*
Natural	++++	++++	?	−
Diagnostic	+++	+++	?	??
Therapeutic	++	++	+	−
Nuclear	+	+	−	−

REFERENCES

Arner, G.B.L., (1908). *Columbia Univ. Studies in History, Economics and Public Law.* 31, 1.

Ash, P., Vennart, J., and Carter, C.O. (1977). *Lancet. i*,
1003.

Brøgger, A. (1977). *In* "Expert Conference on Genetic Damage
in Man Caused by Environmental Agents." Oslo, May 11-13.

Cavalli-Sforza, L.L. and Bodmer, W.F. (1971). "The Genetics of
Human Populations". W.H. Freeman, San Francisco.

Edwards, J.H. (1974). *Progr. Med. Genet. 10*, 1.

Edwards, J.H. and Harnden, D.G. (1977). *Nature 267*, 728.

Fisher, R.A. (1930). "The Genetical Theory of Natural Selection." O.U.P.

Flowers Report (1976). Royal Commission on Environmental
Pollution, 6th Report. Cmnd. paper 6618, HMSO.

Friedman, A.I. and Fraser, G.R. (1967). "The Causes of Blindness in Childhood". Johns Hopkins Univ. Press, Baltimore.

Fraser, G.R. (1976). "The Causes of Profound Deafness in
Childhood". Johns Hopkins Univ. Press, Baltimore.

Haldane, J.B.S. (1937). *Amer. Nat. 71*, 337.

Haldane, J.B.S. (1947). *Ann. Hum. Genet. 14*, 35.

Kochupillai, N., Verma, I.C., Grewal, M.S., and Ramalingswami,
V. (1976). *Nature. 262*, 60.

Morton, N.E., Crow, J.F., and Muller, H.J. (1956). *Proc. Natl.
Acad. Sci. (USA). 42*, 855.

Muller, H.J. (1929). *Proc. Int. Cong. Plant Sci. 1*, 897.

Muller, H.J. (1955a). *J. Hered. 5*, 199.

Muller, H.J. (1955b). *Science. 121*, 837.

Muller, H.J. (1956). *Nat. Acad. Sci. 8*, 207.

Newcombe, H.B. (1971). *Adv. Genet. 16*, 239.

Newcombe, H.B. (1975). *In* "Mutation and the Amount of Human
Ill Health. Radiation Research", pp. 937-946. Academic
Press, New York.

Patil, S.R., Hecht, F., Lubs, H., Kimberling, W., Brown, J.,
Park, S.G., and Summit, R.L. (1977). *Lancet. i*, 151.

Penrose, L.S. (1938). *Special Report Series No. 229*, MRC.

Shaw, M. (1970). *Ann. Rev. Med. 21*, 409.

Uchida, I.A. and Curtis, E.J. (1961). *Lancet. ii*, 848.

Uchida, I.A. and Freeman, C.P.V. (1971). *Nature. 265*, 186.

WHO (1957). "Effect of Radiation on Human Heredity". World
Health Organization, Geneva.

APPENDIX: REPORTS FROM STUDY GROUPS

APPENDIX

REPORTS FROM STUDY GROUPS

During the symposium six study groups composed of experts in different fields prepared statements on their specific areas of competence and interest. The study groups were asked, when applicable, to give a brief presentation of the state of the art concerning standardization and reliability of methods in their areas. They were asked to specifically elaborate on possibilities for the practical application of existing knowledge in monitoring human populations for genetic damage and for the handling of cases of accidental exposure. Finally, the study groups were asked to make recommendations concerning future research, with an emphasis on improving methods which could be utilized in the efforts to combat genetic damage in man.

The reports of the study groups were presented and thoroughly discussed in a plenary session and a round table discussion. In the Appendix, the final reports of five of the study groups are presented. The sixth study group dealt with the possibilities for collaboration between the Nordic countries in combating genetic damage caused by environmental agents. As a result of the deliberations of this study group, a recommendation was sent to the governments of the five Nordic countries (which collaborate in several areas of research and education), to intensify research efforts in the prevention of environmentally caused genetic damage to man.

REPORT FROM STUDY GROUP ON
POINT MUTATIONS IN MAN

Chairmen

Kåre Berg and James V. Neel

We define a point mutation as one which can in principle
be localized to a specific region of a chromosome. In human
genetics at present this term encompasses changes ranging from
alterations of a single nucleotide to small deletions or duplications. It is mutations of this type which, if they occur
in germinal tissues, are particularly apt to persist for many
generations. In contrast to other species, man, because of his
exposure to so very many potential mutagens and because of his
unusually long life span, is especially prone to the induction
of such mutations.

There are four principal approaches available to the
study of *in vivo* point mutations in man. Their occurrence in
germinal tissues may best be investigated either through the
enumeration of sporadic cases of certain dominantly inherited
syndromes, the so-called sentinel phenotypes, or through the
detection of a variant protein present in a child but not in
its parents. Their occurrence in somatic tissue may best be
investigated either by the direct characterization, by histo-
chemical or other means, of cells obtained at biopsy or veni-
puncture or by first cloning such cells, followed by their
characterization. The relative merits of these four approaches,
and the research needed to render them more useful, will be
briefly characterized in the following paragraphs.

I. *Sentinel Phenotypes*

Genetic damage may be manifest either within the next generation or not for at least a century. The relative hazards
of immediate and delayed effects are not known but it seems
likely that agents which induce one will also induce the
other. It is therefore important that an inventory of certain

conditions should be made so that any increase could be noticed and also that a further inventory should be made of the immeidate products of the hereditary material, the protein, so that any increase in some types of mutation which lead to delayed effects can be noticed.

The study of sentinel phenotypes needs the definition of a group which can be recognized with reasonable ease and diagnosed with reasonable precision. Local interests and expertise as well as priorities accepted elsewhere should be considered. Disorders which would seem appropriate to study include dwarfism with normal intellect, fragility of bones, dominant skin disorders, retinoblastoma and Huntington's chorea. Other conditions such as neurofibromatosis, epiloia, or polyposis coli might be considered but difficulties may arise from occasional mild manifestation. Only on the basis of known rates of spontaneous manifestation in any one country, could any opinion be given of the meaning of any additional cases associated with some novel environment. No substantial special investment would be needed in countries where the regular diagnostic services have the monopoly of relevant facilities and expertise, which merely need extending and documenting in a way suited to a regular central audit of incidence. Associated family studies would have to be routine, both to exploit the information on linkage and to confirm apparent paternity. A disadvantage to this approach is that the nature of the ultimate genetic change is in no case understood. Thus, clear reference on the nature of the induced lesion is not possible.

II. *Sentinel Proteins*

The screening of proteins for mutants is the only way of defining the genic mutation rate in man and, since most of these mutants are probably neutral with respect to phenotypic effects when heterozygous and so not subject to early selection, could provide a baseline with greater precision than any procedure based on disease. The proteins to be studied should include some which are well known, such as the hemoglobin and albumin molecules, as well as some on which there is local interest and expertise. The information per unit cost increases with number of tests done per person. There could be advantages in integrating such work with work done on genetically similar groups with distinct environments, and genetically distinct groups with similar environments. This work could also be integrated with screening for variants of medical interest, such as those predisposing to various disorders or industrial diseases, or to difficulties in metabolizing drugs or anaesthetics.

There are considerable opportunities for advanced automated procedures for use in this type of investigation, both for man and experimental organisms. The organization of such screening would be simpler in countries with centralisation of blood banking and maternity services. It would be essential to have a liquid nitrogen facility for the storage of specimens and facilities for coding the information in a uniform way. The technology in this field is developing so rapidly that it is extremely difficult to make exact estimates of costs. However, present experiences suggest that a large-scale screening program comprising 60,000 samples examined annually for 25 proteins may cost between 1 and 3 million dollars. The chief practical problem is to improve the ability to discriminate between true mutation and discrepancies between legal and biological parentage.

The number of mutations found in a ten year study might be small or zero, but this would in no way diminish the value of the work, since only such data could allow the spontaneous genic mutation rate to be defined, or allow the informed documentation of any ecological disaster that might happen.

III. *Point Mutations in Somatic Cells*

In theory, detection of *in vivo* point mutations in somatic cells might permit determination of mutation frequencies in populations of differentiated cells and possibly calculations of mutation rates per cell generation or units of time: however, there is as yet no evidence that this can be achieved and that detection of the *in vivo* somatic mutation can be usefully applied in monitoring environmental mutagenesis.

Point mutations in somatic cells could be depicted with *in vitro* selective systems using cultures of cells with proliferative potentials (lymphocytes, fibroblasts) or with *in vivo* screening of cohorts of easily accessible non-dividing cells (such as red cells, leukocytes, spermatozoa). The first approach is feasible using established methodologies, but techniques are laborious and there is a possibility that new mutations might be induced with the *in vitro* manipulations of cultured cells. Only a few unsuccessful attempts have been made to detect point mutations with *in vivo* cell screening. Progress in this field will require development of reliable methods of detection of structural mutations in individual cells and of direct or indirect approaches for verification of the mutational origin of the abnormal cell phenotypes detected with the *in vivo* screening.

Since it is possible that time or cell division contributes to spontaneous mutations *in vivo*, the frequency of mutations in the terminally differentiated cells of self-renewing tissues may exceed mutation rates per cell generation by several orders of magnitude. Accordingly, somatic cell screening may not be very useful for depicting the *in vivo* effects of short exposures to mutagens. Somatic cell approaches will be more useful for comparisons between populations and studies of individuals with prolonged exposures to industrial mutagenic contaminants.

IV. *General Recommendations*

1) For the screening of human populations, increased effort should be devoted to an approach based on the use of protein variants, with particular reference to decreasing costs and dealing with discrepancies between legal and biological parentage.

2) Parallel studies on protein variation in experimental organisms should make possible much firmer extrapolations from these organisms to the human situation than has hitherto been possible. We recommend that studies on the rate of occurrence of mutations affecting protein structure in experimental organisms be encouraged.

3) The study of mutation rates in somatic cells is in theory much more convenient and less expensive than similar studies on germinal cells. Unfortunately, the relevance of somatic cell rates to germinal cell rates is completely unknown. We strongly recommend that for the immediate future, any study of somatic cell rates be combined if possible with one of germinal cell rates, and *vice versa*. When the relative magnitudes of somatic and germinal rates for the same or similar characteristics have been established, then it should be feasible to extrapolate from somatic to germinal rates in monitoring programs.

4) In principle the monitoring of human populations for changes in mutation rates can proceed either by contrasting the findings in a population with a relatively heavy exposure to one or several mutagens with the findings in a population not so exposed, or by contrasting the findings in successive time periods in a given population or populations. In either event, the necessary studies are so demanding and expensive that we strongly recommend the development of collaborative studies on an international basis, so designed that the results of the various studies can be readily expressed in ways which will allow them to be compared.

The following were members of the Study Group:

Kåre Berg, Oslo
Helge Boman, Oslo
John Edwards, Birmingham
Jan Mohr, Copenhagen
James V. Neel, Ann Arbor
George Stamatoyannopoulos, Seattle

REPORT FROM STUDY GROUP ON
CHROMOSOME DAMAGE AS A MEASURE OF GENETIC HAZARD

Chairmen

Anton Brøgger and Sheldon Wolff

Several cytological tests now exist by which it is possible to demonstrate that chromosomal changes have occurred. These include direct observation of
  a) metaphases for chromosome lesions such as gaps, breaks, rearrangements and sister chromatid exchanges,
  b) anaphases for bridges, lagging chromosomes and fragments,
  c) interphases for micronuclei.

Cells of either man himself or of experimental animals can be studied.

Cytological studies can be used to test the effect of chemicals and to monitor the possible effects in exposed persons.

I. *Testing of Chemicals*

In the past, the standard method for testing the possible chromosomal effect of chemical agents has been the study of chromosome aberrations induced by them.

This test should be continued as part of the battery of short term tests used for chemicals.

In addition, it has now been shown that the induction of sister chromatid exchanges (SCEs) is often a more sensitive indicator of the interaction of chemicals with DNA in mammalian cells. Therefore SCE tests should be included in the battery.

In vitro SCE tests can be used to detect the mutagenic/carcinogenic effect of some compounds that require metabolic activation as well as that of some which act without activation. Additional *in vivo* SCE tests in experimental animals can be used to reveal the effect of those compounds requiring activation but which can not be activated *in vitro*.

II. *Monitoring of Populations*

In order to monitor man for low level effects brought about by chronic exposure to chemicals the standard cytogenic chromosome aberration tests with cultures of lymphocytes from peripheral blood must be continued. The efficacy of SCE testing in this regard has not yet been established.

III. *Accidental Exposures*

Accidental exposures to high levels of mutagenic chemicals could produce the standard chromosome aberrations found in the lymphocytes. Therefore, conventional cytogenetic analysis is recommended in all persons exposed. To make this test more meaningful, the background level of aberration should be established for each individual worker before any possible exposure. Since many of the lesions produced in chromosomes are repaired after some time, this test should be carried out as soon as possible after the exposure.

Studies with SCEs found in peripheral blood cells after treatment with chemicals indicate that the SCE test which is easier and more sensitive than the standard test, should also be used after accidental exposures. Because of repair, this test also must be performed soon after the exposure. Serial testing with both methods should be carried out to follow the residual effects of the exposure.

IV. *Recommendations concerning Future Research*

The following points seem particularly promising for progress in the field of cytogenetics, and some of the developments will have immediate practical applications.

a. Some of the chromosomal effects of chemicals result in the formation of unambiguous aberrations, such as dicentrics, that are scored similarly in laboratories throughout the world. Other aberrations, such as chromatid breaks, however, are often confused with chromatid gaps leading to reports of quantitatively different results from different observers. Work should be carried out to determine what proportion of undisplaced chromatid discontinuities are actual breaks and what proportion are merely gaps. Definitive criteria should be established for discriminating between the two.

b. It is recommended that studies be carried out to determine whether or not SCEs can be used to monitor for low level chronic exposures to chemicals, *i.e.* to determine if there is an accumulation of unrepaired lesions in the genome.

c. SCE and aberration studies should also be carried out to determine if there is a sensitive subpopulation of people.

d. Further studies are recommended to determine the relation between mutations that could affect future generations and the somatic cell damage expressed as either chromosome aberrations or SCEs.

e. Work should be carried out to determine the quantitative relationship between the mutagenicity of a chemical and its carcinogenicity.

f. The possibility of detecting induced chromosome abnormalities at some stages of human germ line cells should be explored. This should be done in both males and females.

g. Because the amount of work required to monitor large (or even small) populations is very great, the development of automatic cell processing and scoring methods should be encouraged. Automation could result in a world-wide standardization of slide preparation and scoring.

h. Problems of possible synergisms from exposure to multiple chemicals should be explored.

i. Attempts should be made to establish dose-response relationships for chemicals with respect to SCEs (biological dosimetry).

j. It is recommended that studies be carried out to determine the quantitative relationships between chemical exposure to mitotic disturbances (turbagenic effects).

The following were members of the Study Group:

Anton Brøgger, Oslo
James German, New York
Carl Birger van der Hageen, Oslo
Olli Halkka, Helsingfors
Inger-Lise Hansteen, Porsgrunn
Preben Bach Holm, Copenhagen
Maj Hultén, Birmingham
Sakari Knuutila, Esbo
Tytti Meretoja, Helsingfors
Nils Nevstad, Oslo
Helga Waksvik, Oslo
Sheldon Wolff, San Fransisco

REPORT FROM STUDY GROUP ON
EPIDEMIOLOGIC APPROACHES FOR THE DETECTION
OF MUTAGENIC AND REPRODUCTIVE HAZARDS

Chairmen

Lars Beckman and Peter F. Infante

Recommendations

To date there has been little emphasis on epidemiologic
investigations of chromosomal aberrations and pregnancy outcome as related to toxic agents. Research and training in
these fields should therefore be promoted.

Specific high risk populations and occupational groups
should be monitored with respect to point mutations, cancer
incidence, chromosomal aberrations, early spontaneous abortions, birth weight, and congenital malformations with high heritability. For studies of congenital malformations, emphasis
should be placed on specific defects where possible and particularly those which have little variability in clinical manifestation. Prospective studies should be conducted to elucidate susceptibility to genetic damage, and cancer risk among
individuals with persistently elevated levels of chromosomal
aberrations.

In cases of emergency, thorough records should be made
and the exposed individuals should be put under life-long surveillance which should include studies of chromosomal aberrations.

Because of concern for the relationship between mutagens,
carcinogens and teratogens, the systematic linking of records
for birth defects, cancer and occupational exposure would
greatly facilitate epidemiological research. To assist in
achieving the latter goal, we recommend that the occupation
of both parents be recorded as part of birth certification.

The following were members of the Study Group:

Lars Beckman, Umeå
Luigi de Carli, Milan
Karl Fredga, Lund
Peter F. Infante, Cincinnati
D. Jack Kilian, Freeport

REPORT FROM STUDY GROUP ON
NON-HUMAN TEST SYSTEMS WITH SPECIAL RELEVANCE TO MAN

Chairmen

Diana Anderson and Claes Ramel

I. Overview of Problem

It has become increasingly clear that as we have been
able to control infectious diseases in the last several decades, the effect of genetic abnormalities, including cancer,
has become one of the major health problems facing our world
population. Present information would suggest that environmental agents including man made chemicals may contribute significantly to the induction of neoplasms and presumably genetic
abnormalities. Since an industrial population, especially
workers employed by chemical companies, in all probability represent a high risk population where genetic abnormalities
may be detected earlier or at a higher frequency than in the
general population, it is particularly important that chemicals that these workers are exposed to are tested by the best
available genetic-toxicological procedures. In addition to
genetical-toxicological studies in non-human systems, it is
imperative to have an ongoing surveillance programme, in our
industrial population to detect potential adverse genetic
effects at the earliest possible time.

II. Evaluation of Chemicals in Non-Human Systems

Fortunately with our increasing awareness of the genetic
hazards of chemicals, a number of procedures have been developed in this area which are as good, if not better, than those
presently available in other areas of toxicology. There is no
single method that can be recommended as the single procedure
to detect potential genetic effects of chemicals, however a
battery of available procedures will in all probability detect the vast majority of mutagenic and therefore presumably
carcinogenic agents.

A. *Criteria for Testing Procedures*. In constructing a meaningful screening programme, the following criteria should be satisfied whenever possible:

1. All genetic lesions, both at the genic and chromosomal level should be detected.

2. The combined tests should have the potential to detect the effect of metabolites produced by the intact host (animals or man).

3. Each of the tests in the battery of tests should be sufficiently reproducible so that the test results are meaningful.

4. In each of the tests the Beta error (the likelihood of a positive response not being detected) should be sufficiently low so that a doubling over background rate is statistically significant at the 5% level.

5. The tests should be of sufficiently short duration and economical so as not to preclude their general use.

B. *Available Testing Methods*. Tables I and II (modified from Committee 17 report: Environmental Mutagenic Hazards, *Science. 187* (1975) 503-514) list genetic lesions and operational characteristics of mutagen screening systems.

C. *Selection of Chemicals to Be Tested*. It should be emphasized that a selection of a battery of tests from the procedures listed in the tables, including *in vivo* as well as *in vitro* procedures represents short term, economical studies when compared to the standard carcinogenicity test. Even though these tests are of relatively short duration, that is can be completed in weeks rather than years, it nevertheless would be useful to have a list of chemicals to ensure an orderly testing of chemicals based on a priority basis. The selection of chemicals for testing can be based on the following criteria:

1. Widespread exposure to the chemical in the industrial or general population.

2. Exposure of individuals for a prolonged period of time.

3. Persistence of the chemical in the environment.

4. Whether the structure of a chemical is related to known carcinogens or mutagens.

Since methods are continually improving in this area and certain chemicals used require specific testing, it is desirable to retain a degree of flexibility in evaluating chemicals rather than adhering to a rigid protocol. A definitive protocol for conducting the various mutagenicity procedures can be found in a soon to be published textbook on mutagenicity test-

TABLE I. Types of Genetic Damage Detected by Currently Employed Mutagen Screening Systems

| Screening system | | Types of damage detected | | | | | | |
| Category | Organism | Chromosome aberrations | | | | Gene mutations | | |
		Dominant lethality	Translocations	Deletions and duplications	Non-disjunctions	Forward or reverse or both	Multiple specific locus	Induced recombination
Bacterial	Salmonella typhimurum#					+		
	Escherichia coli					+		
Fungal	Neurospora crassa#				+	+	+	
	Aspergillus nidulans			+	+	+	+	+
	Yeasts	+			+	+	+	+
Plant	Vicia faba		+	+	+			
	Tradescantia paludosa		+	+	+	+		
	Drosophila melanogaster#	+	+	+	+	+		
	Habrobracon juglandis	+	+			+	+	+
	Bombyx mori	+				+	+	+
Mammalian cell culture	Chinese hamster#		+	+	+	+		
	Mouse lymphoma		+	+	+	+		
Intact mammal	Mouse#	+	+	+	+			
	Rat#	+	+	+	+			
	Man#		+	+	+			

#Most frequently used organisms

TABLE II. Operational Characteristics of Mutagen Screening Systems

Test system	Time to run test	Relative ease of detection	
		Gene mutations	Chromosome aberrations
Microorganisms with metabolic activation:			
Salmonella typhimurium	2 to 3 days	Excellent	
Escherichia coli	2 to 3 days	Excellent	
Yeasts	3 to 5 days	Good	Unknown
Neurospora crassa	1 to 3 weeks	Very good	Good
Cultured mammalian cells with metabolic activation	2 to 5 weeks	Excellent to fair	Unknown
Host-mediated assay with:			
Microorganisms	2 to 7 days	Good	
Mammalian cells	2 to 5 weeks	Unknown	Good
Body fluid analysis	2 days	Excellent	
Plants:			
Vicia faba	3 to 8 days		Relevance unclear
Tradescantia paludosa	2 to 5 weeks	Potentially excellent	
Insects:			
Drosophila melanogaster	2 to 7 weeks	Good to excellent	Good to excellent
Mammals:			
Dominant lethal mutations	2 to 4 months		Unknown
Translocations	5 to 7 months		Potentially very good
Blood or bone marrow cytogenetics	1 to 5 weeks		Potentially good
Specific locus mutations	2 to 3 months	Unknown	

ing methods (Kilbey, B., Legator, M., Ramel, C., and Nichols,
W.,Springer Verlag (in press)).

Recommendations.

1. Based on the criteria set forth in this report, responsible national and international bodies should compile a
priority list of chemicals for mutagenicity testing.

2. These industrial chemicals should be tested for mutagenic and therefore presumably carcinogenic activity by a
battery of the best available testing procedures. The battery
of tests should take into consideration the various criteria
set forth in this report.

3. Considering the large number of chemicals to which man
is exposed, occupationally and otherwise, a complete battery of
tests may not be feasible for chemicals not given a specific
priority. In this situation less complete testing such as the
utilization of *in vitro* procedures with or without microsomic
activation may be necessary. Even for these non-priority chemicals it is desirable to use *in vivo* and *in vitro* tests
where possible.

III. *Human Monitoring*

A. *General Surveillance.* It is now possible to apply several of the procedures that are used in animals for surveillance of human subjects. These procedures, by necessity, are
those which utilize blood or urine for either direct or indirect ascertainment of mutagenic and presumably carcinogenic
potential of specific chemicals. In addition, semen samples
may provide a valuable source for mutagenicity studies when
these samples can be obtained.

The various procedures that can be carried out to monitor
human subjects include:

1. Cytogenetic studies, including conventional metaphase
analysis as well as the selective use of banding techniques
and evaluation for increase in Sister Chromatid Exchange.

2. Analysis of urine and blood for the presence of mutagenic agents by the use of a variety of biological indicators.

3 Determination of increase in alkylation of macromolecules, such as histidine in hemoglobin.

4. Sperm analysis for increase in YY bodies.

Routine metaphase analysis is a practical procedure which
is sufficiently established to warrant its use for industrial
population monitoring. Although there is less experience with
the evaluation of blood or urine for mutagenic substances, it
is sufficiently established to recommend its use in a surveillance programme, especially since recent data have already

shown, with the *Salmonella* system, that mutagenic substances can be detected in a hospital population on specific drug therapy as well as in workers exposed to a specific industrial chemical.

The alkylation of macromolecules, *i.e.* histidine in hemoglobin, by the use of the amino acid analyzer is a technique of great potential for monitoring a population exposed to an important class of mutagenic or carcinogenic agents. If the potential of this procedure is realized we will be able to measure an increase in alkylation that has occurred over an exposure period and achieve quantitative dosimetry but at the present time there is little experience with this technique in industrial surveillance studies. However, this promising procedure may be available in the near future.

If semen samples are available, the examination of sperm for increase in YY bodies is a procedure, that if verified, would be a valuable addition for population monitoring.

B. Monitoring Man's Environment. The above procedures are conducted so as to detect the presence in man of mutagenic action, either directly or indirectly. Another valuable approach may be to monitor the industrial environment for the presence of mutagenic or carcinogenic agents. Although several biological indicators can be used, *Drosophila* may well be the single best system for detecting genetic lesions. The presence of an active microsomal enzyme system makes it well suited for detecting a variety of chemicals. It may be possible to set up *Drosophila* cages in selected areas for subsequent analysis to determine the possible exposure of man to mutagenic/carcinogenic substances.

Recommendations.

1. There should be instituted, where possible, a surveillance programme with workers either on a total employee basis or at least in a selected population where exposure to mutagenic/carcinogenic agents is suspected. Cytogenetic procedures, and the analysis of blood or urine by various indicator systems are suitable for a monitoring programme. The alkylation of protein may be a valuable procedure in the near future. The determination of YY bodies in semen may also be of potential value in this area in future.

2. Investigations should be initiated to determine the practicality of using *Drosophila* to monitor the industrial environment.

3. Whenever possible, industry should make available information of human and non-human studies so as to conserve

resources and avoid unnecessary duplication.

IV. Emergency Procedures. The procedures that are recommended for surveillance of human populations are well suited to be used in the event of an accidental occupational exposure. Special emphasis should be given to the alkylation of macromolecules (histidine in hemoglobin) procedure where exact dosimetry may be obtained.

Recommendations

1. Methods should be available for the estimation of environmental exposure.

2. Responsible individuals such as company or industrial physicians should be made aware of procedures that can be conducted to determine the presence of potential mutagens/carcinogens and secure that adequate blood and urine samples be immediately obtained for appropriate analysis following accidental exposure.

3. Laboratories that can conduct the various procedures should be identified by chemical.

4. International information agencies, such as E.M.I.C. should serve as a repository for information on genetic toxicity studies. The agency should be advised of relevant findings of industrial or environmental exposure at the earliest possible time.

The following were members of the Study Group:

Diana Anderson, Alderley Park
Finn Devik, Oslo
Rune Grubb, Lund
Bo Lambert, Stockholm
Marvin S. Legator, Galveston
Claes Ramel, Stockholm
Marja Sorsa, Helsinki
Øystein Strømnæs, Oslo
Ole Westergaard, Aarhus

REPORT FROM STUDY GROUP ON
PRENATAL DIAGNOSIS

Chairmen

Aubrey Milunsky and John Philip

I. State of Art

In series which now exceed 25.000 patients in the United
States and 6.000 in Europe, the following statement holds:
Transabdominal amniocentesis during the middle trimester by a
competent obstetrician is an acceptable, safe procedure. Tissue culture of fetal cells is predictive in centers with experience (success rates over 95%). The diagnostic accuracy of
chromosomal analysis is better than 99%. Biochemical studies
for inherited metabolic defects are also reliable from both
fetal cell cultures and/or supernatant. Thus, efficiency and
safety are well established.
Open neural tube fusion defects can be reliably detected
by elevated α-fetoprotein in amniotic fluid.
Serial sonographic measurement of fetal biparietal size
and growth curve may confirm neural tybe defects. Localization
of the placenta, diagnosis of multiple fetuses, and confirmation of fetal age are essential to prenatal diagnosis.

II. Applications in Human Populations

Established clinical risk groups are well defined: Increased maternal age (more than 35 years); carrier state for X-linked disease; translocation or metabolic disease; previous trisomic offspring; previous neural tube defect.

III. At Risk Exposures

At present there is no basis to recommend prenatal diagnosis in fetal evaluation in pregnancies exposed to radiation,
viral infection, and/or selected drugs or chemicals.

When, or if a known effect from an implicated mutagen is either chromosomal non-disjunction or an increased rate of neural tube fusion defects - then prenatal diagnosis can effectively exclude these defects - whether random in occurrence or exposure related.

IV. *Recommended Research*

1. Every effort should be made to examine abortal and amniotic fluid material in populations identified to be at high risk to environmental mutagens.

2. Until such time as such high risk groups can be evaluated, antenatal diagnosis has little to offer in the specific defects anticipated in environmental exposures.

The following were members of the Study Group:

Cecil B. Jacobson, Washington D.C.
Aubrey Milunsky, Boston
John Philip, Copenhagen

Index